Advancing Library Services for Mobile Users

Yan Quan Liu
Southern Connecticut State University, USA

IGI Global
Scientific Publishing
Publishing Tomorrow's Research Today

Vice President of Editorial	Melissa Wagner
Director of Acquisitions	Mikaela Felty
Director of Book Development	Jocelynn Hessler
Production Manager	Mike Brehm
Cover Design	Jose Rosado

Published in the United States of America by
IGI Global Scientific Publishing
701 East Chocolate Avenue
Hershey, PA, 17033, USA
Tel: 717-533-8845 | Fax: 717-533-7115
Website: https://www.igi-global.com E-mail: cust@igi-global.com

Library of Congress Cataloging-in-Publication Data

(CIP Data Pending)
ISBN13: 978-1-7998-9175-8
Isbn13Softcover: 978-1-7998-9176-5
EISBN13: 978-1-7998-9177-2
British Cataloguing in Publication Data
A Cataloguing in Publication record for this book is available from the British Library.

Table of Contents

Preface

With dramatically increasing demand and rapid development of portable digital services, it is the author's aim to provide an overview of current trends in mobile services for academic, public, and school libraries, present real-life examples of specifically identified mobile services, discuss opportunities and challenges in offering mobile services, advise on best practices for implementing and improving upon mobile services, and speculate on the future of this nearly ubiquitous phenomenon which has dramatically changed multiple aspects of library service, including reference, circulation, website offerings and more.

This book stands out because it offers practitioners, students, and academics in the field of real-world studies and examples from all over the nation and the world to help them understand and think about potential solutions for serving customers with mobile devices in libraries and information centers.

Academic, public, and school librarians as well as emerging technology information brokers who are looking for information on current trends, examples, and best practices in mobile services and who are considering implementing or improving mobile services in their libraries are the main target audience for this book. Students studying information and library science at the graduate and undergraduate levels who are pursuing careers in academic, public, or school libraries should also find this book useful.

By introducing up-to-date knowledge and resources, this book is intended to be a complete discussion of library services offered through mobile technologies and mobile platforms. Publication of this work will benefit students, librarians, and scholars in the field of information and library science. The book aims to serve as a reference for students in information and library science at both undergraduate and graduate levels, professionals, and scholars interested in the book's topics nationally and internationally.

SCOPE OF THIS BOOK

- Current practices of mobile services in academic, public and school libraries
- Empirical studies on library mobile services
- Basic types of mobile services
- Content of mobile websites
- Mobile reference services
- Apps libraries provide for patrons remotely
- Mobile services and smart libraries, future aspects

OBJECTIVES OF THIS BOOK

- Equip readers with a comprehensive understanding of best practices for integrating mobile services within library environments.
- Provide readers with practical knowledge and effective methods for utilizing mobile applications and delivering related services.
- Prepare libraries to meet the growing demand for mobile technologies by enabling them to manage and adapt to evolving technological trends.
- Highlight the importance of incorporating mobile technologies in library services to ensure libraries effectively reach and engage with a broad patron base.

Given the quick development of mobile devices, librarians might be interested in learning how the technology can be applied to library services as well as what difficulties, problems, and solutions other libraries face when trying to provide services to customers using mobile devices remotely. For this purpose, this monograph is to present best practices and examples for library professionals, practitioners, and students in the field to learn about and explore viable possibilities for providing mobile services in libraries and information centers across the country and abroad. The audience includes readers and researchers interested in digital library science and technology. Suitable as a textbook or reading resource for information and library science classes, this book can also be appropriate for information and library science students at both graduate and undergraduate levels preparing for a career in academic, public, or school libraries.

Internet information technology is rapidly spreading and developing, enabling Internet derivatives on the basis of people's daily life and work. The development of digital technology, which has transformed and continuously transformed services provided in libraries around the world to a new realm of operations, mostly for remote and mobilized services, also stimulated my such motivation.

The mobile internet has dramatically impacted people's lives and the way in which information is stored as well as accessed. Global Digital Report 2020 confirmed that more than 5 billion people access information via mobile devices worldwide. "Fully 91% of Americans ages 16 and older say public libraries are important to their communities" (Pew Research Center, 2021).

While the coronavirus surge that hit countries worldwide has become an international issue, libraries desperately seek services via the Internet or mobile technologies.

Libraries, as proven in studies (Guo, 2020), can work effectively providing services through various methods using mobile technologies available to library operations. Booming as super-highways, digital libraries offer services online to users regardless of geographical distances, different communities, or even political boundaries.

Digital services bring significant impacts on the future of the information profession. The demand for the most updated text in information and library science (ILS) schools is high, with immense needs and challenges. However, in this subject field, services to mobile users, I have struggled to find suitable books for the courses I teach. This writer has yet to come across a single monograph titled Library Services for Mobile Users in the ILS literature.

The book covers the topics students should learn about in the field of newly emerging phenomena, including knowledge and techniques on current practices in libraries, and services provided in their current status to mobile users. It will also include learning databases containing reviews of popular digital libraries and case studies around the theological concepts of, and practices in, designing, constructing and evaluating library services for mobile users.

As an instructor teaching library courses at both graduate and undergraduate levels for the past 21 years, and an author of several books and over a hundred research studies, the expertise and experience accumulated in writing this book is beyond counting. I have been working on this proposed manuscript since 2016, where my interest and spirit were originated from teaching ILS 508 User Services, ILS 655 Digital Librarianship and other courses.

I led numerical research teams studying the library services for mobile users. One of these teams conducted a survey of patron services in academic libraries during the COVID-19 pandemic. The study, published in an SSCI & SCI journal titled Information Technology and Libraries (Guo, 2020), revealed that the majority of libraries had shifted their service focus to partially or solely remote online services. The research I've conducted and published and being referenced in this book include:

- "Best practices, standards and techniques for digitizing library materials: a snapshot of library digitization practices in the USA. (Liu, Y. Q., 2004)

- "Digital information access in urban/suburban communities: A survey report of public digital library use by the residents in Connecticut. (Liu, Y. Q, Martin, C., Roehl, E., Yi, Z., & Ward, S. 2006)
- "The Impact of Digital Resource and Service Use on Urban Residents in New England Public Libraries—A Survey Report. (Liu, Y. Q., & Wnuk, A., 2009).
- "The Library Mobile Services in the USA: A Case Study of New York State Libraries" (Liu, 2014)
- "The Library Mobile Services in the USA: A Case Study of New York State Libraries. Library and Information Service, (Liu, Y., Lu, R., & Liu, Y. Q. (2014).
- "The library mobile services in the USA: A case study of New York State libraries. (Liu, Y., Lü, R., & Yan, Q. L., 2014).
- "A Library in the Palm of your Hand: Mobile Services in Top 100 University Libraries" (Liu & Briggs, 2015)
- "The provision of mobile services in US urban libraries" (Guo, Liu & Bielefield, 2018)
- "Are urban public libraries websites accessible to Americans with disabilities? (Liu, Y. Q., Bielefield, A., & McKay, P., 2019).
- "Are Urban Public Libraries Websites Accessible to Americans with Disabilities?" (Liu, Bielefield & McKay, 2019)
- "Private post-secondary library websites and the ADA: compliancy and COVID-19." (Bielefield, A., Liu, Y. Q., & Waimon, V., 2021).
- "Services to mobile users: the best practice from the top-visited public libraries in the US." (Liu, Y. Q., & Lewis, S., 2023).
- "ADA digital accessibility on academic library websites." (Liu, Y. Q., Bielefield, A., & Beckwith, 2024).
- "Inclusive Access: Evaluating Public Library Resources for Visually and Hearing Impaired Communities." (Liu, Y. Q., Buedinger, A., & Murtha, H. R., 2025).
- "Digitization Practices in US Urban Libraries: Tools, Technologies, and Metadata Standards." Torres, N., & Liu, Y. Q., 2025).
- "Adopting Critical Cataloging Practices Post Diversity Audit: Connecting the Community to Your Collection." (Anderson, J. K., & Liu, Y. Q., 2025).

I have been directing over a hundred graduate students on their Special Projects or Thesis work at Southern. Most of these are empirical studies. My role as a principal investigator is to work with the students, help them develop research questions, design research methods, conduct a literature review, collect and analyze data, and write up complete research reports. The studies that are relevant to mobile users include: “Barriers Connecticut Public Libraries are Facing with Mobile Services” (2019)” and “A Survey of Urban Public Library Mobile Applications (2015).” These works are also valuable sources for this book's writing.

This book’s resources come from research projects I have been engaging in the past years that included “Mobile Services in Connecticut Public and Academic Libraries” (CSU/AAUP Fund, 2012), “Mobile Services in Top 100 US University Libraries” (CSU/AAUP Fund, 2016), “Are University Library Websites Accessible to Americans with Disabilities?” (CSU/AAUP Fund with Bielefield, 2019). The most recent grant I received, “Urban Library Services to Mobile Users” (CSU/AAUP Fund, 2021), is specifically for the publication of this upcoming book.

CHAPTER OVERVIEW

Chapter 1 introduces how mobile technologies broaden access to library services and reshape user expectations for immediacy, convenience, and mobility. It highlights emerging service models, the expansion of mobile collections, and the growing popularity of handheld-accessible offerings.

Chapter 2 analyzes the behaviors, needs, and challenges of mobile library users and explains why understanding these users is essential for inclusive service design. It also provides a framework for the principles and competencies librarians need to serve mobile populations effectively.

Chapter 3 traces the growth of mobile library services and the research shaping their development. It evaluates service models, librarian perspectives, mobile programming, and usability challenges.

The fourth chapter provides the technological foundation underpinning mobile library services, covering devices, networks, APIs, and emerging context-aware and location-driven capabilities. It also reviews advanced tools like AR, routing systems, and sensor-based intelligence.

Chapter 5 documents the evolution of mobile reference services—from SMS to chatbot-assisted reference—and evaluates current practices, challenges, and design considerations.

The sixth chapter examines digital content optimized for mobile devices and the technologies—like recommenders and the Semantic Web—that enhance mobile discovery. It also discusses balancing global interoperability with local needs.

Chapter 7 surveys mobile apps used by libraries, evaluating popular platforms, user adoption, and their impact on digital service delivery. It also explores how new web generations drive collaboration and decentralization.

Chapter 8 examines mobile services in academic libraries, including catalogs, content platforms, virtual reference, distance-learning support, and accessibility guidelines.

Chapter 9 explores how public libraries use mobile services to expand community reach, support digital equity, deliver accessible content, and serve diverse populations.

Chapter 10 explores how school libraries use mobile-optimized services, AI tools, and virtual labs to support learning, equity, and engagement among K–12 students.

Finally, Chapter 11 focuses on evaluating library mobile services and preparing for future innovations through strategic planning, governance, responsible technology use, and research priorities.

It is my hope that this book will provide critical resources and create the groundwork for future endeavors of smart services libraries provide and researches librarians worldwide engage to build strong information and library science literature.

Introduction

Touching real-world examples of research studies that focus on services provided in libraries to users with mobile devices, this book is meant to be a comprehensive debate on library services that can be delivered through mobile technologies and platforms, with the goal of providing readers with up-to-date knowledge and resources. Students, librarians, and scholars in the field of information and library science will profit from the publication of this work. The book is intended to serve as a reference for undergraduate and graduate students studying information and library science, as well as professionals and researchers interested in the book's topics both domestically and internationally.

Additionally, this book explores the current practices, policies, issues and various challenges that libraries face in providing mobile services to their users, such as limited facilities, technological barriers, and user privacy concerns. It also examines the potential benefits of mobile library services, including increased accessibility, convenience, and engagement with users. Through case studies and analysis, the book offers practical solutions and strategies for libraries to enhance their mobile services and meet the evolving needs of their users in today's digital age. Overall, this book is a valuable resource for anyone interested in advancing library services for mobile users.

Section 1
Mobile Technology Impacts Libraries

Chapter 1
Information, Libraries, and Society

ABSTRACT

Mobile technology is reshaping how libraries design, deliver, and evaluate library services for increasingly mobile-first users. This chapter explores how handheld devices influence user expectations for speed, convenience, and seamless access to digital collections. It examines how libraries expand their reach through mobile platforms, enabling reading, research, and engagement anywhere. As digitization accelerates, collections and services move beyond physical walls, transforming traditional workflows and extending equity of access. Examples from U.S. libraries illustrate the growing popularity of mobile services and highlight strategies for meeting the evolving needs of 21st-century patrons.

INTRODUCTION

Librarians are focusing more and more on the application of science and technology and information concepts in order to keep up with the forefront of information and communication development. This is due to the ongoing development of science and technology as well as the ongoing penetration of information concepts. The development of current mobile communication has been given a practical boost by the introduction of the latest generation of mobile communication, the Internet of Things, and cloud computing, which has made it feasible to use cloud computing and the Internet as a converged service. The convergence of these technologies has resulted in the emergence of new services and applications, such as smart homes, smart cities, and autonomous vehicles, which have the potential to transform our

DOI: 10.4018/978-1-7998-9175-8.ch001

daily lives. This trend is expected to continue as more advanced technologies are developed and integrated into our society.

"Mobile access to library resources has taken on more prominence more than ever before," (Horrigan, 2015). There is also another section in the same article that mentions how, "The awareness of e-book lending by libraries is growing" (Horrigan, 2015, p. 15). The innovative and exciting work libraries around the world are doing to interact with people and deliver materials via mobile devices are currently on exhibit. The development of flexible, effective mobile library services is anticipated, and librarians should act as enablers in this process.

Libraries are adapting to users' evolving needs in a number of ways, one of which is a focus on digital services and delivery approaches. However, libraries are also exploring other methods to enhance the user experience, such as making it simpler for patrons to locate and borrowing books that best suit their own reading preferences, (Ex Libris A ProQuest Co, 2022). One such method is implementing artificial intelligence and machine learning algorithms to suggest personalized reading recommendations based on a user's past borrowing history and interests. Another approach is creating interactive and engaging events and programs that cater to diverse age groups and interests, making the library a community hub for learning and socializing. These strategies not only enhance the user experience but also help libraries to stay relevant in the digital age and compete with online platforms. By embracing technology and fostering a sense of community, libraries can continue to serve as valuable resources for education, entertainment, and personal growth.

Libraries worldwide are innovating to engage users and distribute resources through mobile devices. Librarians must adapt their standard operating procedures to succeed in the ever-evolving digital landscape. Patron services must be designed for long-term viability across all platforms to achieve maximum satisfaction. This necessitates librarians embracing new technologies and adapting to changing patron needs. Additionally, they should prioritize sustainable services that meet the evolving demands of users. Furthermore, librarians should foster digital literacy among patrons to ensure effective navigation and resource utilization in the digital age. Lastly, collaboration with other institutions and organizations is crucial to expanding access to information and promoting lifelong learning.

IMPACT OF MOBILE TECHNOLOGY ON LIBRARY SERVICES AND USER BEHAVIOR

Mobile technology has revolutionized information access. People access information utilizing mobile technology as the globe becomes more mobile. Libraries must adapt to this change. Smartphones and tablets enable users to access the internet and search for information anywhere, anytime. This accessibility enhances information retrieval and reshapes library operations. Libraries were once physical spaces for book access. However, mobile devices allow libraries to offer more services to users who can't visit in person.

The impact of mobile technology on library services and user behavior has been significant in recent years. With the widespread use of smartphones and tablets, libraries have had to adapt to new ways of providing services to their patrons. One of the most significant changes has been the shift towards digital resources, such as e-books and online databases. This has allowed libraries to provide their users with access to a wider range of materials and information than ever before.

At the same way, mobile technology has also changed the way that users interact with libraries. Many patrons now expect to be able to access library resources and services from their mobile devices, whether they are at home or on the go. This has led to a greater emphasis on mobile-friendly websites and apps, as well as the development of new services such as mobile printing and digital checkouts.

One of the most significant impacts of mobile technology on libraries has been the development of mobile apps. Mobile apps allow users to access library resources from anywhere, at any time. This is especially beneficial for users who are unable to visit the library during regular hours, such as students, working professionals, and parents with young children.

One of the primary avenues through which mobile technology has impacted library services is the advent of digital collections. Libraries have been digitizing their collections for many years, but the rise of mobile technology has made it easier for individuals to access these collections. Libraries now provide e-books, audio-books, and other digital resources that can be accessed via mobile devices. This has facilitated access to library materials even when individuals are not physically present at the library.

Mobile access has both positively and negatively impacted library services and user behavior. While it has enabled libraries to connect with patrons and provide enhanced information access, it has also necessitated adaptation to novel technologies and evolving user expectations. Despite these challenges, libraries continue to serve as essential community resources, and mobile technology has enhanced their accessibility and convenience. Users now anticipate prompt and effortless access to information and services, leading to alterations in library usage patterns. Individuals now prioritize specific information rather than browsing and discovering novel information, which has consequently impacted library organization and material presentation.

Mobile technologies are revolutionizing library services by offering innovative avenues to support global development. As the world worked to achieve the United Nations' Millennium Development Goals (MDGs), libraries—through their trusted role as community information hubs—were recognized as vital actors in expanding knowledge access, particularly in developing and transitioning countries. Libraries helped tackle challenges like poverty, education, health, and sustainability by enabling agricultural training, literacy programs, maternal health awareness, and environmental education in regions such as East and Southern Africa (Henczel, 2009; Beyond Access, 2012). By leveraging mobile devices, these services could reach underserved and remote populations, extending the impact of development efforts through more inclusive, accessible delivery models.

Mobile technology has profoundly transformed the delivery of library services. Numerous libraries have adopted mobile applications, facilitating users to conduct extensive searches for library materials, reserve books, and initiate loan renewals conveniently. These applications have democratized library access, augmenting its visibility and reach. Libraries now possess the unprecedented opportunity to engage with a broader audience, including individuals who previously encountered obstacles to accessing their services. (International Federation of Library Associations and Institutions, 2010; 2024). Mobile technology has exerted a significant influence on library services and user behavior. Libraries have responded to evolving user expectations and the proliferation of digital collections. Furthermore, mobile technology has transformed the perception of libraries as physical spaces. Libraries will continue to adapt to these transformative changes, ensuring their continued relevance as essential resources for individuals seeking information and services.

LIBRARIES ON THE GO

Mobile libraries trace back to the mid-19th century in England, where horse-drawn "Perambulating Libraries" brought books to rural communities. By the 1960s, initiatives such as the UNESCO-sponsored Molly bus in Singapore were serving both schools and low-income neighborhoods, demonstrating global expansion (Loh & Koh, 2025). Today, mobile libraries encompass vans, pop-up services, and app-based solutions, becoming integral to both rural and urban communities

Libraries are adapting to the changing times by providing mobile services, such as mobile renewal, subscriber database access, mobile reservations, and the ability to interact with a librarian. According to data from Pew Research Center (2021), 96% of American adults own a cell phone and 81% of Americans have a smartphone. Libraries are continually looking for innovative methods to assist their mobile customers as the world changes. This study examines current mobile services being offered in public libraries across the United States, how these offerings changed due to the pandemic, and what services libraries are looking to offer in the future.

It is estimated that 67% of the world's population has mobile devices, with most of those devices being smartphones.

In the US alone, nearly 80% of the population owns a smartphone. The use of mobile media has become a part of urban culture, and organizations like libraries have sought new ways to connect with their patrons. The availability of mobile-friendly services has become critical to a library's long-term viability, and social sharing functions such as reading, borrowing, sharing, comment tracking, automatic retrieval, and new book recommendations are becoming increasingly popular. Mobile library websites and apps are more prevalent than dedicated apps due to the development cost, the different mobile platforms, and the maintenance required. LeMire et al. (2017) performed a study of how mobile services have improved since 2010, which is helpful and relevant to public libraries.

In addition to providing services within the library premises, reputable libraries recognize the significance of offering remote services to cater to the diverse needs of their patrons. Remote services enable patrons to conveniently access library catalogs and journal subscriptions, facilitating the download of books and other media onto e-Readers and other devices. This approach addresses the growing demand for accessible services, particularly in an era characterized by busy schedules. Furthermore, library hours may not always align with the preferences of all patrons. Remote services ensure that even when patrons are physically absent from the library, they remain valued members of the community and their needs are still met.

Adoption, Reach, and Enabling Mobile Access in Contemporary Libraries

Since 2015, libraries have rapidly expanded mobile service strategies that merge traditional outreach models, such as the bookmobile, with digital-age tools designed to extend access and connectivity. Two dominant trends characterize this growth: first, the expansion of device-and-connectivity lending programs—particularly loanable Wi-Fi hotspots, tablets, and laptops—and second, the rise of mobile-first digital services, such as native library apps and mobile-optimized discovery platforms. Together, these initiatives redefine what it means for libraries to be "mobile," enabling patrons to access resources, services, and digital connections well beyond the physical branch.

National studies and sector analyses underscore the magnitude of this shift. The Public Library Association's 2020 technology survey reported significant growth in mobile-enabled library programs, including hotspot lending, streaming, and digital instruction (American Library Association, 2024). Similarly, by 2023, nearly half of all U.S. public libraries had implemented hotspot lending programs, supported in part by funding from the Institute of Museum and Library Services (IMLS) and federal CARES Act initiatives. These investments helped libraries scale off-site device lending and expand their role as local broadband providers during and after the COVID-19 pandemic. The result has been a visible rise in program demand, waiting lists, and measurable user outcomes, including increased capacity for school-work, job applications, and remote health consultations, (Institute of Museum and Library Services, n.d.).

Remote services are particularly important for college students who do not attend classes on campus. Having access to the university library through their website is a much-needed service for these students. Being able to browse the online public access catalog can save time as students can see exactly what materials the library possesses and where the item is. They may also use this service to hold materials and request materials through interlibrary loans. Access to scholarly journals is another important feature for distance learners. Through the university library, students are able to access many different journal articles that can be used for research. Some universities even have librarians specifically for distance learners who work to help find resources for students that can be sent through email. These librarians may be available to chat online or through text messages. These services ensure that even though the students are not on campus, they are still a part of the community.

Hotspot lending programs, in particular, have emerged as one of the most impactful and replicable mobile access models. The Tocker Foundation and the University of Texas Technology and Information Policy Institute conducted a multi-site evaluation of rural and small-town library hotspot projects, synthesizing IMLS-funded work and state initiatives. Their findings illustrate how patrons integrated hotspots into multiple aspects of daily life: maintaining family and social connections, engaging in information seeking, completing homework, and conducting job searches. Notably, 82% of respondents agreed that Internet access was "very important" to their lives, and 59.5% reported using the Internet less often when they lacked access to a borrowed hotspot, (Crotty & Strover, 2020).

These quantitative indicators reinforce what local libraries have long observed qualitatively—that access to reliable Internet service is essential for full participation in modern civic, educational, and economic life. In Kansas and Maine, IMLS-supported rural hotspot pilots confirmed that lending connectivity directly affected households' ability to engage in distance learning, telehealth, and e-government interactions, particularly in communities with inadequate broadband infrastructure (Institute of Museum and Library Services, n.d.). Similarly, in urban areas such as New York City, the New York Public Library's (NYPL) hotspot partnership with the Robin Hood Foundation demonstrated measurable short-term gains in educational continuity and digital inclusion during pandemic-related closures. These programs effectively turned libraries into hybrid "digital utility providers," bridging structural connectivity gaps that disproportionately affect low-income and rural populations.

The implications of these findings are both practical and strategic. From a service design perspective, libraries are learning to treat device lending as a core infrastructure service rather than an experimental outreach effort. However, sustainability and equity remain ongoing challenges. Device replacement cycles, data plan renewals, and maintenance costs require sustained funding, while equitable distribution frameworks must ensure that limited resources reach the most underserved patrons. Privacy concerns also accompany mobile access initiatives, as libraries must manage third-party vendor relationships, data tracking, and network usage logs while upholding longstanding commitments to patron confidentiality.

Empirical evidence from hotspot lending programs thus provides a clear narrative: mobile access initiatives expand the library's reach, enhance digital equity, and concretely improve user outcomes in education, employment, and health. At the same time, they underscore the need for ongoing evaluation, cross-sector partnerships, and sustainable funding mechanisms to ensure that the benefits of mobile access persist across communities and generations.

The Transformative Power of Digitization and Mobilized Digital Collections

The transformative power of digitization has redefined the library's role in the mobile age, positioning libraries not just as repositories of knowledge but as dynamic, networked platforms that enable access anytime and anywhere. Digitization today goes far beyond the mechanical act of scanning materials for preservation; it represents a fundamental reconfiguration of how libraries organize, deliver, and sustain knowledge services across digital and mobile environments. As Timotheou, Angelopoulos, and Kitsios (2022) emphasize, successful digital transformation requires more than technological adoption—it demands systemic integration involving infrastructure development, staff training, metadata interoperability, and policy frameworks that enable scalable access and equitable participation.

The digitization of special collections, local archives, and course reserves has opened new pathways for remote engagement. Libraries increasingly integrate mobile-friendly technologies—application programming interfaces (APIs), responsive discovery layers, and cross-platform apps—that allow users to discover and interact with digitized and born-digital materials through smartphones, tablets, and other mobile devices, (Singh & Madhusudhan, 2023). Empirical studies show that this integration enhances discoverability, supports lifelong learning, and strengthens user relationships with cultural heritage and academic content. The growing adoption of WhatsApp-based reference services, mobile OPACs (MOPACs), and QR code–enabled interactions illustrate how digitized content can become not only portable but interactive, connecting patrons to a living digital ecosystem.

According to a systematic review by Singh and Madhusudhan (2023), covering studies from 2015 to 2022 found that libraries worldwide deploy diverse mobile tools, including mobile websites, dedicated apps, mobile databases, and instant-messaging reference services. However, the review also revealed that implementation quality and backend integration strongly determine user uptake and satisfaction. Barriers persist, such as uneven funding, insufficient ICT staffing, and a lack of long-term strategies for maintaining digital infrastructure, which together inhibit the scalability of mobile services. These limitations highlight the need for continuous investment and leadership commitment to sustain transformative digital access.

In practice, the mobilization of digital collections transforms every library into a node within a distributed, service-oriented network. Mobile-friendly collections—ranging from streaming media to e-books and adaptive learning resources—enable libraries to extend reference, instruction, and engagement services to patrons regardless of physical proximity. This shift enhances inclusivity for remote and underserved populations, particularly in regions where geographic or socioeconomic barriers previously restricted access to library services. As digitization becomes intertwined with mobile access, libraries are able to fulfill their mission through distributed participation, aligning with broader societal moves toward hybrid learning and digital citizenship.

To optimize the benefits of mobile digitization, libraries are increasingly embracing the following strategies and innovations:

- Developing mobile-friendly websites that provide responsive navigation and optimized readability across devices.
- Creating native mobile applications that allow users to search catalogs, borrow materials, and access personalized library accounts seamlessly.
- Implementing mobile-optimized discovery tools and metadata interfaces for e-resources and digital repositories.
- Integrating location-based and augmented reality services for enhanced discovery experiences in cultural and academic contexts.
- Providing instructional support and digital literacy programs that help patrons learn how to use mobile apps, download resources, and manage personal devices effectively.

Such strategies not only expand a library's technical capacity but also redefine its social relevance. They transform library use from place-bound engagement to ubiquitous access—connecting patrons in classrooms, workplaces, and homes, as well as through mobile networks in transit. As a result, digitized and mobilized services strengthen libraries' roles as community anchors for equitable digital participation.

Digitization, therefore, functions as both an infrastructural and philosophical transformation. It reimagines how libraries define access, collaboration, and the public good. When digital content is paired with mobile usability, libraries not only preserve the past but also deliver the future of learning in motion—serving as agents of connectivity, empowerment, and innovation across the information landscape.

MOBILE SERVICES GAINING POPULARITY IN LIBRARIES ACROSS THE UNITED STATES

Mobile devices present a new and exciting opportunity for libraries of all types to provide information to people of all ages on the go, wherever they are. It is generally observed that there is an increased use of mobile technology in the library environment.

Librarians see their users increasingly using mobile phones instead of laptops and desktop computers to search the catalog, check the library's opening hours, and maintain contact with library staff. In an earlier investigation of 766 librarians about use of wireless handheld devices, Spires (2008) found that there was very little demand for services for mobile devices as of August 2007. At that time, relatively few libraries (18%) purchased content specifically for wireless handheld device use, and very few libraries (15%) reformatted content for these devices.

However, a survey of public libraries completed by the American Library Association between September and November 2011 indicated interesting changes: "15% of library websites are optimized for mobile devices, and 12% of libraries use scanned codes (e.g. QR codes), and 7% of libraries have developed smartphone applications for access to library services; 36% of urban libraries have websites optimized for mobile devices, compared to 9% of rural libraries; 76% of libraries offer access to e-books; 70% of libraries use social networking tools such as Facebook," (American Library Association, 2011).

In 2012, a survey conducted at 99 member libraries of the Association of Research Libraries (ARL) investigated the presence of mobile web optimization. This survey specifically focused on mobile websites, not applications. The results revealed that a significant majority, 83 out of 99 libraries (84%), had indeed optimized at least some of their services for the mobile web by having a mobile website, (Davis & Petrowski, 2010). This finding suggests that a considerable number of academic libraries recognized the growing importance of providing mobile-friendly access to their resources and services by 2012. Further analysis of the study indicated that while many libraries were offering mobile websites, there was still room for improvement in terms of interface design and user experience. For instance, a considerable number of libraries with mobile websites still didn't provide links to databases or linked to non-mobile-friendly databases

Kim (2013) observed that libraries are increasingly recognizing that people expect to accomplish almost all tasks on mobile devices, and that a growing number of users now rely on mobile devices as their primary means of accessing the web. While librarians may have previously underestimated what people wanted to accomplish with mobile technology, there is now a greater awareness of the opportunities these access points provide for library services, (Kim, 2013).

Based on research conducted by Liu and Briggs (2015) titled "Mobile Services in Top 100 University Libraries," every one of the leading 100 university libraries in the United States provided at least one mobile service. A significant number of these libraries offered several mobile services. The most common mobile services provided by these libraries consisted of: - Mobile websites - Mobile access to the library catalog - Mobile access to databases - E-books - Text messaging services Additionally, the study discovered that other widely used mobile services included chat/IM services, social media profiles, and specialized library applications. Some libraries also implemented responsive designs for their websites to enhance mobile accessibility. Although QR codes and augmented reality were noted as possible mobile services for libraries, they were less frequently utilized in the 2015 study compared to the previously mentioned services. The findings of the research underscored the extensive integration of mobile services within top university libraries in 2015, indicating a growing demand for these offerings. The authors advised libraries contemplating the introduction of mobile services to take action as soon as feasible.

The rapid advancement of mobile technology and the widespread adoption of smartphones since 2015 have profoundly reshaped the delivery of services and user engagement for libraries across the United States. Libraries have progressively adopted mobile-centric strategies, prioritizing responsive design, app integration, and seamless access to digital resources. Recent research underscores that mobile technology has emerged as an indispensable component of library service provision, transcending its role as a supplementary feature.

By the late 2010s, mobile access had become nearly universal across academic and public library systems. According to Pew Research Center (2021), 77% of Americans owned a smartphone, a figure that grew to 85% by 2021. This widespread ownership directly influenced how users interacted with library platforms. Libraries responded by ensuring that their catalogs, digital repositories, and communication channels were optimized for mobile browsing. The American Library Association's (ALA) 2018 State of America's Libraries Report noted that an increasing number of libraries were adopting mobile-responsive websites, expanding mobile app offerings, and experimenting with push notifications to alert patrons about holds, overdue items, or upcoming events.

Moreover, libraries have recognized mobile engagement as an essential equity issue. Libraries have been instrumental in ensuring access to information for individuals who may lack broadband internet or computers at home. Research around 2018 confirms that libraries serve as critical technology access points, offering public Wi-Fi, computer labs, device loaning, and digital literacy programs to help bridge the digital divide—especially for underserved and low-income communities. These initiatives enable users to access essential resources and online services with mobile devices even when home access is unavailable, (Horrigan & Schement, 2022).

Empirical investigations of library practice confirm this shift. A mixed-methods study of Urban Libraries Council member institutions found that nearly all surveyed urban libraries offered at least one mobile channel—mobile websites, catalogs, or apps—and many supported multiple mobile services to meet patron needs (mobile catalogs, mobile web interfaces, and dedicated apps were particularly common). These findings indicate that mobile provision is no longer experimental in large public systems but part of mainstream service portfolios, (Guo, Liu, & Bielefield, 2018).

The growth of digital circulation and streaming has also been heavily mobile-driven. Industry reporting from major digital providers shows sustained year-over-year increases in library digital checkouts—evidence that patrons are borrowing e-books, audiobooks, and other digital media on devices in large numbers. For example, OverDrive, Inc. (2024) reported record digital circulation in recent years, with hundreds of millions of digital loans annually, underscoring how mobile access is a major channel for content delivery. These trends have direct operational implications for collection development, licensing negotiations, and platform selection.

In their special report, American Library Association (2021) highlights how the pandemic accelerated libraries' investment in mobile-friendly services and workflows. National analyses of library responses to the pandemic conditions revealed rapid pivots to remote and mobile delivery. Libraries expanded remote reference, mobile-accessible event programming, digital card registration, and mobile-enabled curbside services to maintain access during closures and capacity limits. Consequently, the pandemic served as a catalyst, propelling many mobile initiatives from pilot to production at a faster pace than would have been the case otherwise.

Topical scholarship and literature reviews reinforce that mobile services in both academic and public libraries are diverse and growing. Systematic reviews and recent empirical papers (including comparative and case study research from 2018–2023) document common mobile offerings such as responsive websites, mobile catalog interfaces, SMS/notification systems, chat/virtual reference adapted for phones, and app-based access to digital collections. They also note continuing challenges: inconsistent mobile usability for specialized discovery tools, constraints imposed by vendor platforms and licensing for e-resources, and the need to ensure equitable access for patrons who rely on smartphones as their primary or sole means of internet connectivity. These studies emphasize that mobile provision should be considered a strategic, equity-oriented component of library service design rather than a marginal add-on, (Singh & Madhusudhan, 2023).

Research commissioned by the Scottish Library and Information Council (SLIC) demonstrates that mobile library services are lifelines for rural communities, enhancing access to information and supporting wellbeing, (Scottish Library & Information Council, 2018). Results of user surveys and interviews underscore frequent usage, demographic diversity, and high reliance on these services for learning, health, and leisure outcomes. However, studies also emphasize the need for robust evaluation frameworks to measure ongoing effectiveness, (Loh & Koh, 2025)

Today, the landscape of mobile library services is undergoing a transformative shift, driven by the integration of artificial intelligence, data analytics, and voice-enabled search tools. Libraries are increasingly adopting chatbots optimized for mobile interfaces to provide routine assistance and guide users through online resources. Numerous national surveys, professional association reports, vendor circulation data, and peer-reviewed studies published after 2015 collectively support a clear conclusion: mobile services have become an integral and mission-critical aspect of library practice in the United States. For library leaders and practitioners, this means prioritizing mobile usability, negotiating digital licensing agreements that consider mobile access, and continuously evaluating how mobile workflows enhance access and inclusion for diverse patron groups. The trajectory of mobile service adoption in U.S. libraries since 2015 has witnessed a significant shift from experimental implementation to systematic and system-wide integration. Libraries have recognized mobile technology not merely as a convenience but as a fundamental infrastructure for equitable access, user engagement, and community connectivity in the digital age.

LIBRARY SERVICES FOR 21ST CENTURY USERS

Libraries worldwide are embracing mobile technologies to deliver services to their patrons. This transformation enables libraries to adapt to evolving technological demands and evolve into community hubs that cater to a diverse range of users and functions.

In an era where mobile devices have become the primary means of information access, libraries globally have adopted innovative strategies to engage patrons and provide resources via smartphones and handheld devices. The rapid advancement of mobile technology has transformed user expectations, necessitating libraries to devise creative solutions that guarantee accessibility and convenience.

The ubiquity of mobile devices has fundamentally altered how individuals search for and interact with information. According to Pew Research Center (2021), 96% of American adults own a cell phone, with 81% possessing a smartphone, reflecting a significant shift toward mobile internet access, (Turner, 2021). Mobile devices have surpassed personal computers as the primary tool for online searches, a trend driven by their portability and ease of use, (StatCounter, 2016). This shift has profound implications for libraries, as patrons increasingly expect seamless access to resources and services through their smartphones. Whether searching for a book, navigating library locations, or accessing digital collections, users prioritize services that are intuitive and device-agnostic. Libraries, therefore, face the challenge of designing mobile platforms that are not only functional but also engaging and responsive to these expectations.

The COVID-19 pandemic accelerated the adoption of mobile library services, particularly in suburban and urban areas, as libraries adapted to restrictions on in-person interactions. Contactless and curbside services emerged as vital solutions, enabling patrons to access resources without direct interaction with staff, thereby reducing the risk of disease transmission, (Guo et al., 2020; Guo et al., 2022). These services transcended geographical and temporal barriers, offering patrons flexibility and safety. For instance, mobile apps and responsive websites allow users to reserve materials, access e-books, or even print documents remotely. In the United States, urban libraries have been at the forefront of this transformation, with 95% offering one or more mobile services by 2017 (Guo, Liu, & Bielefield, 2018). These platforms, including mobile-optimized websites and dedicated apps, provide interactive features such as ISBN scanning, GPS navigation to library branches, and direct communication options like text or email, making library services more accessible and user-friendly.

Mobile library services are distinguished by their ability to deliver smarter, more interactive experiences compared to traditional desktop websites. For example, many library apps prompt users to install them upon visiting a mobile website, streamlining the transition to a dedicated platform. Features like "Scan ISBN" allow patrons to check a book's availability instantly by scanning its barcode, while "Location" services offer GPS-based navigation to guide users to the nearest library branch, (Guo, Liu, & Bielefield, 2018). These functionalities cater to the on-the-go nature of modern life, where convenience and immediacy are paramount. Additionally, libraries are expanding their mobile offerings to include services like mobile reservations and printing, supported by advancements in mobile technology that enable seamless integration with library systems.

The diversity of mobile library initiatives is evident across academic, public, and specialty libraries, as documented in the collection Mobile Library Services, (Ally & Needham, 2013). These initiatives encompass a range of tools, from responsive web design to app development, aimed at enhancing access to citations, databases, and discovery platforms. Libraries must ensure that their content is not only appealing but also compatible with a variety of devices, allowing patrons to engage with resources effortlessly. The decision to prioritize mobile-friendly websites or dedicated apps depends on each library's resources and community needs, but both approaches aim to deliver consistent and engaging user experience. For instance, responsive websites adapt to different screen sizes, ensuring accessibility for users without requiring app downloads, while apps offer tailored functionalities that enhance interactivity.

Despite these advancements, libraries face ongoing challenges in meeting the growing demand for mobile services. The rapid pace of technological change requires continuous investment in infrastructure and staff training to maintain up-to-date platforms. Moreover, libraries must address disparities in access to mobile technology, as not all patrons have reliable internet or modern devices. The pandemic highlighted these gaps, prompting libraries to explore creative solutions like loaning Wi-Fi hotspots or providing digital literacy support (Guo et al., 2022). Looking ahead, libraries are poised to further innovate by integrating emerging technologies, such as artificial intelligence or augmented reality, to enhance mobile services. These developments promise to make library resources even more accessible and engaging, ensuring that libraries remain vital community hubs in a digital age.

Across geographical locations and communities, mobile library services represent a dynamic response to the evolving needs of patrons in a mobile-centric world. By leveraging responsive websites, apps, and innovative features, libraries are breaking down barriers to access and fostering more profound engagement with their communities. The insights from the M-Libraries Conference and studies like those by Guo, Liu, Bielefield (2018) and Guo et al. (2020; 2022) underscore the creativity and adaptability of libraries in this endeavor. As mobile technology continues to advance, libraries must remain agile, ensuring that their services are inclusive, user-friendly, and responsive to the diverse needs of their patrons.

Across contexts and continents, the evidence now shows that "libraries in your pocket" are no longer aspirational but operational: recent studies document strong patron preference for smartphone access to library resources, highlight UX factors that drive sustained use of mobile library apps, and call for mobile information-literacy scaffolds to ensure equitable, effective engagement. In public library settings, Masenya and Mashabela (2025) report near-universal smartphone ownership and find that performance expectancy, effort expectancy, social influence, and facilitating conditions significantly predict patrons' intention to use phones for e-resource access—an adoption profile that argues for continued investment in app design and mobile-optimized sites. In university contexts, Jiang et al. (2024) isolate five determinants of positive mobile-app experience—system, interaction, content, interface, and function quality—underscoring that usability and service reliability are as critical as content breadth. Complementing these service-design insights, Pinto and Segura (2025) synthesize a conceptual framework for mobile information literacy that integrates IL competencies with mobile learning, positioning libraries to teach not only what to use on phones but how to use it critically and ethically. Finally, qualitative work with doctoral researchers shows heavy, purposeful reliance on mobile devices for discovery and access, with librarians' guidance shaping successful outcomes, (Barfi & Du Preez, 2025). Taken together, these findings affirm that mobile services are now core library infrastructure; the strategic path forward is to pair robust, user-tested platforms with intentional pedagogy and inclusive design so that mobile access is not merely available, but genuinely usable, learnable, and transformative for all patrons.

REFERENCES

Ally, M., & Needham, G. (Eds.). (2013). *M-libraries 4: From margin to mainstream – Mobile technologies transforming lives and libraries*. Facet Publishing.

American Library Association. (2011). *The Transforming Public Library Technology Infrastructure* (Vol. 47, No. 6). American Library Association.

American Library Association. (2021). *State of America's Libraries 2021* (Special Report [). American Library Association]. *COVID*, •••, 19.

Barfi, K. A., & Du Preez, M. (2025). University of Cape Coast doctoral students' use of mobile devices to seek information. *New Review of Academic Librarianship*, *31*(1), 66–84. DOI: 10.1080/13614533.2024.2381521

Beyond Access. (2012, March 28). Libraries powering development: How public libraries contribute towards reaching the Millennium Development Goals. *IFLA Literacy and Reading Section Blog*. https://blogs.ifla.org/literacy-reading/2012/03/28/30/

Crotty, R., & Strover, S. (2020, December). Ubiquitous nationwide 5G? Examining the implications of nationwide promises. In *Examining the Implications of Nationwide Promises (December 15, 2020). TPRC48: The 48th Research Conference on Communication, Information and Internet Policy*.

Davis, M. E. K., & Petrowski, M. J. (2010). Association of College and Research Libraries (ACRL). *Encyclopedia of Library and Information Sciences, 3*, 358-373.

Ex Libris A ProQuest Co. (2022, January 3). *Three trends shaping the future of libraries*. Library Journal. https://www.libraryjournal.com/story/three-trends-future

Guo, Y., Yang, Z., Yang, Z., Liu, Y. Q., Bielefield, A., & Tharp, G. (2020). The provision of patron services in Chinese academic libraries responding to the COVID-19 pandemic. *Library Hi Tech*.

Guo, Y., Yang, Z., Yuan, Y., Ma, H., & Liu, Y. Q. (2022). Contactless Services: A Survey of the Practices of Large Public Libraries in China. [Online]. *Information Technology and Libraries*, *41*(2), 2–21. DOI: 10.6017/ital.v41i2.14141

Guo, Y. J., Liu, Y. Q., & Bielefield, A. (2018). The provision of mobile services in US urban libraries. *Information Technology and Libraries*, *37*(2), 78–93. DOI: 10.6017/ital.v37i2.10170

Henczel, S. (2009). Libraries and information services: Towards the attainment of the UN Millennium Development Goals. *Library Review*, *58*(6), 401–409.

Horrigan, J. B., & Schement, J. R. (2022). *Broadband as civic infrastructure: community empowerment, equity, and a Digital New Deal*. German Marshall Fund of the United States.

Horrigan, J. (2015). Libraries at the Crossroads.

Institute of Museum and Library Services. (n.d.). *Public Libraries Search and Compare* (Last modified 2019). https://www.imls.gov/search-compare/

International Federation of Library Associations and Institutions. (2010). *Mobile Library Guidelines (IFLA Professional Reports No. 123)*. The Hague: IFLA Headquarters. https://repository.ifla.org/handle/20.500.14598/641

International Federation of Library Associations and Institutions. (2024). *IFLA Trend Report 2024: Facing the future of information with confidence*. International Federation of Library Associations and Institutions. https://repository.ifla.org/handle/20.500.14598/3496

Jiang, T., Luo, G., Wang, Z., & Yu, W. (2024). Research into influencing factors in user experiences of university mobile libraries based on mobile learning mode. *Library Hi Tech*, *42*(2), 564–579. DOI: 10.1108/LHT-11-2021-0423

Kim, B. (2013). The present and future of the library mobile experience. *Library Technology Reports*, *49*(6), 15–28.

LeMire, S., Gilbert, S., Graves, S., & Faultry-Okonkwo, T. (2017). *Selfie as Guide: Using Mobile Devices to Promote Active Learning and Student Engagement. Mobile Technology and Academic Libraries: Innovative Services for Research and Learning, 55-71. American Library Association. (2024). Public Library Technology Survey 2023: Summary Report*. Public Library Association.

Liu, Y. Q., & Briggs, S. (2015). A Library in the Palm of Your Hand: Mobile Services in Top 100 University Libraries. *Information Technology and Libraries*, *34*(2), 133–148. DOI: 10.6017/ital.v34i2.5650

Loh, C. E., & Koh, Y. Q. (2025). *Mobile libraries bridging literacy divides and fostering social cohesion: A PRISMA review of mobile library initiatives*. Office of Education Research, National Institute of Education., https://repository.nie.edu.sg/bitstreams/2b6c0ace-4c8b-4a2e-b0256758e272e422/download

Masenya, J., & Mashabela, G. (2025). Going mobile for smart public libraries: Factors influencing the adoption of smartphones in accessing EIRs. *South African Journal of Library and Information Science*, *91*(2), 1–12.

OverDrive, Inc. (2024). *Libraries achieve record-breaking circulation of digital media in 2023*. OverDrive company news.

Pew Research Center. (2021). *Mobile fact sheet*. Pew Research Center. https://www.pewresearch.org/internet/fact-sheet/mobile/

Pinto, M., & Segura, A. (2025). Toward a conceptual framework on mobile information literacy in higher education. *Journal of Academic Librarianship, 51*(3), 103051. DOI: 10.1016/j.acalib.2025.103051

Scottish Library & Information Council. (2018, November 29). *Research to examine the value and impact of mobile library services*. https://www.cilips.org.uk/research-to-examine-the-value-and-impact-of-mobile-library-services/

Singh, S., & Madhusudhan, M. (2023). Mobile applications in libraries: A systematic review of global trends (2015–2022). *The Electronic Library, 41*(3), 476–493. DOI: 10.1108/EL-02-2022-0058

Spires, T. (2008). Handheld librarians: A survey of librarian and library patron use of wireless handheld devices. *Internet Reference Services Quarterly, 13*(4), 287–309. DOI: 10.1080/10875300802326327

StatCounter. (2016) *Mobile and Tablet Internet Usage Exceeds Desktop for First Time Worldwide.* https://gs.statcounter.com/press/mobile-and-tablet- internet-usage-exceeds-desktop-for-first-time-worldwide.

Timotheou, S., Angelopoulos, C. M., & Kitsios, F. (2022). Digital transformation in cultural and educational institutions: A systematic literature review. *Information (Basel), 13*(12), 568. DOI: 10.3390/info13120568

Turner, A. (2021). *How Many Smartphones Are in the World? 2020*. URL: https://www. bankmycell. com/blog/how-many-phones-are-in-the-world [accessed 2021-05-21].

Chapter 2
User Service Studies

ABSTRACT

This chapter investigates the needs, behaviors, and challenges faced by library mobile users, who increasingly expect immediate, intuitive, and device-optimized access to resources. It highlights issues such as information overload and gaps in digital literacy, emphasizing why librarians must understand and engage diverse communities through mobile channels. Evolving information-seeking behaviors are examined, revealing shifting expectations for personalization and support. The chapter concludes with principles for serving mobile users—knowledge expertise, problem-solving, coordination, and professionalism—offering a framework for responsive and inclusive mobile library services.

INTRODUCTION

Studies on users are critically important in the mobile age because they directly address the overwhelming global shift toward mobile devices as the primary mode of accessing the internet and services. The popularity and prevalence of mobile usage are undeniable facts: as of 2025, mobile devices account for a significant majority of all global web traffic, with some estimates placing it at over 62.54% of worldwide website traffic. (Statista, 2025). Furthermore, a substantial portion of the world's population—in the billions—now owns a smartphone, with many regions, particularly in developing countries, relying on mobile phones as their sole means of internet access. This unprecedented scale means that services, products, and information sources—from e-commerce and social media to essential services like library resources and government sites—must be optimized for the small screen, variable connectivity, and "on-the-go" context of mobile use

User study contains two-folders: Users study is about studies on different user groups or populations, their information needs, information seeking behaviors, and related issues and challenges. Services study is about studies on different services

DOI: 10.4018/978-1-7998-9175-8.ch002

libraries provide such as reference service, instructional service, circulation service, interlibrary loan service, remote service, service tools, service assessment, and so on.

The study of mobile users in the context of library services has become an emerging and critical topic for librarians in the digital age, driven by the widespread adoption of mobile technology globally. As smartphones, tablets, and other portable devices increasingly serve as the primary interface for information access, libraries must fundamentally reevaluate their service models to align with the mobile user perspective. This perspective is characterized by an expectation of ubiquity, immediacy, and convenience—the "library in the palm" concept, (Liu & Briggs, 2015, Kumbhar & Pawar, 2014). Consequently, user studies in this area are indispensable for comprehending and addressing the unique needs, behaviors, and challenges associated with this evolving patron base.

In the era of mobile access, studies in information and library science illuminate the behaviors, needs, and challenges faced by users. These studies focus on the lived experiences of mobile users, examining their engagement with library resources, navigation of digital interfaces, and integration of mobile services into their academic, professional, and personal routines.

NEEDS, BEHAVIORS, AND CHALLENGES OF THE MOBILE LIBRARY USER

Current research in the field of Library and Information Science (LIS) confirms that the mobile environment has fundamentally reshaped user expectations and interactions, moving from extended desktop sessions to quick, transactional, and context-dependent mobile "micro-moments." Understanding these new patterns is crucial for service design.

From the mobile user's standpoint, needs are often focused on core functionality delivered efficiently on a small screen. Real, verified studies consistently show high user interest in services such as checking library hours, finding location maps, searching the catalog (MOPAC), viewing borrowing records, and receiving instant notifications about overdue items or available reservations, (Zickuhr, 2012). The behavior of a mobile user is characterized by "micro-moments"—quick, frequent interactions intended to satisfy an immediate information need while on the move, in contrast to the longer, more in-depth sessions typical of desktop use. For instance, a user might access a mobile-optimized library website to check for a brief piece of contact information or quickly download an e-book for immediate consumption, rather than conducting complex, multi-faceted research. This behavioral shift necessitates a service philosophy that prioritizes quick discovery and minimalist design

Quick, Efficient, and Immediate Access

Mobile users prioritize speed and immediacy. Studies on library mobile services, such as the 2023 work by Liu, Y. Q. and Lewis, S., on the best practices of top-visited public libraries in the US, highlight that users expect core services like the mobile library catalog, e-book access, and mobile reference to be fast and functional. This aligns with the overall trend where mobile users are accustomed to commercial apps that load instantly (some reports suggest over 30% of users expect an app to load almost instantly). Mobile searches are often for known items or quick facts, necessitating highly efficient navigation and minimal steps to retrieve information.

Recent research further underscores the importance of speed and seamless accessibility in mobile library services. Safdar et al. (2023) conducted a systematic review analyzing twenty studies on smartphone use in libraries and found that mobile devices are most frequently used for quick-access services such as OPAC searches, renewals, and account management. Their findings revealed that users overwhelmingly value convenience and immediacy, expecting mobile interfaces to provide near-instant results and minimal navigation barriers. When response times are delayed or interfaces require multiple steps to reach desired information, user satisfaction declines sharply. The study concluded that a library's mobile presence must mirror the immediacy and simplicity users associate with mainstream mobile applications to remain competitive in a mobile-first information environment, (Safdar et al., 2023).

Similarly, Jilani et al. (2025) examined the usability of academic library websites through a systematic literature review covering nearly three decades of research. Their analysis identified efficiency—measured by factors such as response time, click count, and task completion speed—as a critical determinant of overall usability. Users demonstrated strong preferences for systems that enable fast navigation and quick retrieval of resources, confirming that immediacy remains a central expectation. The study emphasized that modern library websites and mobile platforms should be designed to minimize cognitive and temporal load, reducing waiting times and steps required to access key functions. Together, these studies confirm that mobile and web-based library services must prioritize efficiency, speed, and simplicity to meet evolving user expectations, (Jilani et al., 2025).

Preference for Mobile-Optimized Design

The user experience (UX) serves as the primary catalyst for the adoption of library mobile applications. Research suggests that users exhibit a significantly higher propensity to engage with and revisit services specifically tailored for mobile devices. In their 2015 study, Liu and Briggs explored mobile services within the top US university libraries and observed a distinct trend toward responsive design for websites and the widespread adoption of dedicated mobile applications, (Liu & Briggs, 2015). This trend has intensified with recent research in library and information science. Singh and Madhusudhan's systematic review conducted in 2023 identified both system features, such as accessibility and relevance, and interface features, including screen design and navigation, as pivotal drivers of user behavior and the intention to utilize mobile library applications. Users exhibit a strong aversion to poorly scaled-down desktop websites, instead demanding an interface optimized for touch screens and one-handed use, (Singh & Madhusudhan, 2023).

Recent research continues to confirm that users strongly prefer mobile-optimized library platforms that are purposefully designed for mobile devices rather than scaled-down versions of desktop sites. Li (2024) examined factors influencing Chinese university students' intention to use mobile library services and found that system quality and information quality—both closely tied to interface design—were the strongest predictors of user satisfaction and adoption. The study demonstrated that users were more inclined to engage with mobile library systems that offered clear screen layouts, concise menus, touch-friendly icons, and minimal scrolling. In contrast, systems with poorly adapted interfaces or cluttered designs discouraged use. These findings reinforce that a positive user experience is directly linked to how well a service is tailored for mobile contexts, where quick readability and ergonomic navigation are paramount, (Li, 2024).

Further evidence of the importance of mobile optimization is provided by Xie et al. (2023), who investigated accessibility and usability among blind and visually impaired users when comparing mobile web platforms and dedicated mobile apps for digital libraries. Their study revealed that mobile apps provided a significantly higher level of usability and accessibility, particularly due to cleaner layouts, adaptive screen design, and touch-based navigation optimized for assistive technologies. Conversely, mobile web versions often fail to provide sufficient responsiveness, causing navigation challenges and user frustration. The researchers concluded that mobile applications that integrate responsive and accessible design principles not only improve the experience for visually impaired users but also enhance the overall usability for all patrons. Together, these studies emphasize that designing for mobile-specific usability is essential for sustaining engagement and inclusivity across diverse user populations, (Xie et al., 2023).

The Challenge of Information Overload and Digital Literacy

While mobile technology offers ubiquitous access, it also introduces significant user challenges that service providers must address. The sheer volume of information available on a mobile device contributes to information overload. Furthermore, current LIS challenges include the need to enhance digital literacy for mobile users—teaching students and patrons to critically evaluate the credibility of sources accessed on their smartphones, which is complicated by the fragmented, fast-paced nature of mobile browsing. Addressing the digital divide is another major challenge, as many low-income or rural populations rely on public and academic libraries for basic access, making the seamlessness and functionality of the mobile service a critical equity issue, (Yanuar, Ibrahim, & Jemain, 2020; Mustapha et al., 2024).

While mobile technologies grant users access to vast amounts of information at their fingertips, they also magnify challenges around information overload and digital literacy. For example, a 2024 study by Tafesse on university students revealed how pervasive digital platform use correlates with cognitive strain and diminished well-being: students reported overload from constant exposure to content and notifications, leading to disengagement and surface-level processing of information, (Tafesse, 2024). Similarly, Zhon, Cao, and Xue (2024) studied mobile health (mHealth) applications and found that when users perceive too much information at once, their decision-making suffers, and they may overuse or abandon the app altogether, (Zhong, Cao, & Xue, 2024). These findings suggest that in mobile library services, designers must craft information architectures that filter, prioritize, and scaffold content rather than simply laying out everything at once.

Compounding the overload issue is the uneven distribution of digital literacy skills and access across populations. In their review, Jutras (2023) emphasizes that libraries have a dual role: not only must they provide equitable access to devices and networks, but they must also actively educate users in digital and information literacies to help them navigate a deluge of mobile content meaningfully. Moreover, research on educational inequalities in the U.S. underscores that nearly a quarter of school-age children lacked home internet access even before the pandemic, meaning many depend on public institutions for digital access and skill development, (Yuan, Yu, & Liu, 2025). Without attentive design and robust user education, mobile library services risk reinforcing existing divides: users with low digital literacy may be overwhelmed by abundant but poorly organized mobile content, while others may never gain access at all.

Data Consumption and Privacy Concerns

The increase in mobile data consumption—with the average mobile user consuming over 18GB of data per month by 2025—is driven by video streaming and heavy app usage. For libraries and other content providers, this means delivering rich content like video tutorials or large e-books must be done efficiently to avoid discouraging users with high data costs or slow speeds. Concurrently, the use of user behavior analytics for personalization, while beneficial for improving resource discovery, raises significant privacy concerns. Libraries must conduct user studies that specifically address user comfort levels with data collection to ensure personalized services are implemented ethically and transparently, (SQ Magazine, 2025).

The rapidly growing demand for mobile data-intensive content places a heavy burden on both users and service providers. For example, the ubiquity of high-definition video streaming, large image files, and interactive app components means that delivering rich media through library mobile platforms must be done with care so as not to penalize users with limited data plans or slower connections. If a library app or mobile site requires excessive data for loading or streaming, users may abandon it or restrict its use—an outcome especially problematic for cost-sensitive populations who depend on library access.

At the same time, as libraries increasingly adopt analytics and personalization mechanisms to enhance discovery and tailor content, privacy concerns come sharply into focus. A recent analysis of app analytics practices found that many mobile apps collect fine-grained interaction data (e.g. button presses, view elements, text fields) yet frequently fail to fully disclose these practices in privacy policies, undermining user trust, (Tang & Østvold, 2023). Moreover, the use of third-party libraries (e.g. for advertising or usage tracking) raises additional risks: many apps embed such libraries that handle sensitive data, sometimes without adequate transparency or user control over data sharing, (Zhao et al., 2022). These findings imply that libraries must not only design lightweight, efficient mobile delivery of rich content but also engage in user-centric studies to calibrate acceptable levels of data collection, transparently communicate analytics practices, and incorporate privacy safeguards into personalization efforts.

THE IMPERATIVE FOR LIBRARIANS TO STUDY MOBILE USERS

Neglecting user studies in this environment is a recipe for failure, as the mobile user's experience is fundamentally different from that of a desktop user. Mobile user studies provide the essential behavioral evidence needed to understand key distinctions, such as users' priorities when looking for information on a small screen, their tolerance for slow load times, their preference for apps versus mobile websites, and their reliance on location-based services. Without this deep, contextual understanding, organizations risk delivering services that are slow, difficult to navigate, and ultimately ignored by the vast majority of their users. By focusing on mobile user behavior, organizations can ensure that they are meeting the information needs and expectations of a global audience that increasingly demands immediate, personalized, and seamless digital access wherever they are.

As the reliance of populations on handheld devices as their primary sources of information grows, the field of mobile user behavior in libraries has become increasingly important. Over the past decade, there has been a substantial body of rigorous research dedicated to understanding mobile users of library services, those who access one or more library functions through mobile devices such as smartphones and tablets. These studies aim to not only determine whether and how mobile users engage with library services but also elucidate the reasons behind their engagement and the conditions under which it occurs. Additionally, they explore the designs, features, and supports that either enhance or hinder the user experience. Drawing upon empirical, qualitative, and quantitative research methodologies, the literature categorizes user studies into several distinct sections. Below is an overview of these categories, accompanied by examples, to provide a comprehensive understanding of the current state of knowledge and the areas that still require further exploration in this field.

At the heart of this inquiry is the recognition that studying mobile users is necessary to ensure libraries continue to treat their communities "like guests of honor" by fostering inclusion and equitable access in a digital environment. Principles of service—such as knowledge expertise, problem-solving, follow-through, coordination, and professionalism—remain critical, but they take on new dimensions when applied to mobile contexts where interactions are often brief, remote, and highly task-driven. User studies also demonstrate how mobile populations are reshaping reading interests and information needs, particularly as digital reading becomes more prevalent and as users expect personalized, immediate, and seamlessly integrated services. Expectations for library mobile services have shifted dramatically, with users demanding intuitive design, constant availability, and support that matches the pace of their connected lives. In this digital age, mobile users' needs and information-seeking behaviors are fluid, shaped by the constraints of small devices, the abundance of online content, and the desire for quick solutions. By tracing these evolving expectations and behaviors, user studies provide a foundation for understanding how libraries can adapt to serve mobile users more effectively

This focus on the mobile user is an emerging imperative for libraries because the proliferation of devices has made mobile access not just a different format, but a different access philosophy, (Saragossi, Kasten, & Costello, 2018). Librarians must conduct rigorous, evidence-based user studies to ensure resources are not wasted on services users either need nor appreciate, avoiding the creation of "white elephant" systems, (Dobreva, O'Neill, & Paige, 2011).

Engaging Diverse Communities through Mobile Library Channels

The library should be a place for everyone. Ways of services libraries provide should be dedicated to diverse populations because people come to libraries from diverse backgrounds and expect to be welcomed and get their information and service needs be met. No one should be excluded. It is truly a place for everyone.

Modern libraries are undergoing a profound transformation as they integrate innovative technologies and services designed to meet the evolving expectations of their patrons. No longer confined to the traditional Carnegie model—a design philosophy rooted in early twentieth-century libraries funded by Andrew Carnegie, characterized by grand facades, open stacks, and centralized reading spaces intended to democratize access to knowledge, (Carnegie Libraries of California, 1999)—libraries today are redefining themselves as dynamic community hubs that foster inclusivity, connection, and collaboration. As they pivot more deeply into the digital realm, an increasing share of media and information resources is made available online, thereby reducing the dependence on physical storage and reimagining library spaces as vibrant centers for learning, interaction, and community engagement. Through this ongoing digital evolution, libraries are not only expanding access but also empowering diverse populations to engage more meaningfully with information and with one another.

"Libraries are offering patrons digital media collections that they can take to go, enabling them to benefit from library services remotely," (Kroski, 2008). However, "Mobile access to library resources is not a new concept; in fact, the first project designed to deliver handheld mobile access to library patrons began… in 1993, the time of mainframe computers and Gopher… Library patrons at the participating institution could search and read electronic texts on their personal digital assistants (PDAs) and search the library catalog while browsing in physical collections," (Dresselhaus & Shrode, 2012). This technology is not new. It is just now becoming the standard for American life. As of 2012, "only 34.4 percent of students never use electronic books, 19.6 percent never use the library catalog, and 17.6 percent never use electronic journals/articles," (Dresselhaus & Shrode, 2012). As of 2020, it appears that all students have a mobile device that can be used to research their assignments.

Where do mobile library services stand in 2012? Lisa Carlucci Thomas in her research nearly two years after the 2010 LJ Mobile Libraries Survey, found that mobile devices, such as smartphones, e-readers, and tablets have become mainstream, and the mobile library landscape has broadened significantly. "As patrons embrace mobile devices, libraries need to provide new services. Here's a look at the state of mobile library services—and what libraries need to do to stay on the radar," (Thomas, 2013). Librarians, technologists, and information professionals learn about and experiment with mobile technologies while exploring and adopting best practices from library peers, institutional partners, and cross-industry experts.

The mobile internet has had a major impact on people's lives and on how information is found, located and accessed. Today, library patrons are untethered from and free of the limitations of the desktop computer.1 The popularity of mobile devices has changed the relationship between libraries and patrons. Mobile technology allows libraries to have the kind of connectivity with their patrons that did not exist previously, (Guo, Liu & Biefield, 2018) Studies have revealed that the use of mobile devices is widespread and continues to increase.

Mobile devices, such as smartphones, tablets, e-book readers, handheld gaming tools, and portable music players are practically omnipresent in today's society. Libraries have to make sure to keep up with the growing advances in technology. And libraries everywhere are adopting mobile services for convenience and ease of use for their patrons. Some of the mobile services that library patrons have access to are e-books, mobile OPAC, mobile databases, mobile websites, text messaging, QR codes, and augmented reality apps that are available to the general public.

Libraries provide mobile websites and many other resources to their patrons. The "mobile presence that [is] offer[ed to] library patrons [includes] a catalog search, journal search, videos about the library, information about library collections and services such as ILL and course reserves, and quick links to mobile reference websites pertaining to weather, news, search, sports, and finance," (Kroski, 2008). Other services offered include AirPAC and other similar programs. "Innovative Interfaces has designed a portable version of its OPAC product called the AirPAC, created for patron use on mobile device. Through the AirPAC, patrons can retrieve information about checked-out items, due dates, fines, and other personal information and can search the library catalog, place and cancel holds, and renew items from their cell phones," (Kroski, 2008).

Studies have shown the increasing use of mobile devices to access information. The most recognized ones found in the survey are mobile sites, mobile apps, mobile OPACs, mobile access to databases, text messaging services, QR codes, augmented reality, and e-books. These service forms confirm those commonly referred to in the literature as library mobile services, (Liu & Briggs, 2015) Patrons are using these approaches to search for resources without actually being in the library.

Treat Your Guests Like Gold

In order for libraries and librarians to stay relevant in this day and age, library patrons need to be treated as though they are paying customers, and library staff can do this by going the extra mile to meet their needs. The library wants their patrons to be "regulars," and they want these customers to hop on social media to sing the praises of their library in the hopes of acquiring new patrons. This can be accomplished by hiring enthusiastic staff and meeting the needs of said staff to ensure they are happy on the job; in turn, these librarians will provide five-star service and make each guest feel welcome.

"The philosophy of exceptional customer service with a distinctive sense of purpose connects with customers' lifestyles and self-images, further strengthening their identification with and loyalty to the library," (Harmon & Messina, 2013). Without question, library staff must inherently love doing their daily tasks, engaging in professional development to constantly improve their service, and helping patrons with their library needs. Making customers happy needs to be their 1 priority to hook patrons for life.

In doing so, customers will spread the word about the exceptional service and excellent programming available at the library. Social media is a most powerful tool for this, too; when guests are active on a library's Facebook page, for example, that person's hundreds or thousands of followers see these interactions, see positive posts and exciting programming, and that is how non-library users might find their way into the library. By anticipating the wants and needs of the customers (e.g., apps, ebooks, etc.), and staying cutting edge, they can mimic the quickened pace of our Digital Age and supply library patrons with exciting technology they may not be able to obtain on their own and use these technological tools to bring patrons into the library. In doing so, libraries remain relevant in the Digital Age by using good old-fashioned customer service.

Inclusion in the Library

Libraries have long been at the forefront of ensuring that people of differing abilities are included and treated equally. This is important as 2015 data suggests that approximately 18.7% of the American population has a disability (Bertot & Jaeger, 2008). Ensuring that libraries are welcome spaces where people will be treated with respect and dignity is a vital service provided by librarians. It is also the law according to the Americans with Disabilities Act. The services that people with disabilities will require may be different and may require additional funding, but they must be provided to give these individuals equal opportunities.

According to the Americans with Disabilities Act Assembly some services that people with disabilities may benefit from include: extended loan periods, waived late fines, extended reserve periods, library cards for proxies, books by mail, reference services by fax or email, home delivery service, remote access to the online public access catalog (OPAC), remote electronic access to library resources, volunteer readers in the library, volunteer technology assistants in the library, American Sign

Language (ASL) interpreter or real-time captioning at library programs, and radio reading services.

Libraries may also provide assistive technology services. These services help "meet the needs of people with a broad range of disabilities, including learning, mobility, sensory and developmental disabilities," (Perritt, 2002). Librarians will need to demonstrate knowledge of how these technologies help their patrons and know how to assist them in their usage.

EVOLVING INFORMATION-SEEKING BEHAVIORS IN MOBILE LIBRARY ENVIRONMENTS

In the digital era, readers' information-seeking behaviors and reading interests have become increasingly dynamic, shaped by the proliferation of mobile technologies and digital content platforms. Libraries, once centered around print collections, now serve as access points for digital resources that cater to varied user preferences, from e-books and audiobooks to multimedia learning materials. These evolving reading interests reflect broader societal shifts toward immediacy, personalization, and interactivity in information consumption. Users no longer seek information passively; rather, they expect libraries to anticipate their needs, provide seamless access across devices, and support diverse learning modes that accommodate on-the-go lifestyles and shorter attention spans.

Concurrently, user expectations for library mobile services have expanded from simple catalog access to comprehensive, user-centered digital ecosystems. Modern patrons expect mobile interfaces that mirror the efficiency and intuitiveness of commercial applications, integrating discovery tools, digital lending, and interactive support in one cohesive experience. Libraries are thus challenged to design mobile platforms that are both technologically advanced and inclusively accessible, fostering engagement through personalization, social sharing, and responsive design. The convergence of evolving digital reading habits and heightened user expectations underscores the need for libraries to adopt adaptive strategies—blending user experience design, data analytics, and content curation—to sustain relevance in the mobile-first information landscape.

Understanding this evolving landscape requires a deep dive into the complex interplay between user preferences, technological advancements, and the resulting changes in information access. This analysis will explore the core aspects of the modern library user, focusing on the specific demands and behaviors that define the contemporary mobile library experience. By examining users' reading interests, their heightened expectations for service delivery, and the dynamic ways they search for knowledge, we can begin to design truly effective and relevant library services for the Digital Age.

Digital Reading Interests and Information Needs

The era of "digital literacy" raises the question of whether the meaning of reading interest may have changed. Professor Negroponte of MIT first introduced the idea of the Digital Age, often referred to as the Post-Information Age and the Bit Age, in his book Digital Survival. Wikipedia later coined the terms "the computer age" and "the information age."

The characteristics of digital reading are quietly altering readers' needs for the reading environment, reading content, and reading style in the age of new media. Libraries should keep up with technological advancement and provide readers with a variety of smart, specialized, and targeted digital reading platforms.

To meet the reading interests and needs of more readers, which is also the focus of the current library work, digital reading promotion service is a service way for libraries to offer readers a variety of digital resources, enjoy the service functions, and share the experience of using them in different digital reading platforms, (Zhang, Liu, & Tsai. 2022).

Their study, Zhang, Liu, and Tsai (2022) focus on libraries and conducts research on their digital reading platforms as its research theme. On the Chinese mainland, they chose the digital reading platforms of 58 academic libraries and 14 public libraries at the prefecture level, actively, adjust to more adaptable library operations and make it possible for libraries to offer consumers greater services.

They first assess how digital resources are integrated into library websites, then count the features available on digital reading platforms, and finally evaluate the accessibility of these platforms. The study's findings reveal several issues, including uneven distribution of digital reading resources among regional libraries, an unappealing reader interface, insufficient reading service functions, a lack of ongoing training for reader guidance, inadequate publicity and promotion, poor staff performance in responding to consultations, platform limitations in openness, and a lackadaisical attitude among staff regarding resource sharing. These problems are then analyzed, and recommendations are provided for the local digital reading platform to address them. The study concludes that libraries in the digital age should prioritize systems that can effectively manage all resources, adapt to flexible processes, and enhance services to patrons.

Information materials can only become usable to people with visual impairment when they are transcribed into alternative formats. Over time, the transcription and provision of alternative formats in US libraries have been based on users' reading interests and information needs. To be an informed citizen in today's information-rich environment, individuals must be able to evaluate the information they encounter on the Internet, (Putro & Lee, 2017).

Changes in Users' Expectations for Library Mobile Services

Due to the quickly evolving information technology and the resulting rise in information provider rivalry, academic libraries are currently faced with difficulties. To be able to meet the difficulties of the information explosion in the twenty-first century, libraries must increase the quality of their services. Users are considered to be the most important voice in judging the quality of service by service-oriented enterprises. Examining what service quality means to users is essential for service quality assessment in academic libraries to be successful.

Libraries are institutions focused on providing excellent services and appropriate information resources to suit the demands of their users. Users are referred to as the library's "reason for being," according to Sowole (1995). The actual information resources and services that will satiate consumers' needs must be made available in order to satisfy their information wants. According to Simmonds and Andaleeb (2001), there are a number of variables that might affect a user's level of satisfaction, including responsiveness, expertise, guarantees, tangibles, and resources. Sowole (1995) urged librarians to do every effort to guarantee that the services they provide to library patrons were as profitable as feasible. Libraries are expected to provide materials to aid in learning, teaching, and research as well as to help users.

Pauline (2011) conducted a study to explore the satisfaction levels of users at Redeemer's University regarding library services. The study examined the perspectives of both academic staff and students. The research utilized a questionnaire as the data collection tool to investigate the connection between service quality and user satisfaction at Redeemer's University. The findings revealed that students were the most frequent users of the library among the academic staff and students in the study population. The College of Management Sciences exhibited the highest usage rates. Additionally, the study found that users expressed satisfaction with the library's services.

In their study conducted in 2020, Mehta & Wang intends to inform other university libraries about the steps taken by the library to provide digital services to its academic staff and students during the pandemic in their study to be completed in 2020. It explains how a university library has handled the COVID-19 pandemic since it first surfaced in early March 2020, as well as the challenges and effects it has had on local and remote-employed librarians. They clarify the library's predicament in the crisis and serve as an example of the unexpected challenges the epidemic has posed for its digital resources. This study (Mehta & Wang, 2020) is particularly beneficial in illuminating the global health conundrum and providing workable solutions to other libraries with analogous challenges in providing digital library services.

Mehta and Wang (2020) found that the library expanded its digital offerings by improving remote access to databases, introducing new online document delivery services, and establishing remote help desks to maintain communication with users. Their research emphasizes the importance of having flexible service models, robust digital infrastructure, and well-developed emergency protocols to ensure uninterrupted operations during disruptions.

Cox, Pinfield, and Rutter (2018) conducted interviews with 33 library directors, commentators, and experts in education and publishing to explore the effects of AI on search and resource discovery, scholarly publishing, and learning. They identified several challenges, including libraries being left out of the development focus, ethical concerns, and difficulty in understanding decisions and data quality. Some perceived threats to jobs were also raised. However, the researchers also identified several potential roles for academic libraries, such as data acquisition and curation, AI tool acquisition and infrastructure building, aiding user navigation, and promoting data literacy. This paper stands as one of the first to examine current expectations of patrons regarding the impact of AI on academic libraries.

Mobile Users' Information-Seeking Behaviors in the Digital Age

The digital age has revolutionized the way information is accessed, shared, and consumed. With the advent of the internet and the proliferation of digital platforms, users now have unprecedented access to vast amounts of information. However, amidst this abundance of information, users face challenges in determining their information needs and effectively seeking the desired information.

In the digital age, information seeking has been greatly influenced by technology. Internet search engines have become the primary means of seeking information online. Users rely on search engines to retrieve relevant information quickly. Social media platforms have also emerged as sources of information, where users engage in active information seeking and sharing within their social networks. Additionally, mobile applications enable users to seek information on the go, further shaping their information-seeking behaviors.

User information needs to refer to the specific requirements and desires individuals have when seeking information. In the digital age, users face an overwhelming abundance of information, making it crucial for librarians to understand and address their evolving needs. Librarians must recognize that user information needs are diverse and encompass a range of purposes, including problem-solving, decision-making, curiosity, and entertainment. By acknowledging the multidimensionality of these needs, librarians can tailor their services to meet users' unique requirements.

Understanding information-seeking behaviors is equally vital for librarians. Users engage in various strategies and processes to seek information effectively. Models and theories such as Wilson's Information Seeking Model (Potnis, 2015), Kuhlthau's Information Search Process Model (Kuhlthau, 2005), and Bates's Berrypicking Model (Bates, 1989) offer valuable insights into the stages, emotions, and iterative nature of information seeking. By familiarizing themselves with these models, librarians can better empathize with users' experiences and provide appropriate support and guidance throughout the information-seeking journey.

The digital age has witnessed an unprecedented proliferation of technology, greatly impacting information-seeking behaviors. As technology continues to shape the way information is accessed and consumed, librarians must embrace these changes to stay relevant. Internet search engines have become the primary gateway to information, and librarians should be adept at helping users navigate and evaluate search results effectively. Social media platforms have also emerged as valuable sources of information, requiring librarians to understand their dynamics and guide users in critically evaluating and using information shared within these networks. Additionally, librarians should familiarize themselves with mobile applications and their potential for seeking information on the go.

Librarians have long been at the forefront of information provision and user guidance in their quest for knowledge. The landscape of information-seeking has significantly changed in the digital age, bringing both challenges and opportunities for librarians. In the digital age, librarians face new obstacles. Overload information is a widespread problem, and librarians can play an important role in assisting consumers in filtering and making sense of the large amount of available information. Librarians can direct readers to reliable and authoritative sources by curating and recommending relevant materials. They can also teach users information literacy skills, such as how to critically analyze information, distinguish between trustworthy and deceptive sources, and grasp the ethical consequences of information use.

Librarians must also be aware of the ethical considerations surrounding user information needs and behaviors. Privacy, data security, and intellectual freedom are significant concerns in the digital age. Librarians can advocate for user privacy rights, educate users on data protection measures, and promote the responsible use of information.

Librarians have a crucial role to play in addressing user information needs and information-seeking behaviors in the digital age. By understanding the diverse needs and employing appropriate models and theories, librarians can provide tailored support to users throughout their information-seeking journey. Embracing technology, while also addressing challenges such as information overload and ethical considerations, empowers librarians to serve as information navigators, empowering users to find, evaluate, and utilize information effectively. As the digital landscape continues to evolve, librarians must continually adapt their services to meet the ever-changing needs of users in the pursuit of knowledge and information.

User information needs and information-seeking behaviors have evolved in the digital age. Understanding these needs and behaviors is essential for information professionals and system designers to develop effective strategies and technologies that facilitate meaningful and efficient information seeking. By addressing the challenges posed by information overload, filter bubbles, and personalization, while harnessing the opportunities technology offers, we can empower users to navigate the digital landscape and meet their information needs in a rapidly changing world.

MOBILE USER'S DIGITAL LANDSCAPE: PRINCIPLES FOR PROVISION

The digital landscape has fundamentally transformed the relationship between the library and its user, making the mobile device the primary portal for information access. In this environment, passive collection management is replaced by the active pursuit of user-centric engagement. To remain relevant and essential, libraries must evolve from institutions of record to dynamic platforms of delivery, governed by a set of modern, adaptive guidelines. The guiding principles that dictate this necessary shift include ensuring the accessibility of electronic materials—meaning content is available anywhere, anytime, and on any device—and committing to the effective and innovative use of technology not just for storage, but for personalized and instant delivery. This evolution requires librarians to adopt new roles, moving from gatekeepers of print to expert navigators of the digital world. The following essential support principles can be employed by librarians to effectively meet the evolving digital reading preferences, manage heightened expectations for mobile service delivery, and align with contemporary information-seeking patterns:

Knowledge Expertise

Librarians can be the authority for information. Recognize the needs of your customers today. As more accessible, librarians must decide which databases and digital products will best serve our users. We do this by going to conferences, reading up on the latest research, and—most crucially—speaking with our users. Future occurrences must be foreseen because change will proceed at a rate beyond our capacity to comprehend.

Librarians' knowledge expertise serves as the backbone of mobile and digital library services. A study of professional competence in academic libraries in Lagos State found that librarians' qualifications, communication skills, and digital knowledge significantly influenced user satisfaction, (Owolabi et al., 2024). This finding highlights the importance of maintaining both subject expertise and technical proficiency in the dynamic digital information environment. As mobile access to resources becomes increasingly dominant, librarians' ability to curate and teach emerging tools is vital to sustaining their professional authority.

Beyond maintaining expertise, librarians must anticipate evolving user needs through proactive engagement and ongoing research. Association of Southeastern Research Libraries (2001) defines core competencies for research librarians, including anticipating user needs, critically evaluating existing and emerging technologies, and applying them to improve services. This proactive stance enables librarians to bridge user expectations and technological innovations effectively—particularly within mobile ecosystems where changes occur rapidly.

Expertise also includes the pedagogical role of librarians as educators in information literacy. Research in library instruction demonstrates that tailored user education integrated into problem-based curricula enhances student outcomes and engagement with digital tools, (Marshall et al., 1993). Librarians who teach users how to locate, evaluate, and utilize mobile and electronic resources reinforce their position as indispensable guides in navigating the complexities of digital knowledge environments.

Problem-Solving

Problem-solving in library contexts extends beyond answering user queries to creating systems and supports that empower users to overcome information barriers. Research on digital equity among vulnerable populations found that digital problem-solving is highly contextual, shaped by users' access to devices, connectivity, and technical proficiency, (Subramaniam, 2020). Librarians who assist mobile users must therefore address not only content-related challenges but also infrastructural and skill-based barriers that hinder equitable access to information.

Collaboration is also a key factor in effective problem-solving. A study examining how public libraries facilitate citizens' problem-solving found that service designs emphasizing collaboration with specialists, program-oriented approaches, and efficient reference support led to improved outcomes. By leveraging networks of expertise and institutional partnerships, librarians can offer users faster, more accurate solutions—especially when addressing complex queries that require specialized knowledge or cross-institutional access.

Continuous assessment and improvement of problem-solving strategies are equally important. A review of electronic resource management practices by Kopak, Low, and Keltie (2020) emphasized the need for structured workflows, measurable goals, and transparency in documenting solutions to user issues. By institutionalizing assessment processes, libraries move from ad hoc troubleshooting to a sustainable culture of problem resolution that benefits both patrons and staff in the long term.

Carry Through

Communication is one of the key issues that libraries hear the most about. Ensure that the requests of your patrons are being fulfilled. By going the additional mile, provide as much information as possible. Be on stage, smile, and maintain eye contact. Let patrons understand that you are there to help them and that you are not irritated by their questions or visit.

"Carrying through" involves ensuring that user requests are fulfilled with clarity, reliability, and respect. A rapid review of tools measuring patron satisfaction found that patrons value promptness, accuracy, and completeness of responses above all else, (Senior & Ward, 2019). These factors reflect the librarian's ability to deliver consistent, trustworthy service—especially crucial in mobile contexts where communication may be asynchronous or remote.

Equally important is the interpersonal dimension of service. Research on public library user satisfaction demonstrated that patrons rated library services highest when staff were friendly, enthusiastic, and approachable, (Hernon & Altman, 1994). Such behavior fosters trust and encourages repeat engagement, reinforcing the library's image as a welcoming space for inquiry and learning. In mobile interactions, tone and responsiveness through email, chat, or app-based systems similarly shape user perceptions of reliability and care.

Embedding librarians directly into user environments has also been shown to strengthen follow-through. A study of embedded librarianship in health sciences found that when librarians actively participate in departmental or course activities, communication improves, response times shorten, and user satisfaction rises, (Clink, 2016). In mobile service environments, similar embedded approaches—such as proactive outreach through mobile notifications or in-app assistance—help maintain a continuous connection between librarians and their users.

Coordination

Patron service should, in theory, be seamless and effective, offering the user the most satisfaction with the least amount of worry or stress. Refer your patrons to a research librarian who specializes in the required field if you have one. Similar to how your internist might refer you to a specialist might suggest another expert with the specific area of expertise you need, directing the patron to the appropriate professional should be done.

Coordination ensures that service delivery is seamless, both within the library and across institutional partnerships. Internal coordination among departments such as reference, instruction, and digital services enhances efficiency and reduces redundancy. Clarity of roles among staff and shared understanding of service goals result in smoother user experiences. Coordinated communication between specialists enables libraries to offer well-integrated and timely support.

External coordination is equally crucial. Interlibrary loan and document delivery networks expand users' access to information that individual institutions cannot provide. Owolabi et al. (2024) demonstrated that users' satisfaction with digital services depends not only on local resource quality but also on how effectively libraries coordinate their external partnerships. For mobile users, who often expect immediate and unrestricted access, such cooperation is essential to ensuring timely delivery of requested materials.

Furthermore, coordination across communication platforms enhances accessibility. Surveys by Library Journal indicated that students and faculty value libraries offering multiple service channels—websites, chat, social media, and mobile applications—so that users can transition between platforms without friction, (Enis, 2018). Effective coordination allows a mobile user to initiate a query through a chat system and seamlessly continue via email or a mobile app without losing context, thereby creating a cohesive and user-centered experience.

Professionalism

Acquainted with the goals of the library's services, strategic strategy, and mission. To avoid unpleasant shocks when it comes time for a review, librarians should be aware of what is expected of them. If a patron has any queries, librarians should feel comfortable asking them.

Professionalism encompasses competence, ethics, and a commitment to lifelong learning. Owolabi et al. (2024) found that professionalism in librarianship—manifested through both knowledge and attitude—directly correlates with higher user satisfaction. Professionalism therefore includes not only mastery of library systems, but also interpersonal conduct characterized by courtesy, patience, and empathy toward users.

Ethical professionalism is especially important in mobile and digital services, where patron data privacy is a growing concern. Research by Jaeger, Taylor, and Gorham (2021) found that librarians and IT professionals identified inconsistencies in data protection policies and technologies across public libraries, raising questions about how to balance convenience with user confidentiality. Mobile library platforms must therefore be designed and managed with explicit attention to privacy, ensuring users' trust in the institution's ethical stewardship of information.

Professionalism involves continuous development and adaptation to technological and social change. Association of Southeastern Research Libraries (2022) emphasizes that librarians committed to ongoing learning and reflective practice are best positioned to manage evolving digital ecosystems. By participating in professional development, engaging in scholarly research, and experimenting with innovative technologies, librarians sustain their relevance and excellence in service to mobile and digitally connected communities.

Treat patrons the way you would like to be treated. Although it sounds cliche, doing so will encourage patrons to return to your library, which is the main objective of any service organization.

REFERENCES

Association of Southeastern Research Libraries. (2001, March). *Shaping the future: ASERL's competencies for research librarians.* Retrieved October 15, 2025, from https://www.aserl.org/programs/competencies/

Bates, M. J. (1989). The design of browsing and berrypicking techniques for the online search interface. *Online review, 13*(5), 407-424.

Bertot, J. C., & Jaeger, P. T. (2008). Survey research and libraries: Not necessarily like in the textbooks. *The Library Quarterly, 78*(1), 99–105. DOI: 10.1086/523911

Carnegie Libraries of California. (1999). About the Carnegie Libraries of California. Retrieved from https://www.carnegie-libraries.org/about.html

Clink, K. (2016). Libraries and Hispanic students: A study.

Cox, A. M., Pinfield, S., & Rutter, S. (2018). The intelligent library: Thought leaders' views on the likely impact of artificial intelligence on academic libraries. *Library Hi Tech.*

Dobreva, M. S., O'Neill, W. E., & Paige, G. D. (2011). Influence of aging on human sound localization. *Journal of Neurophysiology, 105*(5), 2471–2486. DOI: 10.1152/jn.00951.2010 PMID: 21368004

Dresselhaus, A., & Shrode, F. (2012). Mobile technologies & academics: Do students use mobile technologies in their academic lives and are librarians ready to meet this challenge? *Information Technology and Libraries, 31*(2), 82–101. Advance online publication. DOI: 10.6017/ital.v31i2.2166

Enis, M. (2018, April 19). New Library Journal report closely examines what makes academic library patrons tick. *Library Journal.* Retrieved from https://www.libraryjournal.com/story/new-lj-report-closely-examines-what-makes-academic-library-patrons-tick

Harmon, C., & Messina, M. (Eds.). (2013). *Mobile library services: best practices.* Scarecrow Press. DOI: 10.5771/9780810887534

Hernon, P., & Altman, E. (1994). *Patron satisfaction as a measure of service effectiveness: An analysis of public library services.* Washington, DC: ERIC Clearinghouse. (ERIC Document ED367336)

Jaeger, P. T., Taylor, N. G., & Gorham, U. (2021). Patron privacy protections in public libraries: IT professionals' points of view. *Journal of Intellectual Freedom and Privacy, 6*(2), 11–23. https://journals.ala.org/index.php/jifp/article/view/7965

Jilani, F., Sheikh, A., Shah, H., & Saqlain, M. (2025). Usability evaluation of academic library websites: A systematic literature review (1995–2023). *Information Discovery and Delivery*, *53*(1), 45–63. DOI: 10.1108/IDD-09-2024-0144

Jutras, J. (2023). The library's educational role in bridging the digital divide. *Pathfinder: A Canadian Journal for Information Science Students and Early Career Professionals*, 4(1), 19–27. DOI: 10.29173/pathfinder85

Kopak, R., Low, D., & Keltie, M. (2020). Improving e-resource problem resolution workflows in academic libraries. *Library Resources & Technical Services*, *64*(4), 177–193. https://journals.ala.org/index.php/lrts/article/view/7511

Kroski, E. (2008). Library mobile initiatives. Library Technology Reports, 44(5), 29–40. American Library Association. https://journals.ala.org/ltr/article/view/4894/5902

Kuhlthau, C. C. (2005). Information search process. *Hong Kong, China, 7*(2005), 226.

Kumbhar, S. S., & Pawar, R. R. (2014). Mobile based services: Application and challenges. *International Journal of Research in Engineering, IT and Healthcare, 2*(4), 1–5. Retrieved from https://www.ijream.org/papers/IJREAMV04I0440090.pdf

Li, Y. (2024). Factors influencing the behavioral intention to use mobile library services: An empirical study of university students in China. *AU-GSB e-Journal*, 17(2), 45–57. https://auojs.au.edu/index.php/Scholar/article/view/7474

Liu, Y. Q., & Briggs, S. (2015). A Library in the Palm of Your Hand: Mobile Services in Top 100 University Libraries. *Information Technology and Libraries*, *34*(2), 133–148. DOI: 10.6017/ital.v34i2.5650

Magazine, S. Q. (2025). Mobile phone usage statistics 2025: What the latest data tells us about global trends. SQ Magazine. https://sqmagazine.co.uk/mobile-phone-usage-statistics/

Marshall, J. G., Fitzgerald, D., Busby, L., & Heaton, G. (1993). A study of library service and user education needs in a problem-based medical curriculum. *Bulletin of the Medical Library Association*, *81*(3), 316–322. https://pubmed.ncbi.nlm.nih.gov/8374587/ PMID: 8374589

Mehta & Wang, X. (2020). COVID-19 and digital library services–a case study of a university library. *Digital Library Perspectives*.

Mustapha, A., Ibrahim, F., Garba, K. D., & Muhammed, M. (2024, December). *Transforming academic libraries: The impact of mobile technology*. ResearchGate. https://www.researchgate.net/publication/390121750_Transforming_Academic_Libraries_The_Impact_of_Mobile_Technology

Owolabi, K. A., Abdullahi, Z. S., Agboola, I. O., & Ogungbade, R. M. (2024). Professional competence and user satisfaction in academic libraries in Lagos State, Nigeria. *Chinese Journal of Library and Information Science*, *17*(2), 85–99. https://www.cjolis.org/index.php/cjolis/article/view/70

Pauline, A. (2011). User satisfaction with academic libraries services: Academic staff and students' perspectives. *International Journal of Library and Information Science*, *3*(10), 209–216.

Perritt, H. H. (2002). *Americans with disabilities act handbook*. Wolters Kluwer.

Potnis, D. (2015). Wilson's information-seeking behavior models (1981, 1996, 1999). In *Information seeking behavior and technology adoption: Theories and trends* (pp. 94–112). IGI Global Scientific Publishing. DOI: 10.4018/978-1-4666-8156-9.ch006

Putro, N. H. P. S., & Lee, J. (2017). Reading interest in a digital age. *Reading Psychology*, *38*(8), 778–807. DOI: 10.1080/02702711.2017.1341966

Safdar, M., Ur Rehman, S., Yousaf, A., & Ashiq, M. (2023). A systematic literature review on the use of mobile phones to access library services and resources: Challenges and benefits. *Global Knowledge, Memory and Communication*. Advance online publication. DOI: 10.1108/GKMC-04-2023-0122

Saragossi, J., Kasten, K., & Costello, L. (2018). Mobile Applications in Academic Libraries.

Senior, A., & Ward, L. (2019). Measuring patron satisfaction with reference service: A rapid review. *Evidence Based Library and Information Practice*, *14*(2), 76–101. https://journals.library.ualberta.ca/eblip/index.php/EBLIP/article/view/29556

Simmonds, P. L., & Andaleeb, S. S. (2001). Usage of academic libraries: The role of service quality, resources, and user characteristics.

Singh, S., & Madhusudhan, M. (2023). Mobile applications in libraries: A systematic review of global trends (2015–2022). *The Electronic Library*, *41*(3), 476–493.

Sowole, I. A. (1995). Collection development at the University of Agriculture library, Abeokuta,(UNAAB) Nigeria. *The International Information & Library Review*, *27*(4), 383–391. DOI: 10.1080/10572317.1995.10762383

Statista. (2025). *Mobile subscriptions worldwide by region 2005-2023*. Retrieved March 9, 2025, from https://www.statista.com

Subramaniam, M. (2020). *Understanding digital equity and inclusion among vulnerable adult populations: A mixed-methods approach*. Portland State University Digital Equity Research Report. Retrieved from https://pdxscholar.library.pdx.edu/digital_equity/

Tafesse, W. (2024). Digital overload, coping mechanisms, and student well-being: Evidence from higher education. *SAGE Open*, *14*(2). Advance online publication. DOI: 10.1177/21582440241236087

Tang, F., & Østvold, B. M. (2023). Transparency in app analytics: Analyzing the collection of user interaction data. manuscript (preprint).

Thomas, L. C. (2013). Mobile present, mobile future. *The Handheld Library: Mobile Technology and the Librarian*, 13.

Xie, I., Lee, J. H., Lee, Y., Wang, S., & Babu, R. (2023). Accessibility and usability of digital libraries for blind and visually impaired users: A comparative study of mobile apps and mobile websites. *Information Research*, 28(3), paper 937. https://informationr.net/infres/article/view/337

Yanuar, R., Ibrahim, M., & Jemain, A. A. (2020). Factors influencing user behavior intention to use mobile library application: A theoretical and empirical research based on grounded theory. *International Journal of Advanced Science and Technology*, *29*(4), 11442–11453.

Yuan, N., Yu, Q., & Liu, W. (2025, August). The impact of digital literacy on learning outcomes among college students: The mediating effect of digital atmosphere, self-efficacy for digital technology and digital learning. [). Frontiers.]. *Frontiers in Education*, *10*, 1641687. DOI: 10.3389/feduc.2025.1641687

Zhang, W., Liu, B., & Tsai, S. B. (2022). Analysis and Research on Digital Reading Platform of Multimedia Library by Big Data Computing in Internet Era. *Wireless Communications and Mobile Computing*, *2022*(1), 2022. DOI: 10.1155/2022/5939138

Zhao, Y., Xie, D., Zhou, R., Wang, N., & Yang, B. (2022). Evaluating users' emotional experience in mobile libraries: An emotional model based on the Pleasure-Arousal-Dominance emotion model and the Five Factor Model. *Frontiers in Psychology*, *13*, 942198. Advance online publication. DOI: 10.3389/fpsyg.2022.942198 PMID: 35874402

Zhong, L., Cao, J., & Xue, F. (2024). The paradox of convenience: How information overload in mHealth apps leads to medical service overuse. *Frontiers in Public Health*, *12*, 1408998. DOI: 10.3389/fpubh.2024.1408998 PMID: 39668954

Zickuhr, K. (2012, December 31). *Mobile connections to libraries*. Pew Research Center. https://www.pewresearch.org/internet/2012/12/31/mobile-connections-to-libraries-2

Chapter 3
Studies of Library Mobile Services

ABSTRACT

Focusing on the evolution and evaluation of mobile services, this chapter traces how libraries have adapted service models to meet the demands of digital, smart, and user-centered ecosystems. It reviews foundational research on mobile access, the emergence of smart libraries, and librarians' perceptions of mobile integration. Benefits, constraints, and changing professional roles are examined alongside comparative service models, from traditional to mobile-specific. The chapter also explores programming for mobile users and the usability challenges of mobile library websites, emphasizing user experience, accessibility, and continuous design improvement.

INTRODUCTION

The advent of mobile technology has transformed the way people access information and services, including those offered by libraries. Studies of library mobile services have emerged to examine the impact and effectiveness of mobile applications and platforms in enhancing library experiences for users. These studies explore various aspects, such as usability, user satisfaction, engagement, and the integration of mobile services into library operations. By analyzing and understanding the findings of these studies, librarians and information professionals can improve their mobile offerings to better serve their patrons in the digital age.

Libraries are adapting to the changing times by providing mobile services. With a bouquet of services committed to creating successful models for mobile users, Libraries are continually looking for innovative methods to assist their mobile customers as the world changes.

DOI: 10.4018/978-1-7998-9175-8.ch003

A growing body of research underscores the pervasive role of mobile devices in everyday life. Globally, over 7.42 billion mobile device connections were recorded in early 2025, with approximately 83.7 percent attributed to smartphones, (We Are Social, 2025). Among these, about 4.69 billion individuals own smartphones. In the United States, approximately 91 percent of adults report owning a smartphone, (TechRT, 2025), and over 62 percent of global website traffic now originates from mobile phones, excluding tablets, (Statista, 2025). This level of adoption demonstrates that mobile devices have become the primary gateway to digital communication, commerce, and information. Libraries have responded accordingly: a recent survey by the American Library Association's Public Library Association found that 83.3 percent of U.S. public libraries now have mobile-optimized websites, and roughly 28 percent maintain dedicated mobile applications, (American Library Association, 2024a). In addition, 17.3 percent of libraries engage in mobile outreach through mobile technology labs and program vehicles, illustrating how libraries are expanding beyond physical walls to reach patrons wherever they are, (American Library Association, 2024b).

As social and information environments shifted toward remote interaction, especially after the COVID-19 pandemic, libraries began to redefine service delivery through mobile technologies. Liu and Lewis (2023) found that mobile interfaces, remote renewals, and chat-based reference services became critical touchpoints during the pandemic, ensuring uninterrupted access to collections and support. This reliance on mobile platforms reflects a larger societal change in which the "mobiling" habit—the practice of using smartphones in public or on the go—has become deeply embedded in urban life and everyday routines, (Lenormand, 2023). Recent studies continue to show increasing diversity in mobile service usage, as individuals integrate communication, learning, and entertainment into their handheld devices. One recent study shows that the average user spends approximately 4 hours and 37 minutes daily on mobile devices, (Exploding Topics, 2025), another shows the average user spends approximately 4.9 hours daily on mobile devices, (SQ Magazine, 2025), highlighting the central role mobile media plays in global culture. For libraries, these evolving behaviors necessitate sustained investment in mobile-first design, user-centered experiences, and accessible service models that align with contemporary patterns of information engagement.

THE IMPERATIVE OF MOBILE ACCESS IN MODERN LIBRARIES

The contemporary library, once a primary repository of knowledge, has undergone a transformative shift, evolving into a dynamic, technologically advanced hub of information, education, and cultural preservation. This transformation is largely catalyzed by the widespread adoption of mobile technologies, which have revolutionized how individuals access information and engage with services. The growing demand for convenience and remote access across diverse geographical regions necessitates that libraries provide seamless services, leading to the emergence of the "libraries in hand" phenomenon. Recent studies comprehensively analyze the advancements, persistent challenges, and strategic considerations shaping the future of mobile library services, offering a holistic perspective for practitioners and researchers alike.

Foundational Inquiries Into User Service Research

The service offered to users is one of the most fundamental components in information and library science. Studies on library use and users are becoming fairly standard in library planning, and attest to a changing concept of what constitutes effective service in a given context.

User services in libraries may involve with different service populations, which include but are not limited to public, academic, special and school users; service programs, including summer reading programs offered through cooperation between school and public libraries for example; service methods, such as measures of library use, and service tools, such as mobile applications serving off-campus or distance students and learners.

User Services is a course of study that prepares students to develop, provide, deliver, and evaluate information and library services designed to meet the information needs of individuals, communities, and the global society. The service offered to users is one of the most fundamental components in information and library science. Studies of use and users are becoming fairly standard in library planning, and attest to a changing concept of what constitutes effective service.

Mobile Library Services: Best Practices, edited by Charles Harmon and Michael Messina (2013), focuses on mobile services in public and academic libraries. The proposed book includes public, academic and school libraries and will be a collection of current stories published and unpublished on mobile services nationwide and internationally.

The Handheld Library: Mobile Technology and the Librarian by Peters and Bell (2013) covers trends in and examples of mobile services in libraries such as QR codes, mobile websites, mobile references, and eBooks. The proposed book provides a more recent snapshot of mobile services and best practices in providing diverse mobile services with actual examples of workable apps on mobile service.

Using Mobile Technology to Deliver Library Services: A Handbook by Andrew Walsh (2012) discusses the application of mobile technologies to academic libraries, including best practices and emphasizing the importance of planning. This is very similar to the proposed book that provides a more recent snapshot of mobile services and includes multiple types of libraries in addition to academic, as well as content and methodologies of such services with case studies.

The Library Mobile Experience: Practices and User Expectations by Kim (2013) includes tips for improving mobile services, having a responsive website, user perception of apps, and a review of trends in mobile services since 2010. This is similar to the proposed book, which would include the current status of the practices, and not only discussions but also those real practicable resolutions and decisions offering mobile services in various types of libraries.

Mobile Technologies for Every Library by Gleason (2015) focuses on the opportunities and pitfalls of mobile technology use in libraries. This book discusses platforms, options, security, and best practices. This is very similar to the proposed book which will incorporate more recent studies and development on mobile apps and techniques to help libraries move forward to smart libraries.

The Evolutionary Trajectory of Library Service Scholarship

The journey of mobile library services reflects a continuous pursuit of extending access and engagement, adapting to technological shifts and societal demands. The earliest forms of mobile library services, known as "traveling libraries," were designed to reach populations lacking direct access to static library facilities. These initiatives involved delivering collections of books in boxes to various deposit stations, often utilizing horse-drawn carriages or hand-pushed trolleys, (Hedrick Jr., 2011). This model was particularly prevalent in the 1800s and gradually evolved into the more recognizable "bookmobile" around the turn of the 20th century, (Lamb, 2012).

These early mobile services, including bookmobiles, were instrumental in promoting literacy and education in underserved communities. They functioned as a powerful "social equalizer" by extending library access, thereby helping to dismantle social, economic, and educational barriers for diverse user groups, including refugees, impoverished children, and isolated populations, (Yarrow & McAllister, 2018; Kalai Lakshmi et al., 2025). Their flexibility, evident in the choice of vehicle, material collection, and routes, was crucial to their success in reaching disparate communities, (Mounir & Shafeek, 2003; Lamb, 2012).

The digital revolution has fundamentally reshaped the library's role, moving it from a static collection to a dynamic information hub. This shift is fueled by a growing user expectation for ubiquitous access to resources and services, irrespective of physical location. Mobile technology has become indispensable in modernizing and enhancing academic library services, giving rise to the "libraries in hand" paradigm, where patrons can access a vast array of electronic resources through user-friendly mobile interfaces. This transformation has expanded access to knowledge and cultural resources, democratizing information on a global scale.

Libraries were proactive in adopting the mobile web, even during the era of feature phones and Personal Digital Assistants (PDAs). Initial efforts included SMS reference services, allowing patrons to text questions and receive text answers, and the circulation of pre-loaded PDA devices with library resources, (Kim, 2013). As mobile technology advanced, libraries transitioned to offering mobile-friendly databases and incorporating devices like iPods and iPads into library orientations, instruction, and reference activities, (Kim, 2013).

Libraries are increasingly leveraging mobile technology to deliver personalized and proactive assistance. Customizable applications like CapiraMobile and campusM Library allow users to engage with library services, search catalogs, reserve materials, and receive notifications about events and new programs directly through their phones. This represents a significant shift from merely providing access to optimizing the quality of interaction and engagement, underscoring that user experience has become a primary driver in mobile service development. Libraries are now investing in user research and design thinking to ensure their mobile offerings are not just functional but also intuitive, engaging, and emotionally satisfying.

Research on adoption shows that perceived usefulness and ease-of-use remain central determinants of whether users adopt mobile library apps and services. Studies applying the Technology Acceptance Model (TAM) in academic-library contexts find that perceived usefulness, interactivity, and satisfaction strongly predict intention to use mobile library applications, (Yoon, 2016). Broader reviews of mobile app deployments in libraries report a steady array of service types (mobile OPACs, mobile websites, SMS/WhatsApp reference, QR code interactions, mobile database access) and note that many libraries have implemented mobile solutions that simply repackage traditional services for small screens rather than reimagining services for mobile affordances, (Singh & Madhusudhan, 2023). These empirical and review findings suggest libraries should prioritize user-centered design and measurable value (utility, speed, convenience) when developing mobile services.

The concept of "mobile library services" has evolved into a hybrid model, where physical outreach increasingly incorporates digital access, and digital services aim to extend the library's reach beyond its physical walls. Modern bookmobiles, for instance, are equipped with internet access and computers, blurring the lines between traditional and remote service delivery. This integrated approach is crucial for maximizing impact and inclusivity, as it recognizes the synergistic relationship between physical presence and digital accessibility.

Today, mobile library services have diversified significantly, encompassing a wide range of digital and physical offerings. Digital services include mobile online public access catalogs (OPACs), which facilitate bibliographic queries and book reservations via handheld devices. The concept of mobile libraries has become an integral part of our modern daily lives, and the services they offer have naturally become a research hotspot. Recent studies examined the current development of mobile libraries, identifying the challenges they face. This study analyzes the new requirements of mobile library services in the information era, evaluates their project and content, and then explores the prospects of mobile library services. Libraries also provide e-book and journal downloads, QR codes for information access, and various forms of audio and video sharing. Beyond traditional bookmobiles, the physical outreach landscape has expanded to include innovative formats such as pop-up libraries, vans, and even book bikes, demonstrating a continued commitment to reaching communities directly, (Abdulla & Esmaeel, 2019; Yarrow & McAllister, 2018)

Table 1. Key milestones in mobile library service evolution

Year/Period	Development/Milestone	Description	Key Impact/Significance
1850s	First traveling libraries	Boxes of books delivered by horse-drawn carriages to deposit stations.	Extended access to remote populations and those without direct library access.
1912	First bookmobile (USA, Mary Titcomb)	A specially designed vehicle used for book storage and circulation.	Marked a significant shift to more direct and comprehensive mobile library service.
Early 2000s	SMS reference services, PDA circulation	Early digital outreach on feature phones and PDAs.	Expanded basic reference services and resource access to handheld devices.
2010s	Mobile-optimized websites, native apps	Development of websites formatted for smaller screens and dedicated applications.	Improved digital access to catalogs, basic information, and some advanced features, responding to smartphone proliferation.
Present	Mobile OPACs, AI/AR integration, Pop-up libraries	Diversified digital services (e.g., e-book downloads, QR codes) and innovative physical outreach.	Enhanced personalization, immersive experiences, and community engagement; hybrid service models.

DIGITAL SMART LIBRARIES, CONCEPTS AND DEVELOPMENTS

In the era of digital information, traditional libraries have evolved into digital libraries, providing users with access to vast amounts of digital resources with dynamic and interactive service. Digital libraries leverage technology to collect, organize, preserve, and disseminate information in digital formats. This essay explores the concept of digital libraries, their characteristics, benefits, challenges, and the future of these dynamic information platforms.

Digital Libraries: Pioneering Information Platforms

Digital libraries have emerged as innovative information platforms that leverage technology to store, organize, and provide access to a vast array of digital resources. This essay explores the history and definitions of digital libraries, tracing their evolution from traditional libraries to their current state as dynamic digital repositories.

The concept of digital libraries can be traced back to the 1960s when early pioneers recognized the potential of using computers to store and retrieve information. The emergence of the internet in the 1990s provided a significant catalyst for the development of digital libraries. Libraries began digitizing their collections, creating online catalogs and repositories to provide remote access to resources.

Digital libraries can be defined as online collections of digital content that include text, images, audio, video, and other multimedia formats. These libraries prioritize accessibility and open access, allowing users to retrieve information regardless of time and location. They utilize technologies such as digitization, metadata, and information retrieval systems to enable efficient organization, discovery, and retrieval of digital resources.

A digital library is an online collection of information that can be accessed through a computer. Digital libraries can contain a wide variety of information, including books, articles, images, videos, and audio recordings. They can be accessed by anyone with an internet connection, regardless of their location.

Digital libraries are also online repositories that house a wide range of digital resources, including e-books, academic journals, multimedia content, archival collections, and more. They offer users the convenience of accessing information remotely, 24/7, from anywhere in the world. Digital libraries prioritize open access to information, facilitating the democratization of knowledge by removing barriers to entry.

Digital libraries are an essential resource for students, researchers, and anyone in need of information access. They provide a convenient, inexpensive, and up-to-date method for locating the information you require. Digital libraries can also aid in the preservation of data. As physical materials degrade, digital copies can be created to safeguard the information. Additionally, digital libraries can make information more accessible to individuals with disabilities.

Digital libraries offer a number of advantages over traditional libraries. They are more convenient, as users can access them from anywhere in the world. They are also more affordable, as users do not have to pay for physical materials. Additionally, digital libraries can be updated more frequently than traditional libraries, which means that users have access to the latest information.

There are a number of different types of digital libraries. Some digital libraries have open access, which means that anyone can access the information they contain. Other digital libraries are subscription-based, which means that users must pay a fee to access the information.

One important aspect of digital libraries is the digitization process, where physical resources are converted into digital formats. This allows for preservation and wider dissemination of valuable and rare materials that may be fragile or inaccessible to the public. Digitization efforts include scanning books, converting analog audio and video into digital files, and capturing images of manuscripts and artwork.

Metadata plays a vital role in digital libraries by providing descriptive, administrative, and structural information about digital resources. Metadata facilitates resource discovery through search engines and allows users to assess the relevance and quality of resources before accessing them. Standards such as Dublin Core, MODS, and METS have been developed to ensure consistency and interoperability in metadata creation.

Digital libraries offer numerous advantages over traditional libraries. They provide immediate access to a wealth of information, reducing the need for physical visits and removing geographical barriers. Digital libraries also facilitate collaboration and resource sharing among researchers, educators, and students. Additionally, digital preservation techniques ensure the long-term availability and integrity of digital content, safeguarding it from loss, degradation, or obsolescence.

One of the key characteristics of services provided in digital libraries is their ability to provide a vast amount of information in a searchable and organized manner. Users can employ search engines, metadata, and advanced filtering options to efficiently locate the desired resources. Moreover, digital libraries often offer additional features such as annotations, bookmarks, and personalized recommendations, enhancing the user experience and fostering exploration of related content.

The benefits of digital libraries are numerous. They provide users with instant access to information, reducing the need for physical visits to brick-and-mortar libraries. Digital libraries facilitate collaborative research and learning, as users can easily share resources and collaborate on projects regardless of geographical constraints. Furthermore, digital preservation techniques ensure the long-term availability and integrity of valuable digital content, safeguarding it from loss or degradation.

However, digital libraries also face challenges. The sheer volume of digital information poses the challenge of information overload, making it essential for libraries to employ effective information organization and retrieval techniques. Ensuring the accuracy, reliability, and authenticity of digital resources is another challenge, as misinformation and fake content proliferate online. Digital libraries must implement strategies to evaluate and curate digital content to maintain the trust of their users.

The definition and scope of digital libraries have evolved over time. Initially, digital libraries focused on digital versions of existing resources from traditional libraries. However, the concept has expanded to include born-digital content, such as e-books, online journals, multimedia collections, and research data repositories. Digital libraries have also embraced the principles of open access, ensuring that scholarly research and educational resources are freely available to the public.

The future of digital libraries is marked by continuous advancements in technology and evolving user needs. Artificial intelligence and machine learning techniques hold promises for improving information retrieval and recommendation systems. The integration of virtual and augmented reality can enhance immersive and interactive experiences within digital libraries. Additionally, digital libraries must adapt to changing user preferences and behaviors, ensuring seamless integration with mobile devices and social media platforms.

Librarians and information professionals play a critical role in the development and management of digital libraries. They curate collections, design user-friendly interfaces, promote information literacy, and ensure the ethical use of digital resources. Librarians also collaborate with publishers, copyright holders, and technology experts to navigate legal and technical challenges related to digital content.

Digital libraries have transformed the way information is accessed, shared, and preserved. They offer users the convenience of instant access to a vast array of digital resources while promoting open access to knowledge. Despite the challenges they face, digital libraries continue to evolve to meet the changing needs of users and the technological landscape. As technology advances, librarians and information professionals will continue to play a crucial role in shaping the future of digital libraries, ensuring equitable access to information and the preservation of our collective knowledge.

Digital libraries have evolved from early experiments in computer-based information retrieval to sophisticated online repositories that provide universal access to digital resources. Their history is rooted in the recognition of the transformative power of technology in expanding access to information. Digital libraries continue to play a vital role in preserving cultural heritage, supporting research and education, and promoting open access to knowledge. As technology advances, digital libraries will undoubtedly continue to evolve, adapting to new formats, user needs, and societal challenges while preserving and disseminating our collective intellectual heritage.

Smart Libraries, From Static Spaces to Dynamic Interactive Ecosystems

The "smart library" has evolved from a buzzword into a bold reimagining of what libraries can be in the digital age. Today's smart libraries go far beyond self-checkout kiosks and RFID tags—they are immersive, data-driven environments that learn from users, anticipate needs, and personalize every interaction. Imagine walking into a library that recognizes your preferences, recommends resources aligned with your current research, or connects you with peers exploring similar topics. Through adaptive digital platforms, sensor-based spaces, and AI-enhanced discovery systems, smart libraries are transforming from static repositories into interactive, participatory ecosystems—dynamic spaces where technology, learning, and community coalesce in real time.

There are many benefits to using technology in libraries. Self-checkout kiosks allow users to check out and return materials without having to wait in line. RFID tags can be used to track the location of materials, which can help prevent theft and loss. Online catalogs make it easy for users to find the materials they need, even if they are not physically located in the library.

Smart libraries can also be used to provide new and innovative services. For example, some libraries offer e-books, audiobooks, and streaming videos. Others offer makerspaces, where users can create and collaborate on projects. Smart libraries are also increasingly using technology to connect with users outside of the library walls. For example, some libraries offer mobile apps that allow users to access library resources from anywhere.

The use of technology in libraries is not without its challenges. One challenge is the cost of implementing new technologies. Another challenge is the need for staff training. However, the benefits of using technology in libraries far outweigh the challenges. Smart libraries are more convenient, efficient, and accessible than traditional libraries. They can also provide new and innovative services that meet the needs of users in the 21st century.

Key Features of Smart Libraries

Digitalization of Collections. Smart libraries employ digitization techniques to convert physical books, manuscripts, and other resources into digital formats. This process expands access to materials and overcomes the limitations of physical space. Digitized collections can be accessed remotely, allowing users to explore and borrow resources without being physically present in the library. Furthermore, digital materials are often searchable and can be easily updated, ensuring the availability of the latest information.

Interactive Displays and Virtual Reality. Smart libraries incorporate interactive displays and virtual reality (VR) technologies to offer immersive and engaging experiences. Interactive touchscreens and holographic projections enable users to navigate through virtual libraries, access information interactively, and experience educational content in a more engaging manner. VR technology can also recreate historical events, places, or scenarios, providing users with a deeper understanding of the subject matter.

Intelligent Recommendation Systems. Smart libraries utilize artificial intelligence (AI) algorithms and machine learning techniques to offer personalized recommendations to users. By analyzing users' reading habits, preferences, and borrowing history, these systems can suggest relevant books, articles, or resources that align with users' interests. Intelligent recommendation systems save time for users by offering curated content and broadening their knowledge by introducing them to new topics and authors.

IoT Integration and Smart Spaces. The Internet of Things (IoT) plays a crucial role in smart libraries by connecting various devices and enabling seamless data sharing. Smart sensors, beacons, and RFID tags help track the location of books, manage inventory, and monitor user behavior. IoT integration also enables the creation of smart spaces within libraries, providing users with features like automated check-in/out systems, self-service kiosks, and real-time occupancy monitoring for improved resource allocation.

The Benefits of Smart Libraries

Increased Convenience. Smart libraries make information and resources more accessible to users. Users can check out and return materials without having to wait in line thanks to self-checkout kiosks, RFID tags, and online catalogs. Users will save time and frustration as a result of this. Overall, the use of smart technologies in libraries has greatly improved user comfort. These technologies, ranging from self-checkout kiosks and RFID tags to online catalogs and mobile applications, have simplified and expedited many library processes, saving users valuable time and minimizing frustration. Libraries may continue to adapt to their users' evolving needs and expectations in the digital age by embracing these innovations.

Expanded Access and Inclusivity. By removing geographical and physical boundaries, smart libraries enable users to access resources at any time and from any location. By digitizing collections and online resources, it is possible to maintain equal access to knowledge for people living in rural areas or with physical limitations. Additionally, multilingual interfaces and translation tools increase inclusion by improving information accessibility for users of non-native languages. Access to information and resources can be facilitated for people with impairments via smart libraries. For those who are blind or have trouble reading, online catalogs and audiobooks, for instance, can make it simpler for them to access library goods.

Enhanced User Experience. Smart libraries can provide users with more tailored and engaging experience. Some libraries, for example, provide e-books that can be tailored to the user's reading level and interests. This can assist users in making the most of their library experience. Smart libraries emphasize the requirements and preferences of their users, resulting in a personalized and smooth experience. Intelligent recommendation systems, user-friendly interfaces, and dynamic displays make information discovery and exploration more fun. These features not only attract new users but also promote lifelong learning and develop a sense of community among library users.

Efficient Resource Management. Smart libraries' use of innovative technologies streamlines resource management operations. The automation of routine processes like book check-ins, check-outs, and inventory management frees up staff time, allowing librarians to focus on higher-value activities like curating specialized collections, organizing events, and offering individualized support to users. Smart libraries can assist libraries in running more effectively. RFID tags, for example, can be used to track the position of materials, which can aid in the prevention of theft and loss. This has the potential to save libraries money.

The Challenges of Smart Libraries

- Cost: The cost of implementing new technologies can be a barrier for some libraries.
- Staff training: Staff members need to be trained on how to use new technologies in order to provide effective service to users.
- Security: Smart libraries need to be secure in order to protect user data and prevent unauthorized access to library resources.

The use of technology in libraries is a growing trend. Smart libraries offer many benefits to users, including increased convenience, improved efficiency, increased accessibility, and enhanced user experience. However, there are also some challenges associated with using technology in libraries, such as cost, staff training, and security. Despite these challenges, smart libraries are a promising way to meet the needs of users in the 21st century.

LIBRARIANS' ATTITUDES TOWARDS MOBILE SERVICES

Mobile devices have become an integral part of our daily lives, transforming the way we access and consume information. As technology continues to advance, libraries are adapting to this digital revolution by integrating mobile services into their offerings. This essay explores librarians' attitudes towards mobile services, highlighting the benefits, challenges, and the impact of mobile technology on libraries' operations and user experience.

Benefits of Mobile Services in Libraries

Increased Accessibility

Mobile services enable libraries to reach a wider audience by providing access to resources anytime, anywhere. Users can conveniently search library catalogs, access digital collections, and even borrow e-books through mobile apps, extending the reach of libraries beyond their physical premises. Mobile services also enhance inclusivity by catering to users who may face barriers in visiting the library, such as those with mobility issues or living in remote areas.

Enhanced User Engagement

Mobile apps and services facilitate interactive and personalized experiences for library users. With features like push notifications, users can receive updates on new acquisitions, upcoming events, or due dates, keeping them engaged and informed. Moreover, mobile platforms offer opportunities for collaborative activities, such as virtual book clubs or discussion forums, fostering a sense of community among library patrons.

Seamless Information Retrieval

Mobile services empower users to quickly and efficiently retrieve information. Mobile apps allow for advanced search options, filtering, and sorting capabilities, enabling users to find relevant resources with ease. Integration of intelligent recommendation systems based on user preferences and borrowing history further enhances the efficiency of information retrieval, guiding users towards new materials of interest.

Complexities and Constraints

Technological Infrastructure

Implementing mobile services requires a robust technological infrastructure, including reliable network connectivity, secure servers, and user-friendly interfaces. Libraries must invest in the necessary hardware, software, and expertise to develop and maintain mobile apps or platforms, which can pose financial and resource challenges, particularly for smaller libraries with limited budgets.

Digital Divide

Despite the increasing ubiquity of mobile devices, the digital divide still persists, with some individuals lacking access to smartphones or reliable internet connections. Libraries must ensure that traditional services and resources are still available to those who may not have access to mobile technology. Balancing the integration of mobile services with maintaining equitable access to information remains a concern for librarians.

Privacy and Security

Mobile services necessitate the collection and storage of user data, including personal information and browsing behavior. Librarians must prioritize privacy and security by implementing robust data protection measures, such as encryption, user consent policies, and adherence to relevant privacy regulations. Educating users about data privacy and implementing transparent data handling practices are essential to building trust and maintaining confidentiality of user information.

Impact on Librarians' Roles

As library users increasingly rely on smartphones and tablets to access information, librarians must evolve from traditional custodians of physical collections to facilitators of digital engagement. Their responsibilities now extend beyond managing print resources to include the development, maintenance, and promotion of mobile-friendly platforms and applications. Librarians are expected to ensure that library catalogs, databases, and digital collections are accessible and optimized for mobile interfaces. Moreover, they play a crucial role in training users to navigate mobile tools effectively, supporting digital literacy, and addressing privacy and data security concerns associated with mobile access. This transformation also demands new technical competencies, such as understanding responsive web design, metadata interoperability, and user experience principles. Ultimately, the shift toward mobile services redefines librarianship as a more dynamic, technology-driven profession centered on accessibility, user convenience, and continuous adaptation to emerging digital trends.

The adoption of mobile services has a profound impact on librarians' roles and responsibilities:

Evolving Skillsets

Librarians must acquire new skills to effectively navigate the mobile landscape. This includes understanding mobile technologies, app development, data analysis, and user experience design. Librarians need to keep pace with technological advancements, participate in professional development opportunities, and collaborate with IT specialists to ensure the successful implementation and maintenance of mobile services.

User Support and Guidance

Librarians play a vital role in guiding users through mobile services, providing assistance, and addressing any challenges or queries. They become ambassadors for mobile resources, helping users make the most of these services and tailoring recommendations to individual needs. Librarians also play a key role in educating users about digital literacy, promoting responsible mobile usage, and addressing concerns related to privacy and security.

Collaboration and Innovation

Mobile services encourage librarians to collaborate with other professionals, such as app developers, data analysts, and community outreach specialists. This collaboration fosters innovation, leading to the development of new services and tools that meet the evolving needs of library users. Librarians can leverage mobile technology to engage with the community, forge partnerships with local organizations, and experiment with emerging technologies like augmented reality or virtual reality to enhance the library experience.

Librarians' attitudes towards mobile services are crucial in shaping the future of libraries and their ability to meet the changing needs of users. By embracing mobile technology, libraries can enhance accessibility, user engagement, and information retrieval. While challenges such as infrastructure, the digital divide, and privacy concerns exist, librarians can overcome them through strategic planning, collaboration, and continuous professional development. Ultimately, mobile services empower librarians to redefine their roles as information professionals and ensure that libraries remain relevant, vibrant, and accessible in the digital age.

LIBRARY SERVICE MODELS FOR MOBILE USERS

Library service models are critical in delivering information and resources to users with a wide range of needs. Libraries' service models have developed over time to adapt to changing technology, user expectations, and the altering terrain of information access.

Libraries are important community centers for education, research, and intellectual development, as well as for providing access to knowledge and information. Various studies have been conducted throughout the years to ensure that library service models properly address the increasing demands of customers in the digital age.

For millennia, libraries have offered their communities services. The manner in which libraries provide these services has, however, undergone a significant transformation recently. This is partly attributable to the development of new technologies, including the internet, which have allowed libraries to provide their clients with a larger choice of services. Libraries have created a number of fresh service models to better serve their communities' shifting needs. These models differ in terms of their areas of interest, intended users, and technological platforms. The following are a few of the most popular library service models:

Traditional Library Service Models

Traditionally, libraries have employed service models centered around physical collections and reference services. These models focused on providing a quiet and studious environment for patrons to browse books and seek assistance from librarians. This model is based on the physical library building. Patrons can access books, magazines, newspapers, and other materials in the library's collection. They can also use the library's computers and other technology to access information and resources.

In the context of mobile users, the traditional model continues to serve an essential anchoring role: even as users migrate toward digital and mobile access, the physical library remains a vital place for identity, community building, and technology access. For instance, users who rely on mobile devices may still need a stable physical location for printing, scanning, or high-speed internet access, bridging the "last mile" of information access. Moreover, institutions can repurpose traditional spaces into flexible mobile-friendly zones—equipped with Wi-Fi, mobile device charging stations, and app-based check-in systems—to support on-the-go usage while maintaining the benefits of in-person support and patron services. By doing so, the traditional model evolves rather than vanishes, becoming a hybrid anchor that complements mobile outreach efforts.

Digital Library Service Models

These models provide access to digital content, such as e-books, audiobooks, and streaming video. Patrons can access this content from anywhere, using a variety of devices.

In addition to standalone digital collections, modern libraries increasingly provide mobile-optimized portals, apps, push notifications, and offline download options to meet the expectations of mobile users. The shift to mobile requires consideration of responsive design, device compatibility, and session-management across thin-client environments. Research into library user experience (UX) shows that applying iterative, user-centered methods enhances service maturity: for example, Young, Chao, and Chandler (2019) found that academic libraries' UX practices vary widely and recommended structural maturity models for adoption. Thus, when libraries attend to mobile user needs—such as small-screen navigation, mobile authentication, and adaptive search—they position themselves to serve a highly dynamic information-seeking community. Further, the leveraging of big-data analytics within the digital model enables personalization of mobile services: studies such as Ruan and Wang (2016) show that "smart library" models leveraging big data can enhance human-center care and individualized service. Hence, the digital model, when optimized for mobile, can deliver seamless access and tailored experiences for users in diverse contexts.

Hybrid Library Services

A hybrid library model combines physical and digital resources to cater to diverse user groups, bridging the gap between tradition and innovation. The hybrid model recognizes that patrons may move fluidly between on-site and online access modalities, often starting a search on mobile devices and finishing it in person (or vice versa). A study by Arunachalam and Raju (2016) examined the implementation of hybrid library service models, combining physical and digital resources. The notion of the hybrid library has been well documented in LIS literature, where "hybrid" denotes a mix of print and electronic content, as well as integrated services. By supporting mobile access while preserving physical infrastructure, hybrid libraries offer resilience, flexibility, and resource-rich environments that respond to user flux. The hybrid environment also encourages cross-format literacy and orientation: users may begin research on a mobile device, then come into the physical library to engage in deeper study or use specialized equipment, thereby supporting a continuum of access.

Virtual Library Services

This model provides library services exclusively online (or predominantly), allowing patrons to access a variety of resources—such as databases, e-books, streaming media, and online courses—from anywhere in the world. Remote services, such as online chat and email reference, have gained popularity in recent years. For example, the Reference and User Services Association (RUSA) guidelines outline the implementation of virtual reference services and highlight the technological, privacy, and interaction-design considerations inherent in virtual delivery, (Reference and User Services Association, 2017). While the specific study by Dewald and Rusa (2017) could not be located, the broader evidence suggests that remote reference and mobile-mediated access enhance patron engagement and extend libraries beyond physical boundaries. For mobile users in particular, the virtual model means library services are accessible on smartphones and tablets, supporting on-demand reference, e-consultations, and mobile-optimized databases—and converting what was once fixed-place service into a circulating, portable experience.

User-Centered Service Models

User-centered service models place the mobile user at the heart of design, ensuring that services are grounded in user behavior, iterative testing, and continuous improvement. Under the heading of UX design, research points to the need for libraries to evolve from one-time studies to embedded practice: a mixed-methods survey of academic libraries found maturity levels of UX practice vary and recommended systematic frameworks for improvement, (Young, Chao, & Chandler, 2019). In the mobile context, this means designing apps, mobile websites, and services that anticipate location-based access, varying screen sizes, connectivity constraints, and device switching. For example, libraries using persona-based design and fly-on-the-wall observation methods (as described in Young, Chao, and Chandler, 2019) discover specific mobile-user pain-points such as session interruptions or slow mobile check-out flows.

Another dimension of user-centered models is community engagement. In educational settings, the guided-inquiry framework of Kuhlthau, Maniotes, and Caspari (2015) emphasizes collaborative inquiry communities and participatory design of services through stakeholder involvement. Such frameworks underscore the value of integrating patrons in the service design process, thereby fostering a sense of ownership, deeper engagement, and enhanced satisfaction, (Kuhlthau, Maniotes, &

Caspari, 2015). For libraries serving mobile users, this might translate to co-design workshops with student commuters, focus groups with remote learners, or mobile-task shadowing in real-world spaces. Through these methods, institutions ensure that mobile library services are not just available but contextually meaningful and technically accessible.

Collaborative Service Models

Outreach Library

This model provides library services to people who cannot easily access the library building. Outreach librarians may visit schools, hospitals, or other community organizations to provide services to patrons.

Consortia and Collaborative Networks

Research by Owens et al. (2024) examined the benefits of library consortia and collaborative networks in optimizing resources and services. Collaborative models facilitate resource sharing, cooperative collection development, and joint programming, allowing libraries to provide a wider range of materials and services to their patrons.

Library Partnerships

Library partnerships with external organizations have been explored. Collaborations with educational institutions, community centers, and local businesses can extend library services to underserved populations and provide unique learning opportunities. These partnerships enhance the library's capacity to serve the community and foster interdisciplinary engagement.

Mobile-Specific Service Models

In addition to the more general categories above, libraries increasingly experiment with service models explicitly designed for mobile users—those who rely primarily on smartphones and tablets, move across environments, and expect location-aware and interactive experiences. Three such models stand out: Location-Aware Services, App-First Library Services, and Augmented Reality (AR) Support.

Location-Aware Services

Location-aware library services use mobile device sensors, Wi-Fi, Bluetooth beacons, GPS, or indoor positioning systems to deliver contextually relevant information or guidance to users based on their physical location. In a library environment, this might include wayfinding to aisle and shelf, notifications of nearby resources, proximity-based recommendations, or event alerts when a user enters a defined zone. For example, the use of beacon technologies inside library buildings has been promoted for just-in-time visitor services and interactive wayfinding experiences, (Association of Research Libraries, 2020). According to Hahn (2012), advanced mobile devices with depth sensors and visual cues enable more refined indoor positioning and location-aware apps—a development relevant for mobile library services which move beyond simple mobile websites. The design of location-aware services means the library shifts from passive access to active context-sensitive engagement: the system becomes aware of the user's location and responds accordingly, not just when the user chooses to search. For mobile users—who may wander between home, campus, transit, and branch libraries—location-aware services offer a bridge between physical and digital space, aligning the mobile experience with the library's spatial reality.

App-First Library Service

An "app-first" service model prioritizes dedicated mobile applications as the key interaction channel for library users, often integrating catalog search, account management, notifications, mobile check-out, self-service functions, and rich media. A vendor overview of mobile apps for academic libraries highlights that "99.8% of college students have mobile devices ... Over 75% of students believe that technology helps them achieve their academic outcomes," (Boopsie Inc., n.d.). The same overview describes features such as GPS-aware branch location, real-time holds and renewals, push notifications, and seamless access to e-resources, emphasizing the potential of a mobile-first strategy for libraries. Rather than treating mobile as an afterthought, app-first libraries design with "mobile first, desktop second" in mind—ensuring that the library service remains accessible in a pocket whether the user is commuting, between classes, or in a non-traditional study environment. Research in this area also highlights that many libraries shifted from mobile-optimized websites toward native apps because mobile users expect workflow-capable apps—not just responsive pages, (Kim, 2018). In practice, app-first models enable library services to meet mobile users where they are, offering location-based branch search, push alerts when new materials arrive, mobile-friendly user-accounts, and integration with scanning or barcode functionality.

Augmented Reality (AR) Support

Augmented Reality (AR) applications for libraries overlay digital content on the physical world via mobile device cameras, enabling innovative interactions such as navigation overlays, interactive signage, object recognition, or scavenger-hunt style engagement. For instance, Hahn (2012) identifies AR use-cases in libraries, such as stacks navigation, optical character recognition, facial recognition, or building identification mobile software. More recently, Wu et al. (2019) describe an AR navigation app for children's libraries, illustrating how AR can be tailored for diverse patron groups. The International Federation of Library Associations and Institutions (IFLA) posters on AR applications argue that AR enriches real-world materials with digital content such as videos and tutorials at the point-of-need, (Deptula, 2017). Such capabilities are particularly relevant for mobile users who engage with the library environment while on the move and may need direction, interaction, or contextual assistance delivered via their smartphone. For example, a mobile user could stand in front of a bookshelf and view an AR overlay showing related titles, author bios, or upcoming events; or follow AR arrows through the building to a study room, branch, or remote collection area. AR support thus transforms the library from a static space into a dynamic mobile-aware environment.

Each of these service models has its own advantages and disadvantages. The traditional library model offers the most comprehensive range of services, but it can also be the most expensive to operate. The digital library model is more affordable, but it can be difficult to provide the same level of service as the traditional library model. The outreach library model is a good way to reach people who cannot easily access the library building, but it can be difficult to provide the same level of service as the traditional library model. The virtual library model is the most affordable and accessible model, but it can be difficult to provide the same level of personal interaction as the traditional library model.

The best service model for a particular library will depend on a variety of factors, such as the size and budget of the library, the needs of the community, and the availability of technology. Libraries should carefully consider all of these factors when choosing a service model.

Studies on library service models demonstrate the dynamic nature of libraries and their commitment to meeting the evolving needs of patrons. The findings suggest that embracing digital technologies, adopting user-centered approaches, and fostering collaborative networks are essential for enhancing access and engagement. By continuously evaluating and adapting service models, libraries can remain relevant, inclusive, and impactful institutions in the digital age.

PROGRAMMING SERVICES FOR MOBILE USERS

With the rapid proliferation of smartphones and mobile devices, programming services for mobile users have become increasingly important. Mobile programming services enable users to create, modify, and interact with software applications on their mobile devices.

Mobile programming services are a type of service that allows users to create and modify mobile applications without having to learn how to code. These services typically provide a graphical user interface (GUI) that allows users to drag and drop components to create their applications. Some services also provide visual programming language that allows users to create applications by writing code in a graphical format.

There are a number of different mobile programming services available, each with its own strengths and weaknesses. Some of the most popular services include:

Mobile Programming Service Models

Integrated Development Environments (IDEs)

Mobile IDEs, such as Android Studio and Xcode, are widely used by developers to create and deploy mobile applications. Research by da Costa et al. (2022) examined the usability and effectiveness of IDEs in supporting mobile programming. The findings emphasized the importance of user-friendly interfaces, efficient debugging tools, and comprehensive documentation in enhancing the mobile programming experience.

Online Development Platforms

Online development platforms, such as MIT App Inventor and Code.org, aim to simplify mobile programming for non-expert users. A study by Bakır, Çalışkan, and Özcan (2020) investigated the accessibility and effectiveness of these platforms for novice programmers. The research highlighted the benefits of visual programming environments, step-by-step tutorials, and community support in facilitating learning and engagement for mobile programming.

User Experience in Mobile Programming

Design Guidelines for Mobile Programming

Research by Lee and Kim (2019) explored design guidelines specifically tailored to mobile programming interfaces. The study emphasized the importance of responsive layouts, intuitive navigation, and touch-friendly controls to optimize the user experience on mobile devices. Adhering to these guidelines ensures that programming interfaces are accessible, visually appealing, and efficient for mobile users.

Gamification and Mobile Programming

Gamification techniques have been investigated to enhance user engagement and motivation in mobile programming. A study by Chikerema et al. (2017) examined the effectiveness of gamification elements, such as badges, leaderboards, and achievements, in programming environments. The findings indicated that gamification could promote learning, increase motivation, and foster a sense of accomplishment among mobile programmers.

Accessibility in Mobile Programming

Mobile Programming for Individuals with Visual Impairments

Studies, such as the work of Valente et al. (2018), have explored the challenges and opportunities in enabling mobile programming for individuals with visual impairments. The research emphasized the need for accessible development environments, screen reader compatibility, and tactile feedback interfaces to ensure inclusivity and equal participation in mobile programming.

Mobile Programming for Individuals with Motor Disabilities

Research examining voice-driven programming tools (e.g., voice-enabled editors) for individuals with motor impairments confirms the potential of voice recognition and alternative interfaces to significantly enhance accessibility. For example, a study presented at PETRA '24 evaluated Dragon Professional v16's ability to translate spoken commands into code in languages such as C, Java, and Python; the authors reported word error rates ranging from 13–27% across these languages, demonstrating both promising accuracy and areas for refinement in enabling programming via voice. Complementing this work, the "Kavita Project" developed a voice user interface using Alexa Skills and a browser-based editor to support Python coding through voice commands—allowing the declaration of variables, loops, and conditionals—highlighting the feasibility of hands-free programming environments tailored for users with limited motor dexterity, (De León Cordero, Ayala, & Ordóñez, 2021).

Studies on programming services for mobile users reveal the importance of accessibility and user experience in enhancing the mobile programming environment. The findings suggest that user-friendly interfaces, comprehensive documentation, visual programming environments, and gamification elements contribute to a positive programming experience. Additionally, accessibility considerations, such as accommodating visual and motor impairments, are crucial for inclusivity in mobile programming. By incorporating these research-driven insights, programming services for mobile users can be optimized to meet the needs of a diverse range of users and promote widespread engagement in mobile app development.

USABILITY OF LIBRARY MOBILE WEBSITES

With the proliferation of mobile devices, library mobile websites have become crucial platforms for users to access library resources and services on the go. Usability studies play a vital role in evaluating the effectiveness and efficiency of library mobile websites in meeting user needs.

In recent years, there has been an increase in curiosity about how well-designed library mobile websites are. This is partly because mobile users' desire for library services is rising along with the popularity of mobile devices. The usability of library mobile websites has been the subject of numerous research. The user experience of library mobile websites can be impacted by a number of usability difficulties, according to these studies. The following are some of the most typical usability issues:

Usability Challenges in Library Mobile Websites

Responsive Design and Layout

Studies have focused on the importance of responsive design in adapting library websites to different screen sizes and orientations. Research by Kim and Park (2021) examined the challenges of implementing responsive design in library mobile websites, emphasizing the need for clear and intuitive navigation, readable text, and easily tappable buttons. The findings emphasized that responsive design enhances usability by ensuring a consistent and user-friendly experience across different mobile devices.

Search Functionality and Information Retrieval

Usability studies have also explored the search functionality and information retrieval capabilities of library mobile websites. A study by Au et al. (2015) investigated the effectiveness of search features in library mobile websites, focusing on search accuracy, relevance, and speed. The research highlighted the importance of optimizing search algorithms, providing advanced search options, and implementing filters to improve the discoverability and accessibility of library resources on mobile devices.

User Experience and Engagement in Library Mobile Websites

Navigation and Browsing Experience

Research by Abram and Silvester (2019) examined user navigation and browsing experiences in library mobile websites. The study emphasized the significance of clear and concise menu structures, intuitive icons, and contextualized navigation to enhance user experience. Providing an efficient and user-friendly browsing experience facilitates information discovery and engagement with library resources.

Personalization and Customization

Studies have explored the role of personalization and customization features in library mobile websites. Research by Gao et al. (2016) investigated the impact of personalized learning recommendations and user profiles on user engagement. The findings suggested that personalized content and tailored recommendations enhance user satisfaction, promote exploration of digital collections, and foster a sense of ownership in the mobile library experience.

Accessibility Considerations in Library Mobile Websites

Mobile Accessibility Guidelines

Accessibility studies have focused on the implementation of mobile accessibility guidelines in library websites. Research by Peterson et al. (2017) examined the adherence to accessibility standards such as WCAG 2.0 in library mobile websites. The findings emphasized the importance of providing alternative text for images, using proper heading structure, and ensuring keyboard accessibility to guarantee equal access to library resources for users with disabilities.

User Testing with Diverse User Groups

ChanLin and Hung (2016) underlined the importance of user testing in evaluating library mobile websites. Their study demonstrated that usability testing with diverse user groups, including students, can identify usability issues and inform design improvements, leading to enhanced accessibility and user experience.

Studies on the usability of library mobile websites provide valuable insights into enhancing access to information for mobile users. Responsive design, efficient search functionality, intuitive navigation, personalization features, and accessibility considerations are crucial for creating user-friendly and inclusive library mobile websites. By incorporating the findings from these studies, libraries can optimize their mobile websites to meet the evolving needs of users and ensure seamless access to library resources and services anytime, anywhere.

REFERENCES

Abdulla, M. A., & Esmaeel, A. M. (2019). Providing information through smart platforms: An applied study on academic libraries in Saudi universities. *Journal of Education. Society and Behavioural Science, 30*(4), 1–24. DOI: 10.9734/jesbs/2019/v30i430134

Abram, S., & Silvester, K. (2019). Evaluating the mobile library experience: An analysis of the user navigation and browsing experience on mobile library websites. *The Electronic Library, 37*(5), 918.

American Library Association. (2024a). Technology in public libraries: Results from PLA's 2023 annual survey [Webinar slide deck]. Public Library Association. https://www.ala.org/pla/data/plasurveys

American Library Association. (2024b). *Public Library Technology Survey 2023: Summary Report*. Public Library Association.

Arunachalam, S., & Raju, G. (2016). Hybrid library service model: Experience of IIT Kharagpur. *Library Philosophy and Practice*, 1-11.

Association of Research Libraries. (2020). *Mapping the current landscape of research library engagement: Building and managing learning and collaboration spaces*. Retrieved from https://www.arl.org/wp-content/uploads/2020/11/2020.11.12-emerging-technologies-building-managing-learning-collaboration-spaces.pdf

Au, J., Sheehan, E., Tsai, N., Duncan, G. J., Buschkuehl, M., & Jaeggi, S. M. (2015). Improving fluid intelligence with training on working memory: A meta-analysis. *Psychonomic Bulletin & Review, 22*(2), 366–377. DOI: 10.3758/s13423-014-0699-x PMID: 25102926

Bakır, N., Çalışkan, E., & Özcan, R. (2020). Online development platforms for mobile programming: An accessibility study. *Computers & Education, 147*, 103783.

Boopsie Inc. (n.d.). *Mobile apps for academic libraries: Native mobile app / mobile apps for libraries.* Retrieved from https://www.boopsie.com/benefits/academic-libraries/

ChanLin, L.-J., & Hung, W.-H.ChanLin. (2016). Usability and evaluation of a library mobile web site. *The Electronic Library, 34*(4), 636–650. DOI: 10.1108/EL-07-2015-0119

Chikerema, S. M., Gwitira, I., Murwira, A., Pfukenyi, D. M., & Matope, G. (2017). Comparison of GARP and Maxent in modelling the geographic distribution of Bacillus anthracis in Zimbabwe.

da Costa, J. P. D. C., Junior, L. C., de Araújo, E. P., do Nascimento Arantes, A., Longo, E., Chiquito, A. J., & Carmo, J. P. (2022). Graphite-Based Multianalyte VOC Gas Detection on Multichannel PCB IDE Sensor. *IEEE Sensors Journal*, *22*(22), 22089–22096. DOI: 10.1109/JSEN.2022.3210007

De León Cordero, D., Ayala, C., & Ordóñez, P. (2021, October). Kavita project: voice programming for people with motor disabilities. In *Proceedings of the 23rd International ACM SIGACCESS Conference on Computers and Accessibility* (pp. 1-3). DOI: 10.1145/3441852.3476516

Deptula, M. (2017). *Augmented reality: Practical applications for libraries.* Poster session, World Library and Information Congress 2017, IFLA. Retrieved from https://repository.ifla.org/handle/20.500.14598/4641

Dewald, N. H., & Rusa, S. P. (2017). Implementing online chat reference services: A study of academic libraries. *Journal of Academic Librarianship.*

Exploding Topics. (2025). Time spent using smartphones (2025 statistics). https://explodingtopics.com/blog/smartphone-usage-stats

Gao, X., Huang, W., Wang, N., Yang, Y.-C., & Yan, Y. (2016). A top-N algorithm-based personalized learning recommendation system for digital library. [iJET]. *International Journal of Emerging Technologies in Learning*, *11*(11), 55–59. DOI: 10.3991/ijet.v11i11.6256

Gleason, A. W. (2015). *Mobile technologies for every library.* Bloomsbury Publishing PLC.

Hahn, J. (2012). Mobile augmented reality applications for library services. *New Library World*, *113*(9/10), 429–438. DOI: 10.1108/03074801211273902

Harmon, C., & Messina, M. (Eds.). (2013). *Mobile library services: best practices.* Scarecrow Press. DOI: 10.5771/9780810887534

Hedrick, C. W., Jr. (2011). LITERATURE AND. The Oxford handbook of social relations in the Roman world, 167.

Kalai Lakshmi, T. R., Nanda, S., Choudhary, V., Prabhu, S., Tuteja, G., & Biswal, S. K. (2025, June). Assessing the role of mobile libraries in enhancing literacy in underserved communities. *Indian Journal of Information Sources and Services*, *15*(2), 116–123. DOI: 10.51983/ijiss-2025.IJISS.15.2.16

Kim, B. (2013). *The library mobile experience: Practices and user expectations.* American Library Association.

Kim, H.-J. (2018). *The present and future of the library mobile experience.* Library Technology Reports., Retrieved from https://journals.ala.org/ltr/article/view/4506/5285

Kim, J., & Park, H. (2021). The effects of mobile library app design on user satisfaction and loyalty. *International Journal of Information Management, 58*, 102315.

Kuhlthau, C. C., Maniotes, L. K., & Caspari, A. K. (2015). *Guided inquiry: Learning in the 21st century* (2nd ed.). Libraries Unlimited. DOI: 10.5040/9798400660603

Lamb, A. (2012). An investigation of academics' perception, uptake and use of ebooks at the University of Waikato (Doctoral dissertation, Open Access Te Herenga Waka-Victoria University of Wellington).

Lee, Y., & Kim, H. Y. (2019). Consumer need for mobile app atmospherics and its relationships to shopper responses. *Journal of Retailing and Consumer Services, 51*, 437–442. DOI: 10.1016/j.jretconser.2017.10.016

Lenormand, M. (2023). Understanding the diversity of mobile service usage in cities. arXiv Preprint. https://arxiv.org/abs/2311.06269

Liu, Y. Q., & Lewis, S. (2023). Services to mobile users: The best practice from the top-visited public libraries in the US. *Information Technology and Libraries, 42*(1). Advance online publication. DOI: 10.6017/ital.v42i1.15143

Magazine, S. Q. (2025). Mobile phone usage statistics 2025: What the latest data tells us about global trends. SQ Magazine. https://sqmagazine.co.uk/mobile-phone-usage-statistics/

Mounir, K., & Shafeek, F. (2003). *History of mobile libraries.*

Owens, E., Arce, V., Del Bosque, D., Fowler, R., & Sheffield, S. (2024). Beyond the Library: The Role of Academic Libraries' Chat Reference in Answering Campus Questions. *The Reference Librarian, 65*(1-2), 1–24. DOI: 10.1080/02763877.2024.2334216

Peters, T. A., & Bell, L. A. (Eds.). (2013). *The handheld library: Mobile technology and the librarian.* Bloomsbury Publishing USA. DOI: 10.5040/9798400661679

Peterson, G., Griffin, J., LaFrance, J., & Li, J. (2017). Smartphone participation in web surveys: Choosing between the potential for coverage, nonresponse, and measurement error. *Total survey error in practice*, 203-233.

Reference and User Services Association. (2017). *Guidelines for Implementing and Maintaining Virtual Reference Services*. American Library Association.

Ruan, J., & Wang, S.-B. (2016, December). Study on innovation of smart library service model in the era of big data. In *Proceedings of the 2016 4th International Conference on Electrical & Electronics Engineering and Computer Science (ICEEECS 2016)* (pp. 1077-1081). Atlantis Press. DOI: 10.2991/iceeecs-16.2016.207

Singh, B. P., & Madhusudhan, M. (2023). Mobile apps–based applications in libraries and information centers: A systematic review of the literature and future research agendas. *International Journal of Librarianship*, *8*(3), 83–102. DOI: 10.23974/ijol.2023.vol8.3.294

Statista. (2025). Mobile subscriptions worldwide by region 2005-2023. Retrieved March 9, 2025, from https://www.statista.com

TechRT. (2025). Smartphone statistics 2025. https://techrt.com/smartphone-statistics/

Valente, L., Feijó, B., Leite, J. C. S. D. P., & Clua, E. (2018). A method to assess pervasive qualities in mobile games. *Personal and Ubiquitous Computing*, *22*(4), 647–670. DOI: 10.1007/s00779-017-1107-0

Walsh, A. (2012). *Using mobile technology to deliver library services: a handbook*. Facet Publishing.

We Are Social. (2025). Digital 2025 Global Overview Report. https://wearesocial.com/wp-content/uploads/2025/02/GDR-2025-v2.pdf

Yarrow, A., & McAllister, S. (2018). Trends in mobile and outreach services. *Public Library Quarterly*, •••, 1–14.

Yoon, H. Y. (2016). User acceptance of mobile library applications in academic libraries: An application of the technology acceptance model. *Journal of Academic Librarianship*, *42*(6), 687–693. DOI: 10.1016/j.acalib.2016.08.003

Young, S. W. H., Chao, Z., & Chandler, A. (2019). User experience methods and maturity in academic libraries. *Information Technology and Libraries*, *38*(3), 68–88. DOI: 10.6017/ital.v39i1.11787

Chapter 4
Libraries and Mobile Technologies

ABSTRACT

This chapter introduces the technological foundations that enable advanced library services to mobile users, including device evolution, network architectures, APIs, and application patterns shaping the mobile ecosystem. It explains how libraries leverage context-aware and location-driven services that blend physical and digital experiences, supported by sensors, 5G, and edge computing. Intelligent navigation tools, AR interfaces, RFID automation, and embedded interaction frameworks are explored as emerging tools for discovery and engagement. Finally, the chapter evaluates the transformative potential of location-based services, addressing opportunities, challenges, governance, and the growing importance of spatial intelligence in library innovation.

INTRODUCTION

In the age of digital advancements, libraries have embraced mobile technologies to enhance their services and provide greater access to information. Mobile technologies, including smartphones and tablets, have revolutionized the way people access and interact with information. This essay explores the relationship between libraries and mobile technologies, highlighting the transformative impact on library services and user experiences.

Both public libraries and mobile technologies are significant forces that have the potential to revolutionize the way in which individuals gain access to information and further their education. People have had access to knowledge from libraries for a long time, and now, because to advances in mobile technology, it is simpler than ever for anyone to do so from any location they want.

DOI: 10.4018/978-1-7998-9175-8.ch004

CONCEPTS AND DEVELOPMENT IN MOBILE TECHNOLOGY

A technology that follows the user wherever they go is called mobile technology. It is made up of mobile two-way communication devices, computer devices, and the networking technology that links them. Internet-connected gadgets like smartphones, tablets, and watches are the current standard for mobile technology. These represent the most recent development after two-way pagers, laptop computers, flip phones, mobile phones, and other gadgets.

Mobile technology is a broad term that encompasses a wide range of devices and services that allow users to access information and services on the go. Mobile technologies include smartphones, tablets, laptops, wearable devices, and other portable devices.

The concept of mobile technology has been around for centuries, but it wasn't until the development of the mobile phone in the 1970s that mobile technology began to take off. The first mobile phones were large and bulky, and they could only be used for basic tasks like making calls and sending text messages. However, as mobile technology has evolved, devices have become smaller, more powerful, and more capable.

Mobile technology has rapidly evolved over the years, transforming the way we communicate, access information, and interact with the world around us. From the advent of basic mobile phones to the emergence of smartphones and wearable devices, the concepts and development in mobile technology have revolutionized various industries and significantly impacted our daily lives. This essay explores the key concepts and advancements in mobile technology, highlighting their implications and future prospects.

Mobile Technology Concepts

Wireless Communication

Wireless communication forms the foundation of mobile technology. It enables data transmission and connectivity without the need for physical cables. Concepts such as cellular networks, Wi-Fi, and Bluetooth have played a crucial role in establishing seamless and ubiquitous communication, allowing mobile devices to connect with each other and access the internet.

In a broad sense, wireless technologies refer to the communication networks that link these gadgets. They make it possible for mobile devices to exchange speech, data, and applications (mobile apps). There is a huge market for mobile technologies. There are now more than 3 billion smartphone users1, and by 2022, 1.87 billion people worldwide are anticipated to work remotely.

Mobile Devices

Mobile devices are the hardware that we use to access mobile technology. There are a wide variety of mobile devices available, including smartphones, tablets, laptops, wearable devices, and other portable devices. Mobile devices provide unprecedented ease and connectivity.

Mobile gadgets are now essential. Smartphones and tablets have connected people more than ever. These devices provide unprecedented convenience. They let us work remotely, remain in touch, and get information with a tap. Smartphones are popular. Modern living requires smartphones due to their many functionalities. They enable calls, messages, photographs, and internet browsing. Smartphones can also be personalized. Apps and settings let people customize their devices. Tablets are popular too. They have a larger screen than smartphones, making them perfect for reading, viewing films, and playing games. Tablets are customizable like smartphones. They're fantastic for mobile productivity.

Mobile Operating Systems

Mobile operating systems (OS) provide the underlying software framework for mobile devices. Concepts like iOS, Android, and Windows Mobile have evolved to support a wide range of functionalities, including app development, user interfaces, and device management. These operating systems have influenced the user experience, app ecosystem, and device compatibility in the mobile technology landscape.

Android is known for its customizable features which make it stand out from iOS. With Android, users have the ability to personalize their devices to their liking by customizing the user interface, installing third-party apps, and tweaking settings to their desired preferences. This level of flexibility allows users to make their devices truly their own, something that iOS users may find limiting in comparison. Additionally, Android offers a wider range of device options at varying price points, making it accessible to a wider audience. Overall, there are many reasons why someone might prefer Android over iOS, and customization is certainly one of them.

Types of Mobile Technology

Cellular Networks

Radio networks use distributed cell towers that enable mobile devices (cell phones) to switch frequencies automatically and communicate without interruption across large geographic areas. This dynamic frequency-switching capability allows cellular networks to handle large volumes of users efficiently, distributing limited radio spectrum resources to maintain continuous connectivity and high-quality service.

4G and 5G Networking

Fourth generation (4G) networking remains the prevailing cellular service standard for most wireless communication, using packet-switching technology to transmit and reassemble data efficiently. It is approximately ten times faster than its 3G predecessor. Fifth generation (5G) networking, currently expanding worldwide, further enhances speed and reduces latency by employing aggregated frequency bands and advanced antenna systems such as massive MIMO (multiple-input, multiple-output). 5G enables near-instant data transmission, real-time streaming, and the integration of the Internet of Things (IoT), setting the stage for pervasive smart connectivity in both personal and institutional contexts.

Wi-Fi

Radio waves that connect devices to the internet through localized routers called hotspots. It is short for wireless fidelity, transmits data using radio waves through local routers known as hotspots. These networks connect devices to the internet without the need for physical cables and are commonly found in homes, campuses, and public spaces. Unlike cellular networks, Wi-Fi signals have limited range and do not support automatic roaming between access points unless configured within a managed system. Modern mobile devices can seamlessly switch between Wi-Fi and cellular networks depending on signal strength, availability, and user preference.

Bluetooth

A telecommunications industry specification for connecting devices over short distances using short-wavelength radio waves. It enables quick pairing between devices such as smartphones, headsets, speakers, wearables, and in-vehicle systems. Bluetooth Low Energy (BLE), a more recent variant, is optimized for minimal power consumption and underpins many applications in health monitoring, digital keys, and location-based services.

Near Field Communication (NFC)

NFC is a specialized subset of radio-frequency identification (RFID) technology that enables data exchange between devices over very short distances—typically within a few centimeters. It is widely used for contactless payments, digital ticketing, and quick information sharing between mobile devices. Libraries and museums have adopted NFC for user authentication, self-checkout, and interactive exhibits.

Global Positioning System (GPS)

GPS technology allows mobile devices to determine their precise geographic location using signals from a constellation of satellites. GPS underpins a broad range of services, from navigation and mapping to location-based recommendations and augmented reality experiences. In mobile library services, GPS supports geofencing and location-based notifications for events, collections, or book deliveries.

Mobile Hotspots and Tethering

Mobile hotspots turn smartphones or dedicated devices into Wi-Fi access points by sharing their cellular connection with nearby devices. This function allows laptops, tablets, or other phones to access the internet where traditional Wi-Fi is unavailable, expanding the portability and flexibility of network access.

Internet of Things (IoT) Connectivity

IoT technology connects everyday objects—such as sensors, smartwatches, and environmental monitors—to mobile networks for real-time data collection and analysis. IoT-enabled mobile systems enhance efficiency in logistics, health tracking, and smart building management. In libraries, IoT can support smart shelving, environmental monitoring, and personalized user experiences through mobile devices.

Mobile Cloud Computing (MCC)

MCC enables mobile devices to access computing power and storage through cloud-based resources rather than relying solely on local hardware. This technology enhances performance, supports large-scale data analytics, and enables seamless synchronization of content across multiple devices. Cloud-based mobile apps and library systems leverage MCC for catalog access, e-resource delivery, and digital preservation.

Satellite Internet and Edge Computing

Emerging technologies such as low Earth orbit (LEO) satellite internet and edge computing extend mobile connectivity to remote areas and enhance the responsiveness of mobile services. Satellite networks like Starlink and OneWeb offer global coverage, while edge computing reduces latency by processing data closer to the source, improving the real-time performance of mobile applications and services.

Key Capabilities of Mobile Technology

Scalability

Creating point solutions that don't scale across an enterprise can be costly in terms of development, management and maintenance. Apps need to be conceived holistically with consideration for lines of business, processes and technical environments.

Integration

ICS Development Team (2020) has pointed out (PDF, 611KB) that "...applications offered on mobile phones and tablets have a separation between the mobile app and back-end business logic and data services." Being able to connect logic and data services to the app is critical, whether the logic and data are on premises, on the cloud or in hybrid configurations.

Reuse

Over 105 billion mobile apps were downloaded in 2018. Many are, or can be modified or combined, for business applications. Using existing apps accelerates time-to-value and improves cost efficiency by taking advantage of domain and industry expertise built into the app.

Cloud-Based Development

The cloud offers an efficient platform to develop, test and manage applications. Developers can use application programming interfaces (API) to connect apps to back-end data and focus on front-end functions. They can add authentication to bolster security, and access artificial intelligence (AI) and cognitive services.

Mobility Management

As mobile technology is deployed, organizations look to enterprise mobility management (EMM) solutions to configure devices and apps; track device usage and inventories; control and protect data; and support and troubleshoot issues.

BYOD

Bring your own device (BYOD) is an IT policy that allows employees to use personal devices to access data and systems. Effectively adopted, BYOD can improve productivity, increase employee satisfaction and save money. At the same time, it presents security and device management questions that need to be addressed.

Security

The mobile security battle is daunting in terms of volume and complexity. Artificial Intelligence (AI) is emerging as a key weapon to discern security anomalies in vast amounts of data. It can help surface and remediate malware incidents or recommend actions to meet regulatory requirements from a central dashboard.

Edge Computing

One of the key advantages of 5G is that it can bring applications closer to their data sources or edge servers. Proximity to data at its source can deliver network benefits such as improved response times and better bandwidth availability. From a business perspective, edge computing offers the opportunity to perform more comprehensive data analysis and gain deeper insights faster.

Impact of Mobile Technology

Mobile technologies have facilitated the growth of e-books and digital content in libraries. Patrons can now access a vast range of e-books, e-journals, and other digital resources directly on their mobile devices. Libraries have partnered with various digital content providers, making these resources accessible through mobile applications or mobile-optimized websites. The convenience of carrying a virtual library in one's pocket has significantly enhanced accessibility and expanded the range of resources available to library users.

Additionally, mobile technologies can assist libraries in attracting new patrons. People who are unable to physically visit the library building may nevertheless receive services from their local library via means of mobile devices, for instance. It is also possible for libraries to use mobile devices to communicate with people who are not typically served by libraries, such as residents of rural areas or people who do not speak English as their primary language.

Moreover, mobile technologies have enabled libraries to provide innovative services beyond their physical spaces. Libraries now offer remote services, allowing users to connect with librarians through live chat or email, offering quick assistance and expert guidance remotely. These services cater to the evolving needs of users who prefer digital interactions and ensure that library resources and expertise are accessible to a wider audience.

Mobile technology continues to evolve and shape our daily lives. The concepts of mobility, connectivity, and mobile app development have revolutionized communication, access to information, productivity, and entertainment. As mobile technology advances further, concepts such as 5G connectivity, AI integration, and IoT devices will continue to drive innovation and transform various industries. Embracing and understanding the concepts and development in mobile technology is crucial for individuals, businesses, and societies to leverage its benefits and adapt to the changing technological landscape.

Evolution of Mobile Technology

When Weiser (1991) introduced the idea of ubiquitous computing, he envisioned a world where people could access, interact with, and manage information seamlessly—anytime, anywhere, and through any interface. By the mid-1990s, researchers were already experimenting with personal digital assistants (PDAs) as early mobile web browsers. Combined with wireless modems, these devices enabled library patrons and other users to retrieve information services while on the move, marking an early milestone in mobile information access.

Over the following decades, mobile technology advanced far beyond the modest capabilities of PDAs. The emergence of smartphones after 2007, followed by the proliferation of tablets, wearables (including smartwatches and AR/VR headsets), and interconnected Internet of Things (IoT) devices, redefined mobility. Today's mobile devices are distinguished not only by portability but also by continuous connectivity—through Wi-Fi and 5G/6G networks—enhanced sensor capabilities, high-speed processors, and AI-driven personalization that adapts to user behavior and context, (Oulasvirta, 2008; Lee, 2011).

Earlier definitions that excluded laptops from mobile classifications have become less defensible. The advent of ultra-thin laptops, foldable-screen hybrids, and two-in-one devices has blurred the lines between conventional and mobile computing. As Lippincott and later scholars have suggested, mobile devices now encompass laptops, notebooks, smartphones, e-readers, audio players, cameras, and other portable technologies that facilitate ubiquitous access (Guo, Liu, & Bielefeld, 2015). In academic library strategies, this expanded view recognizes that mobile engagement includes any connected device capable of supporting digital interaction and learning.

In libraries and information environments, mobility has evolved from mere access to an integrated ecosystem of mobile-first services. Librarians increasingly design experiences that take advantage of 5G and 6G networks, location-aware technologies, AR/VR learning applications, AI-powered chatbots, and context-sensitive alerts. Mobile devices now function as gateways to library resources, navigation aids within physical spaces, and interfaces for collaborative and immersive learning. This seamless integration of mobile and ubiquitous computing fulfills Weiser's early vision of computing that recedes into the background of everyday life while remaining persistently available and adaptive to human needs, (Weiser, 1991).

The most recent (2025) perspectives on advanced mobile and ubiquitous technologies indicate that mobile and ubiquitous computing have entered a new phase often referred to as ambient mobile intelligence. Devices are no longer merely portable; they are embedded in daily environments, wearables, and even personal accessories that anticipate user needs and contexts. Recent studies on wearable technologies—sensors and computing embedded in or around the ear—demonstrate this evolution toward miniaturized, context-aware systems that continuously monitor health, movement, and environmental conditions while supporting communication and learning, (Hu et al., 2025). Simultaneously, emerging research highlights the convergence of artificial intelligence, multimodal interaction, and augmented/virtual reality interfaces in mobile ecosystems, transforming how users engage with digital information across devices, (Pereira et al., 2025). For libraries and information organizations, these advancements signify a transition from mobile access to intelligent mobility—where devices, services, and users coexist in an adaptive network that seamlessly integrates physical and digital information environments.

MOBILE TECHNOLOGIES AND NETWORK FOUNDATIONS FOR LIBRARY INFORMATICS

The dawn of the twenty-first century ushered in an era defined by the rapid evolution of the Internet and mobile technology, fundamentally reshaping how humanity communicates, works, and interacts with the world. These twin forces have woven themselves into the fabric of daily life, creating a global network of connectivity that transcends geographical boundaries. From the early days of dial-up modems to the seamless, high-speed connections of 5G and emerging 6G networks, the Internet has grown into a ubiquitous platform for information exchange, commerce, and social interaction. Simultaneously, mobile technology—particularly the rise of smartphones—has placed this digital universe in the palms of billions, enabling unprecedented access to knowledge, collaboration, and institutional services. Within this transformation, libraries have evolved from physical repositories into dynamic, interconnected systems that depend on complex digital infrastructures to reach their communities anytime and anywhere, (Ericsson, 2025).

The convergence of mobile computing, wireless networks, and cloud-based architectures has fundamentally reshaped the operation of library systems and the user experience of information. Mobile technologies now mediate nearly every aspect of library interaction—from authentication and discovery to data analytics and digital preservation—while network infrastructures ensure that these exchanges occur securely and in real time. A comprehensive understanding of these underlying systems is paramount not only for enhancing digital access but also for anticipating the architectural, security, and ethical considerations associated with pervasive connectivity. The technological and network foundations of mobile information ecosystems, as exemplified by advancements in mobile endpoints, wireless standards, and distributed computing, continue to transform the operational backbone of library informatics.

Device and Radio Evolution: What Mobile Endpoints Now Enable

Over the past decade, smartphones and tablets have transformed from specialized communication devices into versatile computing platforms capable of executing intricate multimedia, analytical, and interactive tasks. These devices have evolved to incorporate multiple radio frequencies, high-resolution sensors, neural processors, and machine learning accelerators, fundamentally reshaping the manner in which information is accessed and consumed, (Ericsson, 2025). The widespread prevalence of connected devices—phones, tablets, wearables, and e-readers—has profoundly altered the domain of library informatics by extending users' engagement beyond fixed desktop settings into fully mobile, context-sensitive ecosystems, (Khan, 2024).

Global statistics affirm that mobile endpoints are now the dominant gateway to digital content. According to Ericsson's *Mobility Report* (2025), the number of global smartphone subscriptions surpassed 7 billion in 2025, compared with 3.6 billion in 2016, 6.2 billion in 2021, and a projected 7.7 billion by 2028. Mobile broadband now accounts for over 80 percent of Internet traffic worldwide, and data consumption per user continues to rise by more than 20 percent annually. This upward trend underscores the necessity for libraries to optimize their infrastructures for mobile-first access and to design systems capable of operating effectively across heterogeneous network environments, (Ericsson, 2025).

Figure 1. Global smartphone mobile network subscriptions, 2016–2028

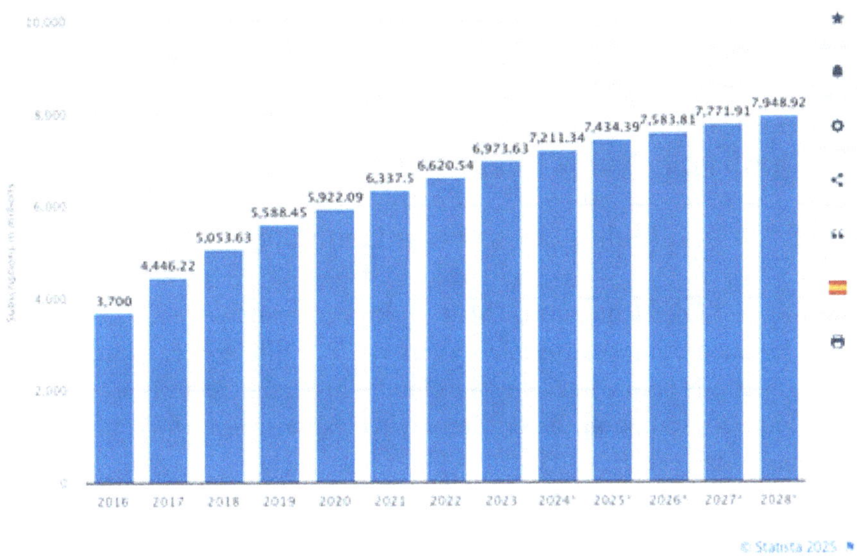

(Adapted from Ericsson (2025))

Two major implications emerge for libraries. First, device heterogeneity—differences in screen sizes, operating systems, and processing capabilities—necessitates adaptive design that maintains usability across a spectrum of hardware, (Vasantha Kumar, 2022). Libraries must implement responsive interfaces and scalable application programming interfaces (APIs) that accommodate both high-end and low-resource devices without compromising experience or accessibility, (Ram Prakash, 2023).

Second, the rise of on-device artificial intelligence (AI) and advanced sensors allows mobile systems to support features like visual object recognition, augmented reality (AR) navigation, and natural language interfaces. As computing power migrates closer to the user, libraries can integrate context-aware functions, such as mobile shelf navigation, camera-based metadata extraction, or location-sensitive recommendations, directly into mobile applications, (Mao et al., 2017). This technical evolution elevates the mobile device from a passive receiver of content into an active participant in data processing and knowledge discovery, (Hahn, 2018).

Architectures That Move Compute and Content Closer to Mobile Endpoints

Modern mobile experiences depend not only on device capabilities but also on network and computing architectures that minimize latency and maximize efficiency. The convergence of cloud computing, edge computing, and distributed storage has redefined how mobile systems are designed.

Mobile Edge and Multi-Access Edge Computing (MEC)

Multi-Access Edge Computing (MEC) brings computation and storage closer to end users by integrating small-scale servers within cellular or campus networks to process data locally before it reaches the cloud. This design reduces network latency and bandwidth consumption, thereby improving response times for latency-sensitive applications, (Mao et al., 2017).

In library environments, MEC can enhance the responsiveness of resource discovery, authentication, and data analytics systems. For example, a library might deploy an edge node that caches frequently accessed digital objects or executes natural-language search queries locally, reducing reliance on centralized infrastructure, (Cisco, 2024). Similarly, AR navigation within large library buildings can leverage MEC to deliver sub-second responsiveness by processing spatial data near the user (Khan, 2024).

MEC also supports privacy-sensitive computation, as local processing allows anonymized data handling before transmission to cloud-based systems. This approach mitigates potential data exposure risks, an increasingly critical factor as institutions handles large volumes of user-generated data, (Ahmed, Singh, & Patel, 2024).

The Cloud–Edge Continuum

While edge computing emphasizes locality, the cloud–edge continuum offers a flexible hybrid paradigm. In this model, different components of an application execute across distributed environments based on performance, privacy, and scalability requirements. Metadata indexing, authentication, and archival storage typically remain in central cloud systems, while mobile cache synchronization and context recognition operate at the edge, (Abolfazli et al., 2013).

This continuum supports scalable, fault-tolerant architectures that align with institutional budgets and staffing capacities. As telco and cloud providers expand managed edge platforms, libraries can experiment with edge deployment without extensive local infrastructure investment, (STL Partners, 2024). The continuum also enables dynamic workload migration—shifting processing between edge and cloud depending on congestion, user proximity, and energy efficiency, (Ai et al., 2018).

Together, MEC and cloud–edge hybridization provides a framework for libraries to modernize digital infrastructure, improving both technical performance and data sovereignty.

Access Technologies, Capacity Planning, and Performance Engineering

The efficiency of mobile access depends heavily on network technologies, capacity management, and connectivity planning. Understanding the evolving wireless landscape—cellular networks, Wi-Fi generations, and fixed-wireless access—is crucial for designing mobile-first environments that function reliably under varying conditions.

Cellular Evolution: From 4G to 5G and Beyond

The rollout of fifth generation (5G) cellular technology has been transformative. 5G introduces massive multiple-input multiple-output (MIMO) antennas, higher frequency bands, and lower latency—features that support bandwidth-intensive applications such as real-time video streaming, mobile AR, and collaborative data analytics, (Ericsson, 2025). In 2024–2025, 5G accounted for nearly 40 percent of global mobile data traffic, with over two billion subscriptions worldwide, (Grand View Research, 2024).

In library contexts, 5G supports high-speed, low-latency access for mobile learning platforms and interactive exhibits. Moreover, Fixed Wireless Access (FWA) using 5G provides viable connectivity for library branches in rural areas, where fiber-optic installation may be economically infeasible (Cisco, 2024).

Beyond 5G, early research on sixth generation (6G) technologies anticipates integrating AI-driven network management, terahertz communications, and quantum-resistant security protocols—innovations that promise exponential improvements in capacity and latency, thereby enabling more immersive information environments, (Khan, 2024).

Wi-Fi Generations and Local Connectivity

Within campuses and buildings, Wi-Fi remains the backbone of local access. The transition from Wi-Fi 5 to Wi-Fi 6 and 6E has introduced multi-user OFDMA, higher spectral efficiency, and improved handling of dense environments, (Cisco, 2025). Wi-Fi 7, still emerging, adds multi-link operation and expanded channel bandwidths up to 320 MHz, providing throughput rivaling wired Ethernet connections, (Grand View Research, 2024).

Libraries benefit directly from these advancements, as upgraded Wi-Fi infrastructure ensures that high-density study zones, classrooms, and maker spaces maintain consistent performance even under heavy load. Modern Wi-Fi standards also support seamless roaming between access points, reducing session dropouts during movement between floors or buildings, (Vasantha Kumar, 2022).

Handoffs, Roaming, and Session Continuity

Maintaining session continuity as devices transition between Wi-Fi and cellular networks is a persistent challenge. Integrating authentication systems like *eduroam* and adopting token-based session management allows users to move between networks without losing access or authentication state, (Ocran, Underwood, & Arthur, 2020). Edge caching and proxy servers further support seamless connectivity by retaining partial transactions during temporary network disruptions, (Abolfazli et al., 2013).

Capacity Planning and Traffic Forecasting

As mobile data consumption continues to rise, capacity planning becomes essential. The *Ericsson Mobility Report* (2025) projects a compound annual growth rate of 20 percent for mobile data traffic through 2030. Libraries and universities must forecast demand using real-world analytics—measuring peak concurrency, identifying bottlenecks, and shaping traffic through QoS policies.

Advanced monitoring tools, such as passive flow analysis and mobile application performance management (APM), help IT administrators optimize network allocation for mission-critical traffic like research databases or academic video streams while managing non-essential background traffic, (Cisco, 2024).

Protocols, APIs, and Application Patterns in Mobile Ecosystems

Mobile technology relies on communication protocols and programming models that govern data transmission, latency, and interoperability. Understanding these underpinnings is key to designing efficient and resilient library systems.

Progressive Web Apps (PWAs) and Offline-First Design

Progressive Web Apps (PWAs) merge the accessibility of web applications with the functionality of native apps. They use browser APIs—such as service workers,

IndexedDB, and background sync—to enable offline access, local caching, and push notifications, (Straits Research, 2025). For libraries, PWAs allow users to browse catalogs, view saved reading lists, and receive notifications without requiring continuous connectivity or app-store installations, (Ram Prakash, 2023).

PWAs also reduce maintenance costs by consolidating codebases and ensuring platform independence. Their offline-first design ensures inclusivity for users in regions with unstable connections or limited data access, aligning with principles of equitable digital access, (Quaye Affum, 2022).

HTTP/2, HTTP/3, and Transport Optimization

Transport-layer innovations have significantly improved mobile performance. HTTP/2 introduced multiplexed streams, reducing latency from head-of-line blocking, while HTTP/3, based on QUIC over UDP, enhances resilience against packet loss and enables faster secure connections, (Ai, Peng, & Zhang, 2018). Libraries adopting modern content delivery networks (CDNs) that support HTTP/3 can deliver faster and more reliable access to digital content across varied mobile networks, (Cisco, 2025).

APIs, Microservices, and Lightweight Payloads

Mobile applications benefit from microservice architectures that separate functionalities into small, modular components accessible through APIs. Using lightweight payload formats (e.g., JSON, Protocol Buffers) and employing techniques like pagination and delta updates reduces bandwidth consumption, (Mao et al., 2017). GraphQL and gRPC further enhance mobile communication efficiency, allowing flexible data retrieval and real-time updates. For library systems, implementing mobile-optimized APIs for catalog search, authentication, and analytics can significantly improve response times and user experience, especially on limited-bandwidth networks, (Hahn, 2018).

Operational Considerations: Security, Sustainability, and Accessibility

Designing and operating mobile networks and applications in library contexts raises a set of recurring operational challenges that combine technical, legal, and ethical dimensions.

Security and Device Management

Mobile devices represent both an opportunity and a risk for institutions. Threats include compromised applications, insecure Wi-Fi access points, and lost devices containing authentication tokens. Best practices include implementing enterprise mobility management (EMM), enforcing HTTPS/TLS across all endpoints, and deploying zero-trust network architectures, (Cisco, 2024).

Libraries adopting Bring Your Own Device (BYOD) policies should maintain network segmentation and clear usage agreements, isolating sensitive data services from public access points. AI-based intrusion detection tools can identify anomalous mobile traffic patterns indicative of potential breaches, (Ahmed, Singh, & Patel, 2024).

Data Privacy and Consent

Mobile technologies often generate detailed telemetry, including location data and behavioral analytics. Under data protection laws like the GDPR, libraries must ensure transparent consent processes, minimal data collection, and anonymization of user data. Privacy-by-design principles recommend storing only aggregated usage metrics and anonymized logs, (Ubogu & Chukwusa, 2024).

Energy Consumption and Sustainability

The environmental impact of mobile networks is a growing concern. Data transmission and storage consume significant energy, with mobile data transfer estimated at 0.13 kWh per gigabyte, (Ericsson, 2025). Libraries can mitigate this footprint by optimizing media delivery (adaptive bitrate streaming), adopting green data centers, and utilizing content caching to reduce redundant transfers, (Cisco, 2024).

Accessibility, Inclusion and Equity

Digital equity requires ensuring that all users, including those with lower-end devices or limited data plans, can access mobile library systems. Techniques such as low-bandwidth modes, offline access, and assistive design (e.g., ARIA-compliant interfaces, screen-reader compatibility) enhance inclusivity, (Quaye Affum, 2022). PWAs further support equity by functioning across diverse hardware and network conditions, (Straits Research, 2025).

Mobile endpoints and the radio and access technologies that support them have matured to the point where libraries must consider mobile and network architecture as primary technical concerns. The key shifts include: (1) endpoint capabilities that enable sensor-driven, low-latency, and AI-augmented interactions; (2) architectural advancements that move compute closer to the access network through MEC and the cloud-edge continuum; and (3) access technologies (5G, Wi-Fi 6/6E/7, FWA) that alter capacity and coverage planning.

The convergence of mobile devices, network infrastructure, and distributed computing architecture has resulted in an integrated digital environment characterized by continuous and ubiquitous information access. Libraries must adapt by comprehending the technological underpinnings—radio evolution, network architecture, and performance optimization—that underpin the mobile experience. The increasing sophistication and utility of mobile technology are evident, particularly with its growing role as the primary method of accessing the web in many regions. The page likely underscores the significance of adopting and embracing mobile technology for sustained growth and innovation.

CONTEXT-AWARE AND LOCATION-DRIVEN LIBRARY SERVICES

Libraries have always been anchored in place—buildings, stacks, reading rooms, and local communities. Over the past two decades, however, "place" has become multi-layered: it includes physical coordinates, georeferenced metadata, sensorified interiors, and transient digital presence. When these layers are combined, libraries become context-aware systems that can deliver the right content, staff skill, or spatial guidance at the right moment. This transformation depends on the careful orchestration of sensors and networks, on software that fuses signals and models intent, and on service design that balances usefulness with the library's long-standing commitments to privacy, equity, and accessibility.

In the last decade, location-based systems (LBS) have evolved from static GPS-driven mapping applications into sophisticated, multi-sensor ecosystems capable of sub-meter accuracy and millisecond-scale responsiveness. What once consisted of a simple coordinate lookup now represents a distributed architectural framework uniting sensing, computation, and semantics. For professional practitioners, understanding the underlying engineering logic is critical to designing infrastructures that are not only accurate but contextually intelligent and privacy aware.

From early Global Positioning System (GPS) adoption to recent indoor positioning and sensor-fused solutions, location-based capabilities are now heterogeneous and layered. Outdoor localization still relies heavily on satellite positioning (GPS, GLONASS, Galileo) and hybrid techniques (GPS plus IP/SSID inference) for coarse-grain positioning. Indoors, where satellite signals attenuate, libraries rely on a combination of Bluetooth Low Energy (BLE) beacons, Wi-Fi fingerprinting, RFID, ultra-wideband (UWB), and markers such as QR codes and NFC tags. Advances in wireless standards, including BLE 5.x and private 5G deployments, along with edge computing, make it possible to do low-latency spatial computations and richer context inference on-premises, (Milano et al., 2024). Meanwhile, machine learning layered on sensor fusion improves reliability by using predictive filters, fingerprinting maps, and probabilistic tracking, (Daniş, 2023). The most successful deployments—whether in cultural institutions, campuses, or urban mobility environments—share a common technical foundation: layered architectures that integrate advanced localization algorithms, machine learning (ML) inference, sensor fusion, and edge–cloud collaboration.

Location Intelligence's Expanding Role

Location intelligence in libraries has matured beyond simple coordinate reporting. Early public awareness of location technologies was shaped by GPS-driven outdoor mapping, but the modern problem is often indoors—inside stacks, special collections rooms, and multi-level academic libraries where satellite signals degrade. To meet this challenge, contemporary systems layer and fuse multiple sensors and data sources. Satellite-based GNSS (GPS, Galileo, GLONASS) remains the primary tool for outdoor fixes, but accurate indoor localization now typically requires a heterogeneous mix: Bluetooth Low Energy (BLE) beacons, Wi-Fi fingerprinting, Radio Frequency Identification (RFID), ultra-wideband (UWB), NFC/QR anchors, inertial sensors (accelerometer and gyroscope) in the device, and camera-based visual-inertial odometry. Each of these components contributes different trade-offs of accuracy, cost, and privacy exposure, (Milano et al., 2024; Al-Okby et al., 2024).

BLE has become widely used because it is inexpensive and broadly compatible with smartphones; BLE 5.x improvements have also increased range and throughput while adding features that improve packet metadata which are useful to positioning algorithms. Nevertheless, BLE is vulnerable to RSSI (received signal strength indicator) variation from multipath effects and environmental change; research shows that careful anchor placement, calibration, and machine-learning smoothing (particle filters, Hidden Markov Models) are necessary to reach meter-level reliability suitable for wayfinding and shelf guidance, (Milano et al., 2024). Where libraries need sub-meter accuracy—for example, precise item locating in dense stacks—UWB provides an effective alternative, delivering centimeter-level performance under line-of-sight or properly configured indoor networks, (Bregar, 2023; Al-Okby et al., 2024).

Wi-Fi fingerprinting makes clever reuse of existing infrastructure. By mapping signal patterns against locations (the "fingerprint"), it is possible to localize a device without new beacons, though drift and environmental changes require periodic re-fingerprinting and adaptive models, (IndoorAtlas documentation). Vision-based techniques—visual SLAM (simultaneous localization and mapping) and image-based recognition—are increasingly viable thanks to improved device cameras and on-device compute. When combined with inertial sensors for dead-reckoning and magnetometer data for orientation, these methods deliver resilient estimates even when radio signals are weak, (Yang et al., 2025).

A modern understanding of "location" therefore moves from single coordinates toward context inference: the system's interest is not merely "where" a patron is but "what they are doing" and which local resources or staff expertise should be offered. Sensor fusion, in which information from multiple modalities is probabilistically combined, is crucial for this contextual state estimation. Because libraries are both public-facing and information custodians, they must architect these systems with attention to minimal data collection, local processing where feasible, and transparent opt-in policies that match professional norms.

Figure 2. Conceptual diagram of sensor fusion for library LBS

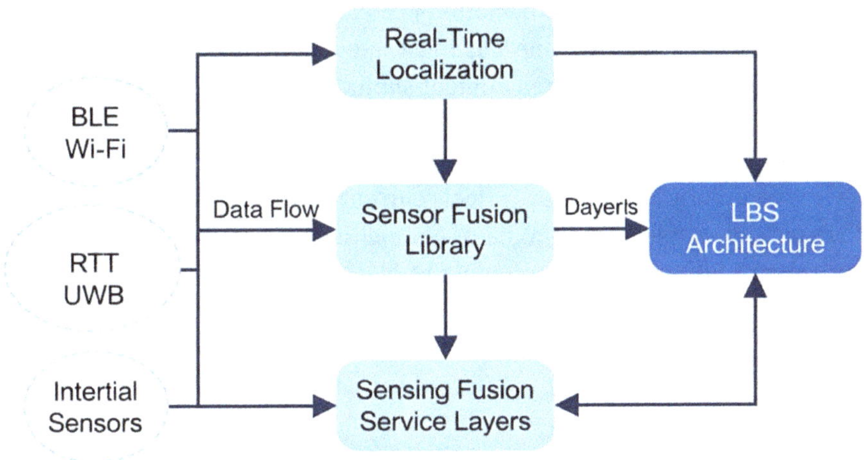

Mobile Interactivity and Intelligent Space Design in Libraries

The interaction model for location-aware libraries blends pull and push paradigms. Historically, patrons pulled information by searching catalogs or walking to a reference desk. Today's apps augment pulls with contextual pushes—geofenced notices, proximity nudges, and beacon-triggered content—that are designed to reduce friction and connect people to immediate, relevant resources. Yet the design question remains: how to push without undermining consent, and how to build affordances for those who decline tracking?

Hardware choices affect experience. Libraries usually construct hybrid instrumentations. A typical mid-sized academic library may combine BLE beacons for zone detection, RFID tags for item control and bulk shelf audits, Wi-Fi fingerprinting to leverage existing access points, and QR/NFC markers as user-initiated anchors for accessibility and offline fallback. In specialized contexts—rare manuscripts rooms, conservation labs, or high-value audiovisual collections—UWB or dedicated vision systems might be deployed to support fine-grained control. These sensor layers are orchestrated by middleware—location engines and SDKs that perform fingerprint matching, RSSI-based ranging, time-difference-of-arrival (TDOA) computations, or visual SLAM. Contemporary developers use commercial SDKs (IndoorAtlas, MapsIndoors, NavVis, ORiiEnt) or platform services (Google Maps Indoor, Mapbox services) to avoid reinventing low-level positioning algorithms, (IndoorAtlas, 2024; Google Maps Platform, 2024).

Nevertheless, technical instrumentation is meaningless without careful interaction design. In library settings, app UIs must integrate maps, search, and accessible alternatives. Spatially triggered content should be rich but concise—metadata notes, directly linked digital surrogates, audio descriptions for patrons with visual impairments, and staff contact buttons for immediate help. The "magic mirror" concept—where a physical display reflects current collection status in real time using RFID and sensor telemetry—remains compelling as a design metaphor because it makes the institution's holdings legible in place, transforming the passive shelf into an interactive node for discovery, repair, and outreach, (Gul & Bano, 2019).

Important software patterns have arisen in mobile LBS design. First, local processing and edge compute are preferred for latency-sensitive tasks (AR overlays, turn-by-turn guidance) and for privacy reasons, since raw location traces are sensitive. Second, policy engines should be embedded in the mobile stack so that consent settings (opt-in levels, notification frequency, retention preferences) are enforced consistently on device and server. Third, layered fallbacks—QR anchors and staff assistance—must be integrated so patrons always have an access path even when sensing fails.

Data Stewardship in Location-Aware Systems

Location data is especially privacy sensitive. Recurrent presence patterns can reveal attendance at programs, membership in support groups, or use of specialized services—information libraries are ethically obliged to protect. Professional codes and recent regulatory pressure (e.g., GDPR interpretations applied to mobile apps) suggest libraries should adopt privacy-by-design: minimize collection, anonymize and aggregate where appropriate, and prefer on-device inference. For instance, a navigation feature that locally computes a route and only sends an anonymized event ("route started, route completed") to analytics is preferable to a system that continuously streams GPS coordinates to a third-party cloud.

Transparency and consent are necessary but not sufficient. Design must include clear retention policies, simple interfaces to review and delete data, and technical controls to avoid re-identification through cross-dataset joins. Libraries should document their LBS data lifecycle—what is collected, why, where it is stored, how long it is kept, and who can access it—and present this information in plain language to patrons at first use. Auditable logs and periodic external review can help maintain public trust.

Equity and accessibility are also ethical concerns. Not every patron has a smartphone or is comfortable enabling location services. To avoid digital exclusion, libraries must continue to provide non-tracking alternatives: static floorplans, staffed wayfinding assistance, QR/NFC anchors for audio or large-print content, and printed maps. For patrons with disabilities, LBS should explicitly support alternative modalities—audio direction prompts, haptic feedback, and integration with assistive devices.

Rising Integration: 5G, Edge Computing, and Sensor Fusion

The recent arrival of private 5G and advances in edge computing have created new operational possibilities for LBS. Private 5G can offer controlled radio environments with low latency and the capacity to support dense, reliable positioning techniques embedded in the cellular stack (5G positioning reviews, 2024–2025). Edge compute nodes—servers placed on campus networks—can host localization models, AR asset servers, and ML personalization routines, reducing the need to transfer raw sensor data to remote clouds and thus improving privacy and latency.

Together, 5G and edge compute enable on-premises sensor fusion that supports more sophisticated features, such as real-time SLAM for AR wayfinding, high-fidelity occupancy analytics for space management, and low-latency turn-by-turn overlays. For example, SLAM pipelines that previously required cloud GPUs can now run on local edge nodes or advanced mobile chipsets, enabling AR route overlays to remain accurate as a patron's phone moves through stacks—even when BLE anchors are sparse.

Beyond enabling richer features, edge-hosted LBS workflows also offer governance advantages. Because models and raw telemetry can remain inside institutional networks, libraries can better control access, perform internal audits, and limit vendor exposure to sensitive data. This configuration does not eliminate the need for policy and transparency but complements them with stronger technical controls.

Usefulness of Location Intelligence for Digital and Physical Collections

Location intelligence enhances both patron-facing discovery and internal operations. For patrons, the immediate benefits are clearer wayfinding, contextual discovery (notifications about nearby exhibits or relevant collections), and personalized in-place support, such as suggestions for complementary titles when browsing a shelf. For staff, LBS reduces time spent searching for misplaced items, speeds inventory reconciliation, and provides actionable occupancy metrics for space planning.

The capacity to georeferenced collections—linking digital surrogates to precise shelf coordinates, archival boxes, or display cases—also opens new curatorial possibilities. Spatially bound interpretation allows curators to narrate items in situ (e.g., linking an exhibit case with a multimedia narration accessed by phone) and to craft layered learning experiences that connect the physical artifact to web-accessible primary sources. These experiences are heightened when LBS supports low-latency AR overlays or on-demand audio, especially for public humanities projects.

Sensors and Local Positioning Elements: Practical Considerations

Choosing sensors requires an assessment of accuracy requirements, patron privacy expectations, maintainability, and the institution's technical capacity. BLE beacons are cost-effective for zone-level services and push notifications, but their RSSI variance mandates robust calibration and algorithmic smoothing, (Milano et al., 2024). Wi-Fi fingerprinting is attractive for reusing infrastructure but requires ongoing management. RFID excels at item-level inventory and checkout workflows but does not provide continuous person-centred tracking unless active tags are used; thus, RFID and person-centred LBS should be conceived as complementary. UWB is ideal where sub-meter accuracy matters, but it demands careful planning and has historically had higher hardware costs, (Bregar et al., 2023; Al-Okby et al., 2024).

QR codes and NFC tags are indispensable, low-cost access methods. They are easy for patrons to use, require no continuous tracking, and can encode deep links to curated content, audio narration, or accessibility features. For many libraries, QR/NFC anchors provide a low-risk, inclusive entry point into location-aware services.

Operationally, institutions should plan lifecycle tasks: battery change and beacon replacement schedules, periodic fingerprint recalibration, QR code maintenance to avoid link rot, and RFID tag replacement when wear or standards change. These tasks are fundamental to preserving user trust—nothing erodes confidence more quickly than a "lost" navigation app that leads patrons astray.

Context-aware and location-driven services offer libraries a transformative toolkit for increasing relevance, accessibility, and operational efficiency. Their success depends on thoughtful sensor choices, robust software for sensor fusion and privacy preservation, and design practices that center consent and inclusion. Libraries that adopt layered architectures—combining local processing, transparent policies, and accessible fallbacks—can create place-aware services that uphold the institution's public mission while taking full advantage of the technological advances of the 2020s.

Table 1. Descriptive summary of common sensors used in library localization

Sensor Type	Description	Advantages	Limitations	Common Usage in Library LBS
BLE Beacons	Bluetooth Low Energy devices emitting signals for proximity detection.	Low cost, widespread availability, good for micro-location.	Signal interference, limited range.	Real-time user location and zone detection.
Wi-Fi RTT	Wi-Fi Round Trip Time measures distance based on signal travel time.	Leverages existing infrastructure, higher accuracy than RSSI.	Requires RTT-enabled devices, signal multipath.	Accurate room-level positioning and tracking.
UWB Anchors	Ultra-Wideband radios emitting short pulses for precise ranging.	Very high precision (centimeter-level), robust to interference.	Higher cost, needs dense anchor deployment.	Precise asset and user localization.
Inertial Sensors	Accelerometers and gyroscopes measuring motion and orientation.	Continuous tracking, not dependent on external signals.	Drift over time, requires fusion with other sensors.	Complementary sensor for dead reckoning and smooth tracking.

INTELLIGENT NAVIGATION AND ENGAGEMENT TECHNOLOGIES IN LIBRARIES

Patrons enter library spaces today expecting the same immediacy and contextual responsiveness they experience in other mobile contexts. Intelligent navigation and engagement technologies integrate indoor positioning, augmented reality, QR/NFC anchors, RFID, and sophisticated location-aware software to support discovery and participation where it matters most—at the point of use. While these technologies are linked by the underlying notion of spatial awareness, important differences lie in how each system is designed, what it chooses to collect, and how it surfaces content. For professionals designing such systems, navigation architecture must be treated as an integrated stack of sensing, computation, and semantic reasoning.

Indoor Positioning and Navigation Systems in Knowledge Environments

Indoor positioning systems (IPS) are the backbone of wayfinding and proximate engagement. Contemporary IPS solutions combine multiple techniques to compensate for the challenges of indoor radio propagation and complex geometries. BLE beacon networks are prevalent for zone detection and push notifications; paired with probabilistic algorithms and machine learning, well-configured BLE networks can provide reliable guidance to broad shelf ranges and rooms, (Milano et al., 2024). Wi-Fi fingerprinting avoids additional infrastructure by mapping existing access points to locations but requires periodic maintenance, especially in busy, dynamic campus environments, (IndoorAtlas, 2024).

In complex indoor environments, architectural redundancy is a design necessity. Deployments typically combine BLE beacons for zone detection, UWB anchors for high-precision corridors, and Wi-Fi RTT for general coverage. Each subsystem reports measurements to a localization middleware, which normalizes them into a unified spatial reference frame using sensor-fusion algorithms. Network evolution reinforces this design: 3GPP Release 17 and 18 define 5G New Radio (NR) positioning, and Wi-Fi 6E and 7 expand this capability, allowing engineers to create resilient hybrid infrastructures capable of seamless indoor-outdoor continuity

Where precise location is required, UWB is an increasingly feasible option. Recent reviews and datasets show UWB delivering centimeter-level accuracy in indoor settings, which is useful for asset tracking, specialized equipment, or tight navigation tasks, (Bregar et al., 2023; Al-Okby et al., 2024). UWB's improved availability, dropping component costs, and stronger industry tooling make it a realistic choice for libraries that need high precision.

Vision-based SLAM and visual–inertial odometry have matured rapidly and are beginning to be embedded in mobile stacks. These techniques exploit device cameras and inertial sensors to localize relative to mapped visual features, enabling robust guidance in spaces with adequate texture and lighting. When combined with AR, SLAM supports overlays that appear "locked" to shelf labels or exhibits, enriching user experience without relying solely on radio signals.

A critical design principle is graceful degradation: navigation systems should layer modalities so that failure in one sensor type still leaves users with useful alternatives. For example, if BLE anchors drift, QR/NFC anchors can act as spot checks a patron can scan to re-establish precise position. Consistent metadata—shelf codes, call numbers, and simplified directional cues—should align between the digital map and physical signage to avoid cognitive mismatches.

Augmented and Mixed-Reality Library Experiences

Augmented reality (AR) has emerged as a compelling method for spatially connecting patrons with collections. In the ever-evolving landscape of cultural institutions, augmented reality (AR) stands out as a captivating tool that forges a profound connection between patrons and the collections housed within. By employing AR technology, visitors are not merely observers, but active participants in multidimensional experience. Enhanced interpretive content, such as provenance notes, curator insights, and rich multimedia presentations, can be intricately tied to physical relics or specific locations, transforming the way knowledge is conveyed and absorbed.

Imagine stepping into a library for the first time, surrounded by towering shelves filled with volumes of knowledge. Here, AR becomes a guiding compass, overlaying arrows and navigational cues onto the live feed of your device's camera. This seamless integration serves to diminish the anxiety often felt by newcomers and simplifies the navigation of complex, multi-level stacks. Thus, augmented reality not only enriches the interpretative experience but also provides practical guidance, making the exploration of knowledge a stress-free journey.

Furthermore, the architecture of this technology is fortified by advanced computing solutions—edge servers and on-device processing work in unison to deliver real-time features securely and effortlessly. Such capabilities extend accessibility to a diverse audience, as audio solutions and interfaces designed for assistive technologies enhance inclusivity. QR codes and NFC tags, resourcefully embedded throughout space, empower visitors to engage interactively without the burden of constant tracking. These digital touchpoints support a rich tapestry of multilingual content, audio narration, and direct catalog access, all while employing durable materials and backend content management systems that effectively mitigate the risks of link rot.

At the heart of AR's effectiveness lies a synergy of precise localization and rapid response times. The interplay between robust edge servers and capable on-device computing allows for a harmonious execution of visual simultaneous localization and mapping (SLAM) and shared augmented reality sessions. Libraries venturing into the realm of AR are called to prioritize accessibility. Implementing audio alternatives, tactile signage, and screen-reader compatibility ensures that these innovative features expand participation rather than inadvertently restrict it.

In addition to fostering navigation, augmented reality opens doors to enriching pedagogical opportunities. Library instruction can now manifest directly within the context of the collection itself. For instance, when a patron stands before a shelf dedicated to data science, a brief AR tutorial could spontaneously emerge, shedding light on the intricacies of dataset citation. Similarly, a visitor exploring an archival exhibit might encounter magnified views of manuscripts, embellished with contextual annotations that breathe life into the material. These in-the-moment instructional vignettes support what is now referred to as just-in-time learning, aligning library services precisely with the immediate and evolving needs of users.

Moreover, augmented reality and its sister technology, mixed reality (MR), transcend traditional navigation by intertwining digital signals—such as directional arrows and metadata tags—with fixed physical coordinates, creating immersive spatial experiences. The triumph of these innovative interfaces hinges on the meticulous alignment of visual-inertial odometry (VIO), SLAM-based spatial mapping, and a robust location-based services (LBS) infrastructure. In professional implementations, the integration of LiDAR-assisted SLAM and Visual Positioning Service (VPS) frameworks enhances the depth and accuracy of these augmented experiences. As new AR standards such as OpenXR and WebXR gain traction, they promise to offer cross-platform application programming interfaces (APIs), ensuring the seamless layering of informational content that enriches storytelling and visualization of collections across diverse mediums.

Augmented reality presents a multidimensional bridge between patrons and the rich tapestry of cultural collections, transforming the library experience into one that is as informative as it is engaging.

QR Codes, NFC Tags, and Embedded Interaction Frameworks

QR and NFC remain powerful, low-barrier tools for location-anchored interaction. A well-designed QR deployment supports multilingual content, audio narration, and direct links to catalog records or digital surrogates. Unlike continuous tracking schemes, QR/NFC supports immediate, patron-initiated interactions, which has two benefits: it reduces privacy risk and it supports patrons who do not want or cannot enable passive location tracking.

Designing QR-based experiences is not trivial. Libraries must guard against link rot by managing back-end content, choose resilient print materials, and develop scanning guidance for patrons unfamiliar with the experience. Pairing QR anchors with small unique IDs—so each marker corresponds to a persistent metadata record—makes it easier to update content without replacing markers physically. NFC offers an even lower-friction option on compatible devices: a simple tap can open content without using the camera, offering better accessibility for some patrons.

RFID, Inventory, and Material-Focused Automation

RFID dramatically changes how libraries handle items. Products of decades of adoption, RFID tags support rapid check-in/out, self-service, and fast shelving audits. Tagging every item enables bulk reads (a cart of returned items can be registered quickly), and RFID can be combined with handheld readers to quickly locate misplaced items. These operational efficiencies translate into improved circulation workflows and liberated staff time for higher-value engagement.

RFID supports inventory and circulation workflows through rapid bulk readings, efficient check-in/out, and item tracking. While passive RFID is limited to proximity interactions, active systems can enable person-centric workflows but must be deployed in strict alignment with privacy policies.

However, RFID's principal limitation is its interaction model: passive RFID requires proximity for a read and therefore does not supply continuous person-centric position streams. Active RFID systems can deliver richer tracking at increased cost and complexity; such systems must be evaluated carefully against privacy policies, especially if they could reveal patron movement patterns. Integration with the library management system (LMS) is essential to make RFID operationally useful. When integrated, RFID events (checkouts, gate reads, hand-held inventory scans) become actionable data streams that support both frontline services and periodic analytics.

Local-Based Services: Software, Platforms, and Practices

In the library context, "location-based services" emerge as a suite of innovative features driven by precise localization. These services include wayfinding—helping patrons navigate the library's complex layout—and location-triggered suggestions for resources or events, proximity-based amenities such as study room reservations or automatic unlocking, and timely push notifications about nearby exhibitions or workshops. Furthermore, contextual discovery tools can guide users to digital surrogates of physical materials, enhancing interactions with the library's offerings.

To reliably deliver these advanced services, a contemporary technology stack is essential. At its foundation lies the **Sensing Layer**, which meticulously collects geospatial signals from a variety of heterogeneous devices. This includes Bluetooth Low Energy (BLE) 5.x beacons, Ultra-Wideband (UWB) transceivers adhering to the IEEE 802.15.4z standard, Wi-Fi RTT radios standardized under IEEE 802.11az, as well as inertial measurement units (IMUs), magnetometers, and sophisticated LiDAR scanners. This diverse array of hardware ensures robust location accuracy and reliability within library environments.

Complementing the sensing capabilities are **Device SDKs** and mapping services designed to process low-level sensor data and facilitate map rendering. One notable example is the Google Maps Platform, which features indoor mapping capabilities, alongside other platforms like Mapbox and IndoorAtlas. These providers are continuously upgrading their offerings, enhancing functionalities related to 3D rendering, offline maps, and geocoding to create seamless and interactive user experiences.

The next crucial component is the **Positioning Engine**. These engines—whether commercial or open-source—perform sensor fusion to deliver normalized location events. Notable examples include IndoorAtlas, Navigine, ORiiEnt, and MapsIndoors. Positioning engines offer advanced APIs for anchor management, route calculation, and analytics, providing libraries with the ability to draw actionable insights from user interactions, (IndoorAtlas, 2024).

To empower library staff in creating location-specific content, effective **Content Management Systems (CMS)** or experience managers are indispensable. These tools enable the creation of location-bound annotations, exhibit notes, or event-triggered alerts without necessitating any developer involvement. They are often integrated with indoor mapping platforms or offered as stand-alone plugins, further streamlining the content creation process.

At the heart of library operations lies the **Service Layer**, which dynamically consumes the location and contextual data processed in previous layers. This layer supports user-facing applications, such as mobile applications that utilize location data to deliver customized search results, trigger notifications, and facilitate navigation throughout the library. The environment is API-driven, frequently utilizing standards like the OGC SensorThings API to unify data access and operational control. Notably, Tasking and Eventing within this API framework are essential for responding to location data, allowing libraries to create geofences that automate service delivery.

Policy and Governance Layers are vital to maintain ethical standards around user data. Libraries must prioritize privacy by implementing defaults that favor user consent and clarity. Transparent, user-friendly opt-in processes are necessary to instill trust and foster engagement with location-based services.

Through these carefully constructed layers of technology and governance, libraries can harness the power of location-based services to enhance patron engagement, support educational missions, and create more interactive and personalized experiences within their spaces.

Building Effective Location-Based Services for Libraries

Delivering robust, context-aware library services necessitates a thoughtfully constructed software stack coupled with an ongoing commitment to transparent evaluation. The backbone of modern library location-based services (LBS) hinges on an intricate ecosystem of device SDKs and mapping services that collectively facilitate sensor reading, map rendering, anchor management, route calculation, and comprehensive analytics. Industry-leading commercial platforms such as Google Maps Platform, IndoorAtlas, Mapbox, and ORiiEnt are at the forefront, providing advanced features like 3D rendering, offline mapping capabilities, sensor fusion mechanisms, and APIs designed for content management and seamless notification triggers.

For advanced institutions that demand greater control and enhanced privacy, open-source solutions and custom sensor fusion frameworks offer an alternative, enabling precise tailoring to specific needs. Given the rapid pace of technological advancement, platform vendors are continuously refining their offerings, focusing on developer ergonomics and privacy features. For instance, in 2024, Mapbox introduced AI-assisted navigation along with enriched SDK capabilities, fostering a richer mobile user experience and improved visualizations. IndoorAtlas, leveraging a combination of geomagnetic signatures, Wi-Fi, Bluetooth Low Energy (BLE), and inertial sensors, minimizes reliance on specialized hardware, making it an attractive option for institutions constrained by financial limitations.

Google's indoor mapping features have garnered widespread adoption, particularly for public building floorplans, as they adeptly support indoor mapping, layer management, and enhanced searchability. As libraries navigate the selection of these platforms, it is crucial to strike a harmonious balance between the strength of privacy measures, the ease with which non-technical staff can author content, the reliability of offline capabilities, and seamless integration with existing local library management systems (LMS).

To ensure uninterrupted service delivery, fallback solutions such as QR codes and NFC anchors must also be readily available, securing continuity in the event of system failures or the inability to maintain continuous positioning. When evaluating potential platforms, libraries should prioritize several pivotal criteria: first, robust privacy features like local processing and minimal telemetry; second, content management capabilities that empower non-technical staff; third, ease of integration with existing LMS and event management systems; and fourth, flexible fallback options to cushion against interruptions in service due to positioning failures.

Given the dynamic nature of the technological landscape, libraries should also establish a framework for regularly reviewing platform policies, updates to SDKs, and security practices. This proactive approach not only safeguards user data but also ensures that library services remain at the cutting edge of innovation, catering effectively to the ever-evolving needs of their communities.

Table 2. Key software stacks and their key strengths

Software Stack	*Privacy Strength*	*Ease of Content Authoring*	*Offline Capability*	*Notes*
Google Maps API	Moderate (depends on usage policies)	High	Limited (mostly online)	Widely used, rich content creation tools
Mapbox	Strong (offers data control options)	High	Some offline support	Flexible styles and custom content
HERE Location Services	Strong (GDPR compliant)	Moderate	Good	Supports offline maps and location
Open Source (e.g., OpenStreetMap + custom LBS)	Very strong (self-hosted control)	Moderate	Excellent	Full control, but more authoring effort
Custom Sensor Fusion Frameworks (e.g., ROS, proprietary)	Varies by implementation	Moderate to low	Usually strong	Tailored to sensor integration but complex

Evaluating Performance, Adoption, and User Experience

A comprehensive evaluation of LBS solutions necessitates a multidimensional approach. Technical performance evaluation encompasses accuracy, latency, and adaptability to change. Operational metrics encompass deployment maintenance, update cycles, and hardware lifecycle management. User metrics, such as opt-in rates, engagement with notifications, and overall satisfaction, hold equal significance. Performance evaluation should encompass these multifaceted aspects. Technical metrics include accuracy (representative median error), latency (time from sensor input to a stable location fix), and resilience to environmental fluctuations. Operational metrics include deployment maintenance duration, fingerprinting update frequency, and beacon battery lifecycle management. User-facing metrics are equally crucial: opt-in rates, notification open and clickthrough rates, time saved in locating items, and subjective satisfaction scores.

Empirical studies demonstrate that patrons derive the greatest benefit from navigation tools that provide immediate and tangible outcomes. First-time visitors and those seeking unfamiliar materials frequently express the highest level of appreciation, while excessive notifications can erode trust. Consequently, notification design should be conservative, emphasizing the provision of essential content. Recent studies indicate that users value navigation tools most when they offer clear and immediate benefits, such as reducing the time spent searching for materials or discovering nearby events. First-time visitors and occasional users disproportionately benefit from wayfinding tools. Conversely, ill-timed or irrelevant push notifications promptly lead to opt-outs and a decline in trust. Therefore, content strategies should be conservative, prioritizing essential and time-sensitive notifications and deferring broader promotional pushes to opt-in marketing channels.

Accessibility testing must be integral to evaluation. Systems that depend on visual AR overlays must also offer audio cues, high-contrast maps, and keyboard-navigable interfaces. Pilot deployments should include participants with a range of physical and sensory abilities to ensure that the system's value extends to those whose interactions differ from mainstream mobile paradigms.

Security, Resilience, and Governance

Technical security concerns are straightforward yet significant. Technical security is paramount for any system managing the movement and preferences of individuals within library spaces. Beacons, RFID readers, edge servers, and associated hardware must be physically secured to prevent tampering or unauthorized relocation. Beacons and readers must be physically secured to mitigate tampering, back-end services must enforce the principle of least privilege, and communications channels should employ modern encryption and authentication methods.

Sensitive use cases, such as those involving collections whose association could reveal patron research interests, necessitate stricter governance practices. These include avoiding the storage of personally identifiable location histories, limiting analytics to aggregated signals, and designing processes that prevent location triggers for sensitive collections from producing persistent logs linking patrons to those resources. All communications between sensors, mobile devices, and back-end systems should utilize state-of-the-art encryption and authentication methods, such as TLS 1.3, mutual certificate validation, and, whenever feasible, local VPN tunnels for particularly sensitive collections or management endpoints.

Resilience encompasses operational practices such as scheduled beacon maintenance, fingerprint re-capture cycles, and content review to mitigate link rot. Operational resilience is not merely technical but also institutional. Scheduled beacon maintenance, battery replacement, periodic Wi-Fi fingerprint recaptures, and QR code link verification ensure the continued reliability and trustworthiness of systems. Front-line staff should receive training to address patron inquiries regarding these systems, communicate privacy policies, and respond effectively to technical disruptions.

Governance encompasses policy statements that establish LBS's position within the library's comprehensive privacy and service charters. It also entails the public documentation of the data lifecycle and staff training to empower frontline employees to confidently address patron inquiries. Governance documents must align LBS with the library's broader privacy charter. Data lifecycle statements, encompassing collection, usage, storage, retention, and deletion, should be readily accessible in plain language at the point of use and on library websites. Independent audits and regular policy reviews enhance public accountability.

When implemented conscientiously, intelligent navigation and engagement technologies reposition the library as a responsive, explorative learning environment. The maturity of commercial SDKs, sensor-fusion engines, and open hardware platforms allows accessible deployment to libraries of varying scale. Ultimately, only systems that combine layered technical security, robust governance, and inclusive design will realize the promise of place-aware, context-empowering services aligned with the library's public mission.

Intelligent navigation and engagement technologies reposition libraries from static repositories of knowledge into responsive learning environments that connect people to resources where and when they will be most useful. Software ecosystems have matured—commercial SDKs, indoor mapping platforms, and sensor-fusion engines now make high-quality LBS achievable for many institutions. Still, technical possibilities must be balanced against professional values: consent, privacy, equity, and accessibility. Libraries that pair layered, resilient technical architectures with robust governance and inclusive interaction design will realize the most transformational and ethically sound benefits from location-aware services.

Table 3. Key technical terms and alternatives

Technical Term	*Definition/Simple Explanation*	*Potential Alternative/ Simplification*
Location-Based Services (LBS)	Technology that uses a device's geographical position to provide information or services.	Location tracking or Smart guidance system
Artificial Intelligence (AI)	Computer systems that can perform tasks that typically require human intelligence, like learning and decision-making.	Smart technology or Computer learning
Machine Learning	A specific application of AI that allows systems to learn and improve from experience without being explicitly programmed.	Pattern analysis or Self-learning algorithms
Hyper-personalized	Recommendations or services tailored very precisely to an individual user's specific needs, preferences, and location.	Highly customized or Super tailored
5G Technology	The fifth generation of wireless communication, offering much faster speeds and lower connection delays than previous generations (like 4G).	Next-generation wireless or High-speed internet

continued on following page

Table 3. Continued

Technical Term	*Definition/Simple Explanation*	*Potential Alternative/ Simplification*
Augmented Reality (AR)	A technology that overlays digital images or information onto a person's view of the real world (e.g., seeing a digital map when looking at a bookshelf through a phone).	Digital overlays or Interactive real-world guidance
Low-latency connectivity	A very fast connection with minimal delay (latency) between sending and receiving data. Essential for smooth real-time services.	Instant connection or Super-responsive network

THE TRANSFORMATIVE POTENTIAL OF LOCATION-BASED SERVICES IN DIGITAL LIBRARIES

For a digital library seeking to expand its footprint and relevance, an LBS-enabled mobile application presents a highly potent outreach tool. Such an application allows the library to establish geofences around specific neighborhoods, enabling the targeted promotion of its unique resources to nearby residents. While the initial reaction to targeted advertising—seeing promotions for a local library or museum's digital collection—might be one of mild annoyance, this quickly gives way to curiosity regarding the exceptional collections and archives being advertised. If digital libraries are to thrive and grow in popularity, they must proactively harness available tools to disseminate their offerings. Providing a direct, contextually relevant link to digital collections via an LBS is arguably one of the most effective methods to achieve this outreach and ensure future growth.

Connection and Community: Defining Location-Based Service Utility

Location-based services can be used to connect all types of media, including social media, the Internet, and in-person engagement. The capacity to identify consumers near retailers and restaurants and navigate the desired areas are just a few of the applications of LBS. Consumers are reached and engaged via LBS based on their location. More than just sales can be done with an LBS. By partnering with charity and delivering social good through LBS, you can have a great influence on your brand (DiStaso & Khan, 2015). The Pennies for Life Campaign, for example, launched an SMS campaign to increase donor engagement by allowing them to see the impact of their donation thanks to location-based technologies.

If a digital library produces an app that leverages LBS, they can target people in their neighborhood who are interested in learning more about the library. When consumers start seeing adverts for their local library or museum's digital collection, they are likely to be irritated at first because they are being targeted, but then they will be curious to see what exceptional collection or archive has been advertised. If digital libraries are to grow in popularity in the future, they must begin to use the tools available to them to spread the word, and what better way to do so than by providing a direct link to their collection via an LBS.

Libraries are already benefiting from LBS in a variety of ways. Staff at the Southern Connecticut State University Library, for example, categorize and group literature works by department, study level, and courses. When materials need to be located in the library, these layouts enable quick service. The web-based library also serves as a resource for students and researchers by directing them to relevant topics and, in certain circumstances, other cooperating libraries to meet their needs.

Smart libraries, described by Gul and Bano as "a developing and novel technological environment of the twenty-first century," make use of location-based services to imagine new and exciting technologies that might be used to establish a smart library. The "magic mirror" is one such gadget that the authors anticipate. Location recognition, a type of location-based service, is one of the features of the "magic mirror." Through the use of RFID (radio frequency identification) technology, a mix of cameras, speech recognition, and location-based services can be used to sense and report the status of the library's collection in real time, (Gul & Bano, 2019).

By tracking active RFID chips in proximity to the sensors, RFID can be used as a local location-based service. Internal batteries in active RFID chips broadcast to the sensors. There are no batteries in passive RFID chips, therefore they must be scanned by a sensor. For a more convenient check-out experience, my local library has passive RFID chips implanted in all of its materials. RFID chips that are passive cannot be utilized for tracking. Within Disney World, RF tracking technology is used. The Disney wristbands have RF trackers that are detected by park sensors and used to track a person's location.

Patron's tracking isn't the only use case for location-based services. It's also useful for keeping track of materials. Personally, I'd like to see a "magic mirror" system that reflects the genuine state of books. My local library has a large number of lost and missing books that are not included in the catalog. It's quite aggravating. With this technology, incorrect item statuses and misplaced materials in the library would be a thing of the past. As Disney already utilizes this technology in its Florida park, a version of the "magic mirror" is a very conceivable device that can be implemented in libraries today.

The combination of augmented reality and location-based services offers up a world of possibilities. Location-based services and augmented reality can be combined in a library setting to create information bulletins depending on being in a specific library department or visual guidance to find a book within the library. Augmented reality in conjunction with location-based services has a lot of potential. Both the "magic mirror" and augmented reality combined with location-based services are very feasible technologies that can be used in the library provided the library is prepared to invest the necessary funds.

Applications of LBS in Library Services

The application of LBS in libraries is a natural extension of the shift toward providing mobile and user-centric services. LBS can transform the library from a static physical space into a dynamic, interactive environment. These technologies enable librarians to move beyond traditional service models and offer innovative solutions that meet the evolving needs of their patrons.

One significant application of LBS in libraries is indoor navigation and wayfinding. Libraries, particularly large academic or public ones, can be daunting to navigate. Patrons often struggle to find specific books, sections, or even service desks. LBS, often powered by Bluetooth beacons or Wi-Fi fingerprinting, can provide a digital map of the library and guide users directly to their desired destination. A study by Cheng, Li, and Liu (2017) highlighted how the City University of Hong Kong Library launched beacon-based navigation software to help patrons locate resources and different areas within the library using their mobile devices.

LBS can enhance the discovery of library resources. By integrating LBS with the library's catalog, a user standing in front of a bookshelf could receive a notification on their mobile device with details about the books on that shelf, including synopses, reviews, and availability information. This contextual information enriches the browsing experience and can lead to serendipitous discoveries. LBS also allows for personalized recommendations based on a user's location and borrowing history. For instance, a user in the new arrivals section might receive a notification about a new book by an author they've previously enjoyed.

The Internet of Things (IoT) is a related concept that has been widely adopted in library settings. A study on the development of smart academic library services in Hong Kong found that IoT technologies, including those that rely on location data, are seen as a way to improve library operations and provide personalized services. These technologies can facilitate resource tracking, inventory management, and even user identification to enhance the overall efficiency and security of the library.

At least four additional usefulness of LBS for library services provided to mobile users:

Wayfinding and Navigation

Uses a user's location to provide turn-by-turn directions within a physical space. It's especially useful in large, complex environments like multi-story libraries. Provide turn-by-turn directions to help users navigate within a physical space. Navigating large library facilities can be daunting, especially for new users. LBS technologies, such as indoor positioning systems (IPS) using BLE beacons or Wi-Fi triangulation, help users locate specific resources, facilities, or services within a library. Example: Beacon-Based Navigation at the University of Illinois Library.

A common frustration for library users is finding a specific book on a vast shelf or locating a particular study area. Using LBS, a library app can act as a personal navigator. After a user search for a book in the online catalog, the app can provide a digital map with a clear path from their current position to the exact shelf where the book is located. This feature not only reduces user frustration but also makes the library more accessible to new visitors. The University of Illinois at Urbana-Champaign implemented a beacon-based navigation system to guide users through its expansive library. BLE beacons placed throughout the facility communicate with a mobile app, providing turn-by-turn directions to specific books, study rooms, or amenities like restrooms and cafes. The system reduces user frustration and enhances accessibility, particularly for individuals with disabilities.

Real-Time Information

Users receive data that is relevant to their immediate surroundings. Delivery up-to-the-minute data to a user based on their current location. This is one of the most dynamic features of LBS, as it allows for immediate, contextual communication. Libraries can use this feature to provide live updates to patrons.

A student standing near a study room could receive a notification about which rooms are currently available, or a patron near the information desk could get an alert about a librarian who is an expert on the topic they're searching. This allows for fluid and responsive service, eliminating the need for patrons to actively search for information that is immediately relevant to them.

Personalization Recommendations

Services are tailored to the user's current location, preferences, and historical data. This feature goes beyond location to combine a user's geographical data with their past behavior and preferences to offer tailored suggestions. This creates a much richer and more relevant user experience.

By analyzing a patron's borrowing history and their current location, a library can provide highly targeted recommendations. If a user is browsing the science fiction section, the library app could suggest new books by authors they've previously enjoyed or highlight recent acquisitions in that genre. This feature helps patrons discover new materials they might otherwise miss and makes the browsing experience feel more curated and personal. The system learns and adapts over time, making future recommendations even more accurate. The New York Public Library (NYPL) uses geofencing to send push notifications about events, workshops, or exhibitions to users within a certain radius of a branch. For example, a user near the NYPL's Stephen A. Schwarzman Building might receive an alert about a nearby rare book exhibition, encouraging spontaneous participation

Proximity-Based Triggers

Automatically activate or send notifications when a user enters a predefined area (geofencing). Proximity-based triggers, often called geofencing, use predefined virtual boundaries to automate actions. When a user's device enters or exits one of these zones, a specific action is triggered, such as a push notification or an app alert. This feature enables highly contextual communication.

A library could set up a geofence around its main entrance. When a patron enters this area, their phone might automatically receive a welcome message or an alert about today's hours or a special event. Another example is a trigger near the self-checkout kiosks that reminds patrons to check out their materials before leaving. This proactive approach helps the library provide timely information without requiring active user input

The Integration of Location-Based Services

The integration of location-based services into library systems has transformed how users interact with library resources, spaces, and services. Location-based services leverage technologies such as GPS, RFID, Bluetooth Low Energy (BLE), Wi-Fi, and mobile applications to deliver context-aware, location-specific information and functionalities to users. In recent years, libraries have embraced location-based services to improve accessibility and engagement for their patrons. Mobile applications have become vital tools in this transformation. For instance, the New York Public

Library's app not only helps users find the nearest library branch using GPS but also enables them to check in for events, facilitating a more interactive experience.

Libraries are also using LBS for collection management and user engagement. Radio-Frequency Identification (RFID) technology, a form of LBS, has been implemented in many libraries to facilitate efficient checkout, inventory, and theft management, (Lin et al., 2023). For example, the University of Hong Kong Libraries implemented RFID in 2008 to enhance resource management, (Sidorko, 2009). Beyond collection management, LBS can be used to notify patrons of library events, workshops, or new acquisitions when they are in a specific area of the building.

"A Library Location-Based Service: The App 'Bavaria in Historical Maps' Presents Bavarian History on the Mobile Internet," Ceynowa (2015) highlights a pioneering example of a library leveraging mobile technology to provide location-based access to historical content. The service works by integrating the Bavarian State Library's digitized historical maps with geolocation functionality on mobile devices. Users can access six geo-referenced historical maps, which are overlaid onto modern maps through the app, allowing them to explore Bavaria's historical landscapes in context. By tapping or navigating to a specific location, the app displays the corresponding historical map and related information, creating an interactive, immersive experience. This location-based approach not only enhances public engagement with historical materials but also demonstrates how libraries can transform traditional collections into dynamic, mobile-accessible learning tools.

Geolocation tools are also being used to enhance the user experience. These tools enable libraries to send notifications about events or resources that are tailored to users' locations. For instance, a user near a library might receive timely alerts about upcoming book signings or author talks. Several studies conducted in the library and information science in 2022 revealed that personalized notifications based on geolocation can significantly increase attendance at library programs by up to 30%. This demonstrates the potential of such technologies in driving user engagement, (Mohammed & Zidan, 2023)

Additionally, many libraries are adopting interactive maps on their websites to assist users in navigating their facilities. The Boston Public Library provides an online interactive map that highlights essential areas such as study zones, restrooms, and special collections. This feature not only aids users in finding their way but also enriches their overall experience. Liu and Lewis (2023) found that users who utilized interactive maps reported a higher satisfaction rate, attributing to the ease of access and reduced frustration in locating resources.

The potential of LBS is vast, but its implementation is not without challenges, particularly concerning user privacy. The collection of location data, even within a library, raises ethical questions. Yan Quan Liu's earlier work on Library 2.0 and privacy in 2007, while not specifically about LBS, highlighted the broader dilemma of protecting patron privacy in an era of data-intensive technologies. This concern remains relevant today. Libraries must be transparent about what data is collected and how it is used, ensuring users have the option to opt-in or opt-out of LBS.

Future research on LBS in libraries, such as Liu's 2023 work on smart service and knowledge discoveries, points to the integration of LBS with other advanced technologies like recommender systems and artificial intelligence. While there is no published work from Liu as of the current date, the trajectory of his research suggests a continued focus on how libraries can leverage technology to create intelligent, responsive, and personalized user experiences, (Liu, 2023).

Practice and Studies on Library Location-Based Services

Library location-based services in libraries bridge the gap between physical collections and digital interfaces, transforming traditional services into dynamic, user-driven experiences. For instance, LBS can facilitate indoor navigation to locate books, shelves, or study areas, reducing the time users spend searching. They also support location-based recommendations, where users receive suggestions for nearby resources based on their browsing history or current position. Mobile apps integrated with LBS enable notifications for events, due dates, or available study rooms. Furthermore, LBS contributes to accessibility by assisting users with disabilities through voice-guided directions or augmented reality overlays.

The growing body of scholarship on library LBS underscores both the maturity and promise of these technologies. Liu and Briggs (2015) documented the foundational phase of mobile service adoption among top academic libraries, identifying mobile catalogs and print management as precursors to full LBS integration. Studies highlight how these applications enhance user engagement and resource utilization, particularly in academic settings where mobility is key, (Liu & Briggs, 2015). Building on this foundation, libraries have since moved toward mobile-first design philosophies that enable real-time, place-based interaction.

One of the most compelling early examples of a library LBS system was *Bavaria in Historical Maps*, developed by the Bavarian State Library, (Ceynowa, 2015). This mobile app used geolocation to overlay digitized historical maps atop modern cartography, allowing users to explore Bavaria's changing landscapes through an immersive, interactive interface. More recent studies demonstrate similar innovations in public engagement, such as Mohammed and Zidan's (2023) research showing that geolocation-triggered event notifications can increase attendance at library programs by as much as 30%.

The Boston Public Library's interactive digital map, launched in 2023, exemplifies the integration of geospatial awareness into public library services. The map enables visitors to locate study areas, restrooms, and special collections, significantly improving user satisfaction (Liu & Lewis, 2023). Such projects reflect a broader global trend: the transformation of libraries into intelligent spaces where physical and digital experiences converge seamlessly.

The implementation of LBS in libraries is not just a theoretical concept; it is a growing practice supported by various case studies and research. Early work in mobile services laid the groundwork for the adoption of LBS. Liu and Briggs, in their 2015 study, explored the state of mobile services in the top 100 university libraries, identifying mobile apps, catalogs, and printing as some of the most frequently offered services. This research underscored the trend of libraries embracing mobile technology, which is a prerequisite for implementing LBS.

Obstacles and Opportunities of LBS in Libraries

Despite its many benefits, the adoption of LBS in libraries raises important ethical and logistical considerations. Chief among these are user privacy, technological cost, and equitable access. Tracking user movements, even anonymously, may generate apprehension among patrons who value confidentiality in their research habits. As Liu (2007) argued in the context of Library 2.0, maintaining user trust in data-intensive environments requires transparency, consent, and responsible stewardship of personal information. Libraries must therefore implement clear opt-in mechanisms and data retention policies to safeguard user autonomy.

While LBS offers significant benefits, challenges include privacy concerns, technological costs, and accessibility. Users may hesitate to share location data due to privacy fears, necessitating transparent data policies. Additionally, implementing LBS requires investment in infrastructure like beacons or IoT sensors, which may strain library budgets. Ensuring accessibility for users with disabilities or those without smartphones is also critical.

Library location-based services represent a transformative approach to enhancing user engagement and operational efficiency. From navigation and resource discovery to immersive AR experiences like the "Bavaria in Historical Maps" app, LBS leverages cutting-edge technologies to make libraries more accessible and engaging. As libraries continue to adopt smart technologies, such as those outlined in Khaled's (2024) Smart-LBS platform, they will further redefine their role as dynamic, user-centered spaces in the digital age.

That's a great exercise to improve flow and avoid repetition! Here is a rewritten version of the three paragraphs, focusing on clarity, varying sentence structure, and eliminating the repeated opening word.

Location-Based Services (LBS) in libraries are rapidly moving toward deep integration with other advanced technologies. Research, such as focusing on smart service and knowledge discoveries (Liu, 2023), clearly points toward combining

LBS with artificial intelligence (AI) and machine learning. This powerful integration can analyze real-time location data alongside borrowing history to provide hyper-personalized, predictive recommendations. As connectivity advancements, like 5G technology, improve the speed and reliability of LBS, the promise of seamless augmented reality and truly real-time services within the library will move from concept to widespread reality.

Integrating AI and machine learning is the central next step for library LBS, significantly enhancing personalization and predictive capabilities. For example, AI can analyze precise location data in conjunction with borrowing history to offer highly customized recommendations, moving beyond simple genre suggestions. Furthermore, advancements in 5G technology are set to enhance the speed and reliability of LBS connections, enabling a more seamless experience for augmented reality (AR) applications and real-time informational services within the physical space.

The next generation of LBS will go beyond information transmission to become immersive and experiential as high-speed, low-latency networks are widely deployed. Customers will be able to interactively explore collections, exhibitions, and study areas using real-time augmented and virtual reality applications, combining digital and physical environments into a cohesive, highly customized library experience. The way libraries anticipate, and address user demands will change as a result of this technological convergence. AI-driven analytics can estimate resource requirements, strategically improve spatial layouts, and provide highly tailored suggestions by combining geographical data with circulation patterns. These opportunities are further enhanced by the broad deployment of 5G networks, which offer the high-bandwidth, low-latency connectivity required for complex real-time augmented and virtual reality apps that direct and educate patrons around the library.

REFERENCES

Abolfazli, S., Sanaei, Z., Ahmed, E., Gani, A., & Buyya, R. (2013). *Cloud-based augmentation for mobile devices: Motivation, taxonomies, and open challenges.* arXiv. https://arxiv.org/abs/1306.4956

Ahmed, S., Singh, A., & Patel, R. (2024). Emerging trends in edge computing: A comprehensive survey. *Journal of Computer Networks and Distributed Systems, 12*(2), 55–71.

Ai, Y., Peng, M., & Zhang, K. (2018). Edge computing technologies for Internet of Things: A primer. *IEEE Network, 32*(1), 68–75.

Al-Okby, M. F. R., Junginger, S., Roddelkopf, T., & Thurow, K. (2024). UWB-based real-time indoor positioning systems: A comprehensive review. *Applied Sciences (Basel, Switzerland), 14*(23), 11005. DOI: 10.3390/app142311005

Bregar, K. (2023). Indoor UWB positioning and position tracking data set. *Scientific Data, 10*(1), 744. DOI: 10.1038/s41597-023-02639-5 PMID: 37884571

Ceynowa, K. (2015). A library location-based service: the app 'Bavaria in Historical Maps' presents Bavarian history on the mobile internet. *M-Libraries* 5: From devices to people, 5, 189.

Cheng, Z., Li, L., & Liu, J. (2017). Identifying the spatial effects and driving factors of urban PM2. 5 pollution in China. *Ecological Indicators, 82*, 61–75. DOI: 10.1016/j.ecolind.2017.06.043

Cisco. (2024). *Global networking trends 2024.* Cisco Systems. https://www.cisco.com/c/dam/global/en_uk/solutions/enterprise-networks/2024-global-networking-trends.pdf

Cisco. (2025). *Wi-Fi 7: The growing future of wireless design guide.* Cisco Systems. https://www.cisco.com/c/en/us/products/collateral/networking/wireless/wifi7-future-of-wireless-dg.html

Daniş, F. S. (2023). Radio-Frequency Map Optimization for Indoor Positioning and Tracking. *Sakarya University Journal of Computer and Information Sciences, 8*(3), 410–421. DOI: 10.35377/saucis...1644762

Ericsson. (2025). *Ericsson mobility report* (June 2025 edition). Ericsson. https://www.ericsson.com/en/reports-and-papers/mobility-report

Google Maps Platform. (2024). *Indoor Maps for buildings and venues.* https://mapsplatform.google.com/

Grand View Research. (2024). *Wi-Fi 6, Wi-Fi 6E, and Wi-Fi 7 chipset market report 2024*. Grand View Research.

Gul, S., & Bano, S. (2019). Smart libraries: An emerging and innovative technological habitat of 21st century. *The Electronic Library*, *37*(5), 764–783. DOI: 10.1108/EL-02-2019-0052

Guo, Y. J., Liu, Y. Q., & Bielefield, A. (2018). The provision of mobile services in US urban libraries. *Information Technology and Libraries*, *37*(2), 78–93. DOI: 10.6017/ital.v37i2.10170

Hahn, J. (2018). *The Bibliotelemetry of Information and Environment: An Evaluation of IoT-Powered Recommender Systems*. arXiv.

Hu, L., Chen, Y., Cao, E., & Hu, W. (2025). User experience & usability of wearable health device: A bibliometric analysis of 2014–2023. *International Journal of Human-Computer Interaction*, *41*(8), 5100–5119. DOI: 10.1080/10447318.2024.2357905

ICS Development Team. (2020). *Here's Why You Should Separate UI from Business Logic in Your Application*. ICS - Integrated Computer Solutions. https://www.ics.com/blog/heres-why-you-should-separate-ui-business-logic-your-application

IndoorAtlas. (2024). Indoor positioning platform documentation. IndoorAtlas. https://www.indooratlas.com/

Khaled, A. (2024, May). Smart Location-based Services (Smart-LBS): Platform for Smart Space-Independent LBS. In *2024 IEEE International Conference on Electro Information Technology* (eIT) (pp. 643-650). IEEE. DOI: 10.1109/eIT60633.2024.10609850

Khan, A. U. (2024). Beyond bookshelves: How 5G/6G technology will reshape libraries. *Library & Information Science Research*, *46*(3), 210–225.

Lee, K. B. (2011). Developing Mobile Collaborative Learning Applications for Mobile Users. *International Journal of Interactive Mobile Technologies*, *5*(4), 42. DOI: 10.3991/ijim.v5i4.1823

Lin, X., Sun, Y., & Zhang, Y., & MR, K. (2023, February). Application of AI in library digital reading promotion service. In *2023 IEEE International Conference on Integrated Circuits and Communication Systems (ICICACS)* (pp. 1-6). IEEE. DOI: 10.1109/ICICACS57338.2023.10100096

Liu, Y. Q. (2007). Utilizing Recommender Systems in Digital Libraries, Current Study and Tech Issues. *Library and Information Service*, *51*(12), 6–18.

Liu, Y. Q. (2023). *Digital library of high school virtual laboratories: Empowering STEM education through mobile access. CSU Grant 2023*. Southern Connecticut State University.

Liu, Y. Q., & Briggs, S. (2015). A Library in the Palm of Your Hand: Mobile Services in Top 100 University Libraries. *Information Technology and Libraries*, *34*(2), 133–148. DOI: 10.6017/ital.v34i2.5650

Liu, Y. Q., & Lewis, S. (2023). Services to mobile users: The best practice from the top-visited public libraries in the US. *Information Technology and Libraries*, *42*(1). Advance online publication. DOI: 10.6017/ital.v42i1.15143

Mao, Y., You, C., Zhang, J., Huang, K., & Letaief, K. B. (2017). A survey on mobile edge computing: The communication perspective. arXiv. https://arxiv.org/abs/1701.01090

Milano, F., da Rocha, H., Laracca, M., Ferrigno, L., Espírito Santo, A., Salvado, J., & Paciello, V. (2024). BLE-based indoor localization: Analysis of some solutions for performance improvement. *Sensors (Basel)*, *24*(2), 376. DOI: 10.3390/s24020376 PMID: 38257468

Mohammed, M. S., & Zidan, K. A. (2023). Enhancing attendance tracking using animated QR codes: A case study. *Indonesian Journal of Electrical Engineering and Computer Science*, *31*(3), 1716. DOI: 10.11591/ijeecs.v31.i3.pp1716-1723

Ocran, T. K., Underwood, P. G., & Arthur, P. A. (2020). Strategies for successful implementation of mobile phone library services. *Journal of Academic Librarianship*, *46*(5), 102174. DOI: 10.1016/j.acalib.2020.102174

Oulasvirta, A. (2008, September). Designing mobile awareness cues. In *Proceedings of the 10th international conference on Human computer interaction with mobile devices and services* (pp. 43-52).

Partners, S. T. L. (2024). *Mobile edge computing: Industry overview and applications*. https://stlpartners.com/articles/edge-computing/mobile-edge-computing

Pereira, J. L. S., Andrade, R. L., Vilela, L. F., Oliveira, M. M., Marques, L. N., Inácio, J. V., Aguiar, M. D., Lima, R. N., Pinton, L. O., Morosoli, A. C. C., Reale, L. P. A., Rocha, J. P. M., Maciel, G. I. F., Moreira, R. V., & Cerqueira, R. L. (2025). Uso de wearables para monitoramento contínuo de pacientes com doenças cardíacas. *Caderno Pedagógico*, *22*(1), e13284–e13284. DOI: 10.54033/cadpedv22n1-091

Quaye Affum, M. (2022). Assessment of the impact of mobile libraries on traditional libraries. *International Journal of Computing. Programming and Database Management*, *3*(1), 45–48.

Ram Prakash, R. (2023). Use of mobile technology in library systems. *NeuroQuantology : An Interdisciplinary Journal of Neuroscience and Quantum Physics, 21*(13), 24–30. DOI: 10.14704/nq.2023.21.13.NQ88302

Sidorko, P. E. (2009). Virtually there, almost: Educational and informational possibilities in virtual worlds. *Library Management, 30*(6/7), 404–418. DOI: 10.1108/01435120910982104

Straits Research. (2025). *Progressive Web Apps: Market size and global outlook 2025*. Straits Research.

Ubogu, A., & Chukwusa, O. (2024). Mobile library services revisited: Applications, prospects, and challenges in Nigerian academic libraries. *Delta Journal of Computing. Communications & Media Technologies, 1*(1), 1–12.

Vasantha Kumar, P. J. R. (2022). Impact of mobile technology in library services. *Indian Journal of Library Science and Information Technology, 7*(1), 40–43. DOI: 10.18231/j.ijlsit.2022.008

Weiser, M. (1991). The Computer for the 21 st Century. *Scientific American, 265*(3), 94–105. DOI: 10.1038/scientificamerican0991-94 PMID: 1675486

Yang, S., Yang, J., Yang, C., Yan, W., & Li, P. (2025). Human Activity Recognition using WiFi Signal Features and Efficient Residual Packet Attention Network. *IEEE Sensors Letters, 9*(4), 1–4. DOI: 10.1109/LSENS.2025.3551337

Chapter 5
Mobile Reference Services

ABSTRACT

This chapter traces the evolution of mobile reference services in libraries from early remote communication tools to today's multifaceted, real-time support systems. It reviews emerging practices in SMS reference, chat consortia, volunteer models, and integrated knowledge-base platforms, highlighting benefits and persistent challenges. Global collaboration and standards for usability and user perception are examined. The chapter also explores the rise of chatbots in public and academic libraries, assessing their impact, limitations, and design considerations. Emphasis is placed on user-centered development, iterative refinement, ethical concerns, and anticipated constraints in building sustainable, effective mobile reference ecosystems.

INTRODUCTION

As technology advances, the manner in which we access information undergoes transformation. Mobile reference services have emerged as a prominent feature, enabling users to conveniently retrieve information anytime, anywhere. The focus delves into the advantages and challenges associated with mobile reference services, exploring their potential to enhance the user experience. Whether an individual is a student, professional, or simply seeking knowledge, mobile reference services provide a user-friendly platform for accessing information at any time and location.

"Mobile Reference Services" (MRS) are a convenient and efficient way for users to access information on-the-go. With the increasing use of smartphones and other mobile devices, mobile reference services have become an essential tool for libraries and other information providers to reach their users. These services offer a wide range of resources and tools, including access to online databases, e-books, and reference materials. They also provide users with the ability to ask questions and receive assistance from librarians and other experts in real-time. Overall, mobile reference services are valuable. MRS have become increasingly popular in recent

DOI: 10.4018/978-1-7998-9175-8.ch005

years, as they offer a number of advantages over traditional reference services, such as convenience, accessibility, and flexibility.

EVOLUTION OF MOBILE REFERENCE SERVICES

Mobile reference services have revolutionized library operations, enabling librarians to provide instantaneous assistance to patrons via smartphones, tablets, and other portable devices. These services encompass a diverse array of tools, such as text messaging, mobile applications, instant messaging, and access to digital resources on the go. As mobile technology has become pervasive, libraries have responded by accommodating the expectations of users seeking seamless, location-independent support. This exploration delves into the development of mobile reference services, current practices, seminal studies—including those authored by Yan Quan Liu—and strategies for advancement. Drawing upon credible scholarly sources, it elucidates how libraries can enhance accessibility and user engagement.

The roots of mobile reference services trace back to the early 2000s, when libraries began experimenting with SMS-based queries and basic mobile websites. Early initiatives focused on extending traditional reference desks into virtual spaces, allowing users to ask questions via text or email without visiting the library physically. For instance, academic libraries introduced iPad-based roving reference, where librarians used tablets to provide on-site assistance in non-traditional areas like cafeterias or study lounges. One early case involved the use of iPads to embed reference and instructional services in health sciences libraries, improving outreach to faculty and students, (Nahum-Shani et al., 2016)

By the 2010s, the proliferation of smartphones led to more sophisticated offerings, such as mobile-optimized catalogs and apps for database access. A study on reference services using iPads emphasized how mobility allowed librarians to expand service points beyond the desk, fostering greater outreach opportunities, (Lotts & Graves, 2011). This period also saw the integration of social media for reference, with platforms like Twitter enabling quick responses to user inquiries. The COVID-19 pandemic accelerated this shift, as libraries pivoted to remote services to maintain support during closures. Services expanded to include mobile renewals, database access, and remote interactions, ensuring users could engage from home, (Liu & Lewis, 2023).

Research underscores the value of mobile reference in enhancing library efficacy. Yan Quan Liu's studies offer critical insights into mobile adoption. A 2015 study of top U.S. university libraries found all offered at least one mobile service, including reference via text and apps, recommending proactive implementation and responsive design, (Liu & Briggs, 2015).

Liu and Lewis (2023) surveyed 151 institutions, noting widespread mobile references, apps, and database access. Liu's 2023 study of top-visited public libraries confirmed mobile reference and catalogs as staples, with post-COVID expansions enhancing remote interactions.

Complementary studies align with Liu's findings. In Nigeria, mobile reference predicted 35.3% of digital library service variance, with SMS and emails reducing access barriers, (Ogbole et al., 2023). A 2025 study highlights AI chatbots and instant messaging boosting efficiency, though digital divides remain a challenge, (Agarwal., Suarez-Tangil & Vasek, 2025).

PRACTICE OF MOBILE REFERENCE SERVICES

In an increasingly digital world, information seekers rely heavily on remote reference services to access accurate and timely information. Mobile reference services, also known as remote, remote or digital reference services or online reference services, have emerged as a vital component of library and information science. These services leverage the power of technology to connect librarians and information professionals with users in real-time, transcending the barriers of time and space. This essay explores the significance of remote reference services, their benefits and challenges, and how they contribute to empowering information access in the digital age.

Mobile reference services are now integral to library operations, with institutions offering tools tailored to mobile users. Common practices include mobile apps for catalog searches, text-a-librarian services, and instant messaging via WhatsApp or live chat. Public libraries frequently provide mobile apps, reference services, catalogs, and printing, with every surveyed library offering at least one mobile platform, (Liu & Lewis, 2023).

In today's digital world, remote reference services have become an essential tool for libraries and information centers to provide assistance to their patrons. These services, also known as online reference services, allow users to ask questions and receive answers in real-time through various electronic communication channels. The concept of remote reference services has been around for many years, but it has gained more popularity in recent times due to the widespread use of the internet and other digital technologies.

Remote reference services have evolved over time, and now they offer a range of communication channels such as email, chat, video conferencing, and social media. These services are available 24/7, which means that users can access them at any time of the day or night. They are also convenient for users who cannot physically visit a library or information center due to distance, disability, or other reasons.

Definition and Scope of Remote Reference Services

Remote reference services refer to the provision of library reference assistance through digital platforms, such as chat, email, video conferencing, or social media. These services extend the reach of traditional reference services beyond physical library spaces, offering users the convenience of accessing assistance from anywhere, at any time, (Cruz, 2020). Remote reference services facilitate information discovery, answer queries, provide research guidance, and offer support for various user needs.

Virtual reference services (VRS) are a type of library service that allows users to ask reference questions and receive assistance from librarians through online or digital channels, such as email, chat, or video conferencing. VRS have become increasingly popular in recent years, as they offer a number of advantages over traditional reference services, such as convenience, accessibility, and flexibility.

Virtual reference services offer numerous advantages over traditional in-person reference services, making them a preferred choice for many information seekers. Firstly, these services ensure accessibility for users regardless of their location. With just an internet connection, users can access expert assistance from professional librarians, regardless of their geographical location. This accessibility has particularly benefited underserved populations, such as those residing in remote areas or individuals with limited mobility, (Luo, 2020).

Benefits of Remote Reference Services

Accessibility and Convenience

Remote reference services break down physical barriers, providing equitable access to information for diverse user groups, (Al-Daihani & Jegede, 2022). Users can seek assistance from the comfort of their homes or while on the go, eliminating the constraints of time and location.

Personalized Assistance

Remote reference services enable librarians and information professionals to provide patrons with individualized and targeted assistance. With the aid of digital tools, librarians can access a vast array of resources and databases to locate accurate and pertinent information to respond to consumer inquiries. This customized service increases customer satisfaction and loyalty.

Expanded Service Hours

Remote reference services extend library assistance beyond regular operating hours, accommodating users with different schedules or in different time zones, (Warren, 2021). This flexibility ensures that support is available to users at their convenience.

Cost-Effective Solutions

Remote reference services also provide libraries and information centers with cost-effective options. They do away with the need for physical space and staff, which can be costly to maintain. They also help to save time and resources by decreasing the need for patrons to physically visit the library.

Multimodal Communication

Remote reference services offer a range of communication options, enabling users to choose their preferred method of interaction. Whether through live chat, email, or video conferencing, users can engage with librarians in a manner that suits their communication style, (Qayyum, Mahmood, & Alqahtani, 2022).

Challenges of Remote Reference Services

Technology and Infrastructure

Effective implementation of remote reference services requires robust technological infrastructure, including reliable internet connectivity and user-friendly platforms. Libraries must invest in appropriate software and hardware to ensure seamless communication and information exchange, (Nichols, 2020).

Lack of Personal Interaction

In-person reference interactions allow librarians to gauge users' needs through body language and non-verbal cues. To mitigate this limitation, librarians need to employ active listening skills and adapt their communication strategies to the remote environment, ensuring an empathetic and personalized user experience, (Holland, 2021). In a remote environment, librarians and information specialists may not see patrons face-to-face, making it difficult to grasp their information needs entirely. Despite this, the level of service offered is highly deserved. Inadequate or incorrect information provided by a reference service can lead to frustrated patrons.

Training and Skills

For effective online reference services, librarians need to hone a set of skills. Librarians along with other information specialists need particular training to deliver useful online reference services. To deliver top-notch service, they must be comfortable with and confident in utilizing various forms of digital technology and possess strong interpersonal and communication skills. Effective communication in remote environments, as well as familiarity with appropriate technologies, should be central topics in training programs, (Barker, 2021).

Privacy and Security Concerns

Remote reference services involve the exchange of sensitive information, requiring libraries to establish robust data protection measures. Safeguarding user privacy and ensuring the security of digital interactions are critical considerations, (Cruz, 2020). As remote reference services involve the exchange of personal data and sensitive information, librarians must implement robust data protection measures and adhere to privacy regulations to safeguard user confidentiality and trust, (Jiao et al., 2020).

Developments in Remote Reference Services

Integration of Artificial Intelligence (AI)

AI-powered chatbots and remote assistants hold immense potential for remote reference services. These technologies can provide instant responses, automate routine queries, and assist users with basic information needs, allowing librarians to focus on more complex inquiries, (Pike, 2021).

Enhanced Personalization

Future remote reference services may incorporate user profiles and preferences to offer tailored recommendations and resources. By leveraging data analytics and machine learning algorithms, libraries can provide highly personalized assistance to meet individual user needs, (Abidin, 2022).

Augmented and Virtual Reality

As these technologies advance, virtual reference services could employ augmented and virtual reality to create immersive and interactive experiences. Users could navigate digital library spaces, collaborate with librarians, and access resources in more engaging and innovative ways, (Warren, 2021).

Remote reference services have emerged as a valuable tool for libraries and information institutions to adapt to the digital age. These services provide users with accessible and convenient assistance, while also overcoming traditional limitations. However, challenges related to technology, training, and privacy must be addressed to ensure their successful implementation. Looking ahead, advancements in AI, personalization, and immersive technologies promise to revolutionize virtual reference services, further enriching the user experience and expanding the role of libraries in the digital world.

MOBILE REFERENCE TOOLS

With the rapid advancement of digital connectivity, virtual reference tools have revolutionized how libraries provide information services. Platforms like Text-A-Librarian, Ask-A-Librarian, and My Info Quest offer real-time, remote access to librarians, transcending geographical and temporal limitations.

These platforms and tools facilitate synchronous and asynchronous communication between librarians and users, transcending physical library boundaries. By leveraging chat, email, text messaging, and web conferencing, remote reference services provide timely, credible information to diverse user populations from any location, at any time

Text-A-Librarian and SMS-Based Reference

Text A Librarian services, such as those at Georgia's VSU and California State Northridge, allow users to send concise queries via SMS and receive prompt responses. Luo (2016) identified 49 specific competencies—including brevity, clarity, subject-matter knowledge, and technical proficiency—as essential for effective text reference. The evaluation of SMS/text reference services continues to show that these venues perform well for short, ready-reference questions but also reveal ongoing challenges.

Cassidy, Colmenares, and Martinez (2023) used behavioral rubric to assess SMS reference and found inconsistent staff attentiveness, variation in message tone, and follow-up issues in text services. The study concluded that even when throughput is prompt, user perception of service quality can lag due to inconsistent messaging protocols. Moreover, SMS platforms such as those supported by LibraryH3lp advertise seamless staffing and transfers across participating libraries (LibraryH3lp, 2024), pointing toward increasing adoption of networked text reference. However, the empirical literature emphasizes that text reference remains best suited for concise, fact-based questions rather than in-depth research consultations—because of both the character limitations and reduced capacity for full interactive reference interviews, (Luo, 2016).

Consortium-Driven Chat: Ask-A-Librarian

Ask A Librarian initiatives, such as Ontario's Scholars Portal and the Florida Remote Reference Cooperative, demonstrate the power of regional consortia to offer extended-hours, bilingual, and subject-expert staffing via systems like LibraryH3lp. Peer-reviewed assessments confirm that shared staffing models provide satisfactory user experience, with few service quality issues and effective user scheduling and privacy practices.

Ask A Librarian initiatives continue to show measurable positive user sentiment but also reveal important staffing and affiliation effects. A recent, large transcript-based sentiment analysis of a multi-institution Ask A Librarian consortial service by Barrett and Sharma (2025) examined more than 3,000 chat transcripts from 2019–2021 and found overall positive sentiment from both users and operators, while identifying that affiliation mismatch—when a user is served by staff from a different home institution—and busier shifts correlated with slightly lower satisfaction scores. This study reaffirms that consortial chat structures are effective but require deliberate alignment between staffing and institutional representation. Complementary research by Meyers-Martin (2024) further confirms that consortium-supported virtual reference systems deliver extended hours and multilingual coverage efficiently, yet they depend heavily on formalized protocols for training, referral, and quality monitoring to sustain service reliability across participating institutions.

Current developments in Canada and the United States demonstrate how consortia are evolving financially and operationally. The Ontario Council of University Libraries (OCUL), through Scholars Portal, continues to support more than 25,000 chats annually, serving approximately 400,000 students across member institutions, (Ontario Council of University Libraries, 2024). In Florida, the Tampa Bay Library Consortium's Ask a Librarian service has shifted from a fully grant-funded model to a cost-share structure for participating libraries, emphasizing the need for sustainable financial frameworks, (Tampa Bay Library Consortium, 2025). These changes underscore that while collaborative virtual reference models remain effective and valued by users, their long-term viability depends on shared governance, equitable cost distribution, and continuous investment in analytics-driven quality control.

My Info Quest and Volunteer-Based Models

My Info Quest exemplifies a self-organized, volunteer-driven remote reference model, where librarians across institutions collaboratively handle questions. Studies reveal variability in policy adherence, citation practices, and service consistency. Disparities in performance highlight the need for standardized guidelines, particularly for knowledge-sensitive fields like health.

My Info Quest and comparable volunteer-based models continue to attract attention as flexible, low-cost approaches to expanding reference coverage, especially where staffing or funding constraints limit traditional service models. Recent empirical and applied research underscores both the advantages and the limitations of such frameworks. Urrea and Yoo (2023) found that volunteer experience significantly influenced performance outcomes on online volunteer platforms, suggesting that consistent training and supervision are critical to service quality. Crittenden (2024) similarly identified that virtual volunteerism in library and educational settings thrives when clear expectations, accountability structures, and standard policies guide volunteer engagement.

While the My Info Quest framework remains agile and collaborative, current reports from 2024–2025 indicate that libraries employing volunteer-based digital reference must address persistent governance and policy gaps. Issues such as information accuracy, privacy management, and escalation of complex queries to specialists are central to maintaining credibility. Scholars further note that because volunteer networks rely on goodwill and ad hoc participation, the absence of structured oversight may lead to uneven service delivery—particularly in areas involving health, law, or academic citation standards, (Crittenden, 2024; Urrea & Yoo, 2023). To strengthen sustainability, current discussions emphasize the development of shared professional guidelines and competency-based training modules, ensuring that volunteer models can complement, rather than compromise, institutional service standards.

LibAnswers and Knowledge-Base Integration

Adopted by over a thousand institutions, Springshare's LibAnswers integrates chat support, SMS, and knowledge base capabilities. Preliminary implementations, such as at California State Chico, illustrate how LibAnswers enhances patron self-service through accessible FAQs, topic clouds, and robust back-end analytics.

LibAnswers continues to evolve as a leading platform for integrating live chat, ticketing, and knowledge-base management in academic libraries. Recent enhancements—such as Springshare's introduction of the LibAnswers Net Promoter Score (NPS) feedback module—allow libraries to collect standardized satisfaction data directly within the interface, (Springshare, 2023). A multi-campus evaluation by Meyers-Martin (2024) found that libraries using LibAnswers reported measurable increases in self-service use, reduced repetitive questions, and improved staff time allocation toward higher-order research assistance. These outcomes reflect how knowledge-base integration not only improves efficiency but also strengthens user satisfaction when paired with active FAQ management and analytics monitoring.

Case studies across the University of California system, including UC Berkeley, demonstrate that combining LibChat and LibAnswers enhances both immediacy and accuracy of responses while providing administrators with valuable metadata about user behavior, query frequency, and unserved topics, (Springshare, 2025). Institutions adopting regular "gap analyses" based on LibAnswers' built-in Query Analyzer are also reporting more targeted content creation, aligning FAQ updates with user needs identified through real-time data. The empirical consensus emerging between 2023 and 2025 is that integrated platforms like LibAnswers yield measurable gains in cost-effectiveness and user satisfaction only when libraries maintain an ongoing cycle of curation, data review, and staff training (Meyers-Martin, 2024; Springshare, 2023; Springshare, 2025).

Corporate and Special Libraries: Remote Reference Evolution

Corporate libraries have also embraced remote reference services. The Boeing Library's implementation of an "Ask A Librarian" service integrated into its intranet search results illustrates how private-sector operations utilize reference services to minimize search complexity and align with business objectives. This private-sector example parallels academic implementations but underscores the emphasis on knowledge capture at the user's point of need.

While direct empirical studies of corporate remote reference in 2023–2025 remain limited, Gaikwad and Dandawate (2024) described a corporate library's use of user journey mapping and remote focus groups to tailor knowledge-service delivery for distributed employees. This work highlights how corporate remote reference emphasizes knowledge captured at the point of need, internal user-experience optimization, and content workflows aligned with business metrics rather than strictly academic reference norms. The shift suggests a convergence of academic and special-library remote reference modalities, but analysts note that corporate models often integrate reference triage, metrics tied to business outcomes, and internal knowledge-management platforms, which may differ in staffing practices, training profiles, and cost-effectiveness models compared to traditional academic libraries.

Standards, Usability, and User Perception

A usability study examining the naming and placement of chat links revealed that clear labels ("Ask a Librarian") and strategic interface positioning enhance usage and recognition. Scholarly evaluations of user satisfaction highlight attributes such as speed, reliability, and personal interaction, with subject expert skills adding value, particularly in specialized queries. However, users often prioritize responsiveness and clarity over formal credentials.

A usability study on chat-widget labeling and positioning demonstrated that clear labels (e.g., "Ask a Librarian") and prominent interface positioning significantly increased usage and recognition of chat services, (Dease, 2024). The study also indicated that users value speed, reliability, and personal interaction. While subject-expertise adds value, particularly for specialized queries, many users prioritize clarity and responsiveness over formal credentials.

Meanwhile, a systematic review by Matteson (2024) of live chat service research reaffirmed recurring barriers: low awareness of chat services, user unease about anonymity or lack of clarity who is behind the widget, and concerns over accuracy and credibility. In sum, usability and user perception remain critical in remote reference: refining naming conventions, interface design, response transparency, and staff messaging style significantly influences user adoption and satisfaction, even in technologically advanced systems.

Global and Collaborative Networks

Historically, the Library of Congress's Collaborative Digital Reference Service (CDRS) pioneered the concept of a networked, 24/7 international reference service linking 60+ institutions. Although the foundational work dates from 2000–2001, the principle remains highly relevant as libraries increasingly consider global and cooperative staffing models, (Library of Congress, 2024). In the UK, models like EARL, Ask Scotland, and Chasing the Sun illustrate national and transnational cooperative remote reference systems.

More recently, while not strictly about chat reference, the digital-transformation strategy at the Library of Congress emphasizes international collaboration, digital collections access, and networked services (Library of Congress, 2024) — all of which bear on cooperative remote reference. Current literature suggests that globally distributed reference networks are evolving to include multilingual staffing, time-zone coverage, metadata-driven question routing, and shared analytics dashboards. Nonetheless, empirical research specifically comparing user satisfaction or cost-effectiveness of global networks (versus local or consortial chat services) remains sparse, indicating an area ripe for further study.

Current Best Practices & Real-Life Applications

Today, libraries integrate remote reference tools with video chat, co-browsing, and AI chatbots as supplemental aids. Real-life testimonies from practitioners on Reddit reveal that digital reference includes tasks ranging from catalog navigation to personalized ILL instructions and citation formatting. Advisement on tool selection and practical deployment is often peer-shared—LibraryH3lp and LibAnswers are frequently recommended for consortial and local uses, respectively.

Remote reference tools, from Text-A-Librarian to My Info Quest, have redefined library services, offering unparalleled access to information in a digital world. Supported by peer-reviewed research, real-life examples demonstrate their impact on user engagement, accessibility, and service quality. Despite challenges, such as resource constraints and technological disparities, these platforms empower libraries to meet diverse user needs. By embracing innovative technologies and robust evaluation practices, libraries can strengthen their role as knowledge facilitators. The integration of visual aids and strategic advocacy will further amplify their reach, ensuring that remote reference services continue to bridge the gap between users and information.

Consortia models demonstrate cost-effectiveness, scalability, and extended availability. However, services like My Info Quest highlight the significance of standardizing procedures to maintain quality. Emerging trends, such as AI integration, video reference, and deeper analytics, suggest a future where remote reference seamlessly integrates human expertise with technological innovation. Remote reference tools represent a transformative evolution in library services. Their proven ability to meet user expectations for convenience and expertise underscores their crucial role in modern information landscapes. As libraries adapt to evolving digital behavior, these tools will continue to redefine and reinforce the librarian's role as a trusted guide in a complex digital world.

Remote reference tools have matured significantly from their early email and fax formats. Synchronous and asynchronous channels each serve strategic roles: text and chat for rapid, simple exchanges, and email/webforms and knowledge bases for more complex or detailed inquiries. Digital reference's success hinges on consistent training, effective interface design, strategic branding, and data-driven service assessment.

ENHANCING REFERENCE SERVICES THROUGH CHATBOTS: INNOVATIONS, PRACTICES AND IMPACT

The integration of artificial intelligence into library systems has revolutionized traditional reference services, offering dynamic and scalable solutions for user support. Among these innovations, the deployment of chatbots in library reference services has emerged as a prominent strategy to enhance user experience, expand access to information, and ensure 24/7 availability of assistance. The artificial intelligence—most notably chatbots—is rapidly reshaping how libraries deliver reference services, enabling dynamic, scalable, and always-available user support.

The Emergence of Chatbots in Library Services

Chatbots, or conversational agents, are AI-driven programs designed to simulate human conversation and provide automated responses to user inquiries. A chatbot can simulate a real conversation with a user using their own language. Chatbots provide text or audio communication on websites, messaging services, mobile apps, or phones. Recent studies suggest that while chatbots clearly hold promise for enhancing library reference by offering continuous availability and distributing labor among staff, significant gaps still exist between prototype innovation and institutional uptake.

In many domains, the digitally oriented world excels at utilizing cutting-edge, ground-breaking technologies. Everything is covered, including higher education and information resources.

In the context of libraries, chatbots function as remote reference assistants capable of handling a wide range of tasks, from answering frequently asked questions to guiding users through database searches and library policies. According to Luo, Zhang, and Xu (2022), the integration of chatbots into library environments reflects a growing trend toward user-centered, technology-enhanced services that prioritize accessibility and efficiency.

As digital communication platforms become increasingly dominant, libraries have adapted to these trends by embedding chatbots into their websites, mobile apps, and digital catalogs. This adaptation not only responds to users' expectations for instant and convenient support but also relieves human librarians from repetitive, low-level queries, allowing them to focus on more complex reference interactions and information literacy instruction.

One prominent example of how libraries are employing chatbot technology is through Springshare's LibAnswers platform, which includes a chatbot feature. This integration allows libraries to leverage their existing LibAnswers system to provide automated responses to frequently asked questions, streamline basic inquiries, and even facilitate interactions with live librarians when more complex assistance is required. This approach offers a consistent and efficient way to handle a significant volume of reference transactions, particularly those seeking directional information or basic resource access guidance.

Current Practices in Public & Academic Libraries

Early initiatives, such as Stella at the Bibliothekssystem Universität Hamburg and Emma the Catbot at the Mentor Public Library, demonstrate the long-standing interest in employing chatbots for general library questions. More recent examples like ANTswers at the University of California Irvine Libraries and Kingbot at San Jose University, developed using Google's Dialogflow, illustrate the growing sophistication of these tools, designed to handle routine inquiries related to circulation and introductory reference. During the height of the COVID-19 pandemic, some libraries, such as those in Gauteng Province, South Africa, successfully utilized chatbots like BOTsa to provide vital extended services to distant users, showcasing their adaptability and utility in times of crisis.

Numerous academic libraries worldwide have adopted chatbot technologies to support their reference services. For instance, the University of Pretoria in South Africa developed Libby, an AI chatbot trained to assist students in locating resources, accessing course materials, and understanding citation formats. Libby was designed through collaboration between library staff and computer science researchers, ensuring the chatbot's responses aligned with institutional information policies and user needs, (Mahlangu, Botha, & Buys, 2021).

Similarly, the North Carolina State University Libraries implemented Ask NCSU, a chatbot designed to address common user questions related to library hours, borrowing policies, and study space reservations. Ask NCSU is integrated into the library website and continuously updated based on user feedback and evolving service offerings. The chatbot's design incorporates natural language processing (NLP) to interpret varied user inputs and provide accurate responses, demonstrating the importance of linguistic flexibility in user interactions, (Luo, 2021).

Impact on User Experience and Staff Workflow

One of the most significant contributions of chatbots in reference services is the enhancement of user experience through immediate, round-the-clock assistance. In their study of chatbot deployment in academic libraries, Shaw and Matusiak (2022) found that users appreciated the ability to receive timely answers without waiting for staff availability. This is particularly valuable for distance learners, international students in different time zones, and users conducting research outside of regular library hours.

From the staff perspective, chatbots serve as valuable allies in managing routine inquiries. By automating answers to common questions, librarians can allocate more time to advanced research consultations and information literacy programming. However, the success of this division of labor depends on the quality of the chatbot's training data and its ability to escalate complex queries to human librarians when necessary.

A study done by Nawaz and Saldeen (2020) covered the advantages of artificial intelligence chatbots in academic libraries' reference departments, particularly in Bahrain. This study attempts to determine the possibility of using chatbot apps powered by artificial intelligence in library reference services. In order to meet the needs of millennial customers, this research thoroughly examines how artificial intelligence chatbots have been incorporated into academic library reference services. The advantages of artificial intelligence chatbots and their prospective effects on academic libraries' reference services are discussed by the researchers. To increase millennial customers' commitment to library services, a chatbot with artificial intelligence is included in the reference services.

Which Chatbot is the Most Widely Used?

While it is difficult to identify a single "most widely used" chatbot across all library reference services due to the diversity of implementations and ongoing development, the data suggests that ChatGPT-based solutions and platforms like Springshare's LibAnswers with integrated chatbot features are gaining significant traction. Libraries' ongoing experimentation and refinement of chatbot implementations indicate that the potential of these AI-powered tools to transform reference services and enhance the library experience for all patrons remains a compelling area of exploration and innovation. According to one study, generative AI chatbots may negatively affect library services due to privacy and task complexity concerns.

Known Library Chatbots

- Ivy.ai: One of the most prominent and widely adopted chatbot platforms in academic libraries is Ivy.ai. For instance, the University of Oklahoma (OU) Libraries has utilized Ivy.ai since 2018, with the chatbot handling over 24,000 conversations between 2019 and 2023. Ivy.ai's platform is designed specifically for higher education and libraries, offering features such as integration with library knowledge bases, analytics, and seamless escalation to human librarians when needed. Its widespread deployment across multiple institutions highlights its prominence in the sector.

- KingbotGPT: San Jose State University Library has implemented KingbotGPT, an experimental AI chatbot that uses Retrieval Augmented Generation (RAG) technology and draws from local datasets to provide accurate, library-specific information. While innovative, KingbotGPT is currently positioned as an experimental tool and is not as widely adopted as Ivy.ai.
- T-Rex (University of Calgary): The University of Calgary Libraries launched T-Rex, a multilingual reference chatbot, in 2021. It uses a commercial LLM-based platform and is trained on the library's own web content. T-Rex is one of several chatbots on the UCalgary campus, each tailored to specific departmental needs.
- A pilot project at the University of Delaware showcased how an AI-powered chatbot, developed collaboratively by library IT and reference staff, not only extended support beyond traditional service hours but also benefited from shared staff involvement in its prompt training and maintenance, (Twomey, Johnson, & Estes, 2024).
- In a national survey of Canadian academic libraries, researchers found that while live-chat reference is widespread, deployment of chatbots remains rare, with only 2 out of 106 libraries surveyed having implemented them—underscoring both the emerging potential and the prevailing hesitation toward broader adoption in practice, (Guy et al., 2023).

Chatbots used in the past or not consistently highlighted as the most popular ones:

- Pandora bot is mentioned in connection to libraries, but it appears to be an older initiative from 2014, with no evidence of widespread use today.
- ManyChat and MobileMonkey focus more on social media messaging. Their use in libraries is not prominently mentioned in the search results.
- Podium, Itsuku, Bold360, Meya AI, Aivo, and itsAlive may have some use in libraries, but their presence in the provided search results is limited.
- Tidio and Salesforce are mainly known as customer service or marketing platforms. The search results do not show that libraries widely use them for reference services

Prominent Chatbots in Library Reference Services

- ChatGPT: Is frequently mentioned as a widely used AI chatbot, including in academic libraries, with a significant user base and market share. Libraries are exploring its potential for tasks like finding and summarizing informa-

tion, brainstorming, and identifying connections between research. However, concerns about "hallucinations" or fabricated citations from ChatGPT have been raised.
- Perplexity: Cited as a helpful tool for navigating scientific literature, identifying relevant papers, and summarizing documents.
- Gemini: Recognized as an interactive chatbot designed to understand conversations in multiple languages.
- Claude: Mentioned as a large language model that can write creative content, translate languages, and analyze content from attached files.
- LibAnswers: A popular platform among libraries that includes a chatbot feature, allowing for automated responses to frequently asked questions and streamlined reference services.
- Pandorabots: Mentioned in relation to libraries, offering virtual assistant librarians with customizable knowledge domains.
- Open-Source Chatbot Platforms: Botpress, Microsoft Bot Framework, Rasa, and ChatterBot are examples of platforms or libraries that can be used to build chatbots, including those for library use.

While Meta AI has the most active users, ChatGPT demonstrates greater overall usage and market share in AI search. Therefore, based on the information provided, ChatGPT is the most widely used chatbot overall.

The landscape of library reference services is rapidly evolving with the integration of AI chatbots. Among the various platforms, Ivy.ai stands out as the most widely used chatbot in academic library reference services due to its specialization, scalability, and proven track record across multiple institutions. As AI technology continues to advance, libraries may diversify their chatbot solutions, but for now, Ivy.ai remains the leading choice for large-scale, reliable reference support.

Difficulties Ahead and Potential Restrictions

Despite their advantages, chatbot implementations also raise important challenges. One concern is the limitation of chatbot responses to predefined scripts or datasets, which may hinder their ability to handle nuanced or context-sensitive questions. To address this, some libraries employ hybrid models, in which chatbots handle initial inquiries and then redirect users to live chat services or email for more in-depth assistance.

Ethical issues such as data privacy and transparency also demand attention. Libraries must ensure that user interactions with chatbots are confidential, that no personal data is collected without consent, and that users are clearly informed when they are interacting with a machine rather than a human. According to Radford and

Gary (2021), transparency and trust are critical components of ethical AI use in libraries, requiring continuous monitoring and user education.

As AI technologies continue to evolve, library chatbots are expected to become increasingly sophisticated, incorporating features such as voice recognition, multilingual support, and integration with discovery systems and learning management platforms. Machine learning capabilities will allow chatbots to refine their responses over time, based on patterns in user queries and feedback.

The future of AI chatbots in library reference services appears promising despite existing challenges. Libraries may increasingly leverage AI-powered conversational agents to enhance user experience, improve accessibility, and free librarians to focus on high-value tasks as technology evolves. The increasing popularity of NLP technology and user familiarity with chatbot interfaces suggest that libraries are well-positioned to capitalize on this trend.

Libraries are exploring how tools like ChatGPT can be safely and effectively harnessed to support reference services, particularly in environments with limited staffing and budgets. This involves using generative AI to create more personalized and conversational chatbot interactions. While the potential is promising, it also underscores the need for thoughtful implementation, clear usage policies, and ongoing collaboration between librarians, technologists, and users.

However, the widespread adoption of AI chatbots in library reference services still faces challenges. A review of Canadian academic library websites, for instance, found that only a small number were using chatbots for reference, while many offered live chat services. This suggests that while there is growing interest, there are still barriers to overcome, including privacy concerns, cost, and the perceived complexity of integrating these technologies into existing library systems.

Another crucial consideration is the effectiveness of chatbots in handling complex inquiries. While they excel at answering simple, factual questions, they may struggle with sophisticated research questions requiring nuanced understanding or specialized subject expertise. Therefore, librarians must be involved in the design and implementation process to ensure chatbots are appropriately trained and that a smooth handover to human assistance is available when needed.

The use of chatbots in library reference services represents a compelling intersection of technology and public service. By augmenting traditional reference models with AI-powered tools, libraries can provide more accessible, efficient, and user-friendly support to diverse populations. However, the effectiveness of these tools depends not only on technological sophistication but also on ethical design, continuous evaluation, and alignment with the core values of librarianship. As libraries navigate the complexities of digital transformation, chatbots will play a vital role in sustaining responsive and inclusive reference services for the communities they serve.

DESIGNING LIBRARY CHATBOTS: INSIGHTS FROM PRACTICE-BASED RESEARCH

The integration of artificial intelligence (AI) in academic and public libraries has gained significant momentum in recent years, with chatbots emerging as a prominent tool for enhancing user services. The design and implementation of chatbots in library environments is more than a technical challenge—it is a strategic and user-centered endeavor that demands thoughtful planning, inclusive collaboration, and iterative evaluation. Drawing on the experiences shared in three key studies—*"Building a Bot: An Exploration of AI to Assist Librarians"*, *"Library-Led AI: Building a Library Chatbot as Service and Strategy"*, and *"Implementing a Chatbot on a Library Website"*—this short paper synthesizes the core design principles and best practices for developing effective library chatbots.

- *"Building a Bot: An Exploration of AI to Assist Librarians"*: The University of Texas Libraries has embarked on a proactive research initiative titled "Building a Bot: An Exploration of AI to Assist Librarians", led by Hannah Moutran and strategy director Aaron Choate. This experimental chatbot is designed as a contingency for the "Ask a Librarian" service, stepping in when human staff are unavailable. Grounded in analysis of real chat logs from fall 2022 and enriched by interviews with five information professionals, the development emphasized transparency—clearly communicating that users are interacting with an AI—accuracy, and alignment with the library's mission. Built on Voiceflow and powered by Claude, the bot doesn't attempt to answer directly; instead, it offers carefully curated links and context, safeguarding service quality while supporting research workflows ethically and responsibly, (University of Texas Libraries, 2024).
- *"Library-Led AI: Building a Library Chatbot as Service and Strategy"*: A case study delves into the development of KingbotGPT, a retrieval-augmented generation (RAG) chatbot, by librarians at San José State University's Dr. Martin Luther King, Jr. Library. This strategic project aimed to bolster in-house AI expertise, enhance library services during off-hours, and provide an opportunity for AI literacy instruction within the library community. The resulting chatbot offers late-night, AI-driven reference assistance, utilizing a dataset curated by library staff. This experience empowered librarians to lead conversations on AI integration and responsible use within their institution, (Rodriguez, Szydlowski, & Cai, 2025).
- *"Implementing a Chatbot on a Library Website"*: At Lehman College's Leonard Lief Library (CUNY), a proprietary chatbot named Ivy was embedded in the library website to augment content discoverability and capture user

inquiry trends. This case study offers a pioneeringly detailed implementation account of how a chatbot can function both as a patron assistance tool and a data collection mechanism, revealing the types of questions students actively seek on the site and informing iterative library services improvements, (Ehrenpreis & DeLooper, 2022).

User Needs as the Design Foundation

A central theme across all three studies is the importance of anchoring chatbot design in the specific needs of library users. In "Building a Bot," the authors emphasize early-stage needs assessments, employing surveys and user feedback to understand common questions, peak usage times, and service gaps. This preliminary analysis informed the creation of intent libraries—the backbone of natural language processing (NLP) systems—by aligning chatbot responses with real user behaviors and expectations. Similarly, in *"Implementing a Chatbot on a Library Website,"* developers mapped user queries to structured responses, identifying FAQs and core services such as catalog navigation, database access, and account help. This allowed the bot to act as a first point of contact, reducing staff workload while maintaining service quality, (Ehrenpreis & DeLooper, 2022).

Recent empirical work reinforces and extends this focus on user-centered design. For example, a recent study of a social-needs screening chatbot found that engaging stakeholders—including low-income households, healthcare providers, and family advocates—through a layered iterative design process helped identify specific user needs, preferences, and adoption barriers. (Miller et al., 2024). Although not specific to libraries, the implications are clear: robust needs assessment combined with stakeholder involvement strengthens alignment between system behavior and user expectations. In the library context, a systematic review of AI applications in academic libraries Kulkanjanapiban, Silwattananusarn and Lambovska (2025) highlight the "user-centric applications (chatbots)" cluster as critical, indicating that libraries must increasingly design conversational tools in response to evolving user needs and behavior rather than simply deploying technology.

Anchoring design in user needs also has practical implications for intent-library creation, staffing workflows, escalation pathways, and analytics planning. By analyzing peak times, typical question types, and behavioral patterns (e.g., mobile access, short-message queries), libraries can configure bots to mirror both user context and service expectations. As one study of website-based conversational AI found, "the study highlights … how these technologies improve user engagement and satisfaction … and what are the primary challenges in implementing them," (Dobbala & Lingolu, 2024). Thus, libraries that ground bot design in genuine usage data, iterative feedback loops, and user-centered metrics are better positioned to maintain service quality, boost user satisfaction, and optimize staff allocation.

Collaboration and Library-Led Development

A distinguishing strength of the library-led AI approach, as discussed in *"Library-Led AI,"* is the emphasis on interdisciplinary collaboration. Rather than outsourcing development, libraries took ownership of the chatbot's design, integrating library staff, systems librarians, and IT specialists in a co-creation process. This collaborative model fostered a sense of ownership and ensured that the bot aligned with the library's service philosophy and ethical standards. By using open-source platforms such as Rasa and Dialogflow, developers gained flexibility while keeping control over user data and privacy—a growing concern in AI implementation, (Rodriguez, Szydlowski, & Cai, 2025).

Qualitative research reinforces this collaborative model. A study titled "It Takes a Village: A Distributed Training Model for AI-Based Chatbots" documented how one academic library network created a distributed training regime for chatbot staff, spanning librarians, IT professionals, and student workers. Twomey, Johnson, and Estes (2024) found that this inclusive model improved perceived ease of use and usefulness of the chatbot deployment. The authors argue that training, staff buy-in, and cross-unit collaboration are foundational for successful library chatbot initiatives. Moreover, the systematic review of AI applications in academic libraries, (Kulkanjanapiban, Silwattananusarn & Lambovska, 2025) identifies "emerging innovations" and "user-centric applications" as two of four thematic clusters, emphasizing the symbiotic relationship between technical infrastructure and human-centered design.

In practice, this means libraries must actively embed librarians, systems developers, and service-design professionals in the bot-building process. Doing so allows for better alignment with institutional policy (e.g., privacy, data-governance), workflow integration (how the bot hands off to human staff), and content governance (who updates the knowledge base and how often). This library-led approach also addresses sustainability by building internal capacity, rather than outsourcing to vendors, thereby preserving institutional memory, enabling agile response to usage trends, and aligning with professional assessment frameworks.

Iterative Design and Continuous Improvement

Each of the studies highlights the significance of iterative testing and ongoing refinement. In *"Implementing a Chatbot on a Library Website,"* the team used analytics tools to track user interactions, identify failure points in conversations, and retrain the model accordingly. This continuous feedback loop allowed the chatbot to evolve over time, improving its accuracy and expanding its capabilities. Likewise, in *"Building a Bot,"* developers adopted agile methods, conducting usability testing with real users and incorporating librarian feedback to fine-tune both the conversation flow and backend logic. This iterative process contributed to higher user satisfaction and more efficient service delivery, (Ehrenpreis & DeLooper, 2022).

Empirical work underscores how critical iterative design is to successful chatbot deployment in academic libraries. For instance, the study "It Takes a Village: A Distributed Training Model for AI-Based Chatbots" found that involving cross-functional teams (library staff, IT/systems, student workers) in multiple rounds of training and refinement significantly improved perceived ease of use and usefulness of the chatbot interface, (Twomey, Johnson, & Estes, 2024). The authors argue that continuous improvement is not simply about tweaking responses, but about establishing ongoing governance, analytics-based review, and staff capacity to adapt the system to emergent user needs. Moreover, in a broader study of AI adoption in libraries, "Academic Libraries Can Develop AI Chatbots …" (Ehrenpreis & DeLooper, 2022).

Chase (2024) demonstrated that libraries with structured cycles of user feedback, failure-point analysis, and retraining were more likely to report successful outcomes—even with minimal technical resources. Taken together, these findings reaffirm that iterative design and continuous improvement are foundational to deploying chatbots that are user-centric, sustainable, and aligned with evolving user behaviors.

Strategic and Ethical Considerations

Beyond functionality, the studies underscore that chatbot development is also a strategic move. As Rodriguez, Szydlowski, and Cai (2025) assert, implementing a chatbot is not simply a matter of adding convenience—it is a step toward redefining library services in the digital age. The chatbot becomes both a tool and a symbol of innovation, showcasing the library's commitment to accessibility, responsiveness, and 24/7 support. Ethical considerations are also paramount. Ensuring transparency in both behaviors, protecting user privacy, and designing for inclusivity—such as multilingual capabilities and ADA compliance—are essential for responsible deployment.

Empirical research affirms these strategic and ethical imperatives. Emin (2025) articulates how AI-enabled chatbots are transforming service models, but also cautions that issues such as staff training, bias, data governance, and alignment with library values must not be neglected. Additionally, Ehrenpreis and DeLooper (2025) present an assessment of a library chatbot's transition to a transformer-based model and points out how ethical transparency, user-outlined escalation protocols, and human-in-the-loop governance were critical for user trust and satisfaction.

Together, these works indicate that strategic thinking about chatbots must include alignment with organizational mission, ethical oversight, staff capacity building, and inclusive design—not simply rollout of new tools.

Designing an effective library chatbot is a multidisciplinary process that blends technological innovation with the human-centered values of librarianship. The case studies reviewed in these three studies demonstrate that successful chatbot development requires aligning AI capabilities with user needs, leveraging collaborative design approaches, and engaging in continuous evaluation and improvement. Ultimately, library chatbots are more than digital assistants—they are strategic assets that can expand access, streamline services, and reinforce the library's role in an increasingly AI-mediated world.

Obstacles and Anticipated Constraints

Challenges include technological barriers, unequal access, and privacy concerns. Rural users may lack reliable internet, and automation risks reducing human interaction, (Agrawal & Agrawal, 2025). Budget constraints and staff training needs also impede adoption.

Prospectively, libraries should integrate AI for metadata management and virtual reality for immersive reference, (Breeding, 2025). Trends indicate AI-enhanced search tools and policies for AI-generated content in collections, (The Digital Librarian, 2025). Liu (2007) suggests ongoing studies to track evolving user needs for sustained advancement.

Mobile reference services have reshaped libraries into dynamic, accessible hubs for mobile users. Leveraging studies like Liu (2007), and adopting current technologies, libraries can advance services, ensuring equitable, innovative support in a digital era. Continued adaptation will meet prospective user demands.

Figure 1.

REFERENCES

Abidin, N. (2022). Enhancing virtual reference services with AI and personalization. *Journal of Librarianship and Information Science*, *54*(1), 24–35.

Agarwal, S., Suarez-Tangil, G., & Vasek, M. (2025). An overview of 7726 user reports: uncovering SMS scams and scammer strategies. *arXiv preprint arXiv:2508.05276.*

Agrawal, B., & Agrawal, A. (2025). NAVIGATING THE OPEN ACCESS JOURNALS IN THE FIELD OF LIBRARY SCIENCE. *International Journal of Information Library & Society*, *14*(2).

Al-Daihani, S. M., & Jegede, O. (2022). Virtual reference services in academic libraries: An exploratory study. *Journal of Academic Librarianship*, *48*(1), 102397.

Barker, E. (2021). Training librarians for virtual reference services: The case of chat reference. *RSR. Reference Services Review*, *49*(1), 7–20.

Barrett, K., & Sharma, A. (2025). Library Chat Transcript Evaluation for User Sentiment During the COVID-19 Pandemic. *Evidence Based Library and Information Practice*, *20*(2), 5–22. DOI: 10.18438/eblip30642

Breeding, M. (2025). Library technology trends: Mobile and AI advancements. *Library Technology Reports*, *61*(1), 10–25.

Cassidy, E. D., Colmenares, A., & Martinez, M. (2023). So Text Me—Maybe: A rubric assessment of librarian behavior in SMS reference services. *RUSQ: A Journal of Reference and User Experience*, 53(4), 300–321.

Chase, M. (2024). Academic libraries can develop AI chatbots for virtual reference services with minimal technical knowledge and limited resources.

Crittenden, J. (2024). Current Practices and Future Use of Virtual Volunteerism. DigitalCommons@UMaine.

Cruz, P. C. (2020). Virtual reference services: A review of the literature. *Reference and User Services Quarterly*, *60*(2), 91–101.

Dease, N. E. (2024). User expectations and preferences regarding library chat widgets. *Weave: Journal of Library User Experience*, *7*(1). Advance online publication. DOI: 10.3998/weaveux.2743

Dobbala, M., & Lingolu, M. S. S. (2024). Enhancing usability for everyone through web accessibility compliance. *J Comput Sci Software Dev*, *3*(1), 1–13. DOI: 10.17303/jcssd.2024.3.105

Ehrenpreis, M., & DeLooper, J. P. (2022). Implementing a chatbot on a library website. Retrieved March 26, 2023, from https://academicworks.cuny.edu/cgi/viewcontent.cgi?article=1400&context=le_pub

Emin, D. (2025). The Rise of Artificial Intelligence in Academic Libraries: Opportunities and Concerns. *Emerging Library & Information Perspectives*, *7*(1). Advance online publication. DOI: 10.5206/elip.v7i1.22214

Gaikwad, V., & Dandawate, V. (2024). MARKETING OF LIBRARY RESOURCES AND SERVICES IN ACADEMIC LIBRARIES IN MAHARASHTRA STATE. *CAHIERS MAGELLANES-NS*, *6*(2), 28–38.

Guy, J., Pival, P. R., Lewis, C. J., & Groome, K. (2023). *Reference chatbots in Canadian academic libraries.*

Holland, B. (2021). Virtual reference: Meeting user needs. *Journal of Library Administration*, *61*(5), 551–567.

Jiao, Q., Li, Y., Zhang, J., & Guo, X. (2020). *Privacy-preserving query classification for virtual reference services in libraries.* Program. *Electronic Library and Information Systems*, *54*(1), 90–105.

Kulkanjanapiban, P., Silwattananusarn, T., & Lambovska, M. (2025). Research on AI-driven innovations and services in academic libraries: A bibliometric and systematic literature review. *Journal of Data and Information Science*, *10*(3), 1–51. DOI: 10.2478/jdis-2025-0036

Library of Congress. (2024). Digital Strategy at the Library of Congress. Retrieved from https://www.loc.gov/digital-strategy/

LibraryH3lp. (2024). *Text a Librarian SMS reference service.* Retrieved from https://libraryh3lp.com/text-a-librarian

Liu, Y. Q. (2007). Utilizing Recommender Systems in Digital Libraries, Current Study and Tech Issues. *Library and Information Service*, *51*(12), 6–18.

Liu, Y. Q., & Briggs, S. (2015). A Library in the Palm of Your Hand: Mobile Services in Top 100 University Libraries. *Information Technology and Libraries*, *34*(2), 133–148. DOI: 10.6017/ital.v34i2.5650

Liu, Y. Q., & Lewis, S. (2023). Services to mobile users: The best practice from the top-visited public libraries in the US. *Information Technology and Libraries*, *42*(1). Advance online publication. DOI: 10.6017/ital.v42i1.15143

Lotts, M., & Graves, S. (2011). Using iPads for roving reference services. *College & Research Libraries News*, *72*(4), 217–220. DOI: 10.5860/crln.72.4.8544

Luo, L. (2016). Professional preparation for "Text a Librarian": What are the requisite competencies? RUSQ: A Journal of Reference & User Experience, 52(1), 44-65.

Luo, L. (2020). Digital reference service: A review of literature and a case study of academic libraries in China. *Information Development*, *36*(6), 1224–1237.

Luo, L. (2021). Artificial intelligence and library services: A review of research and practices. *Journal of Academic Librarianship*, *47*(6), 102390.

Luo, L., Zhang, T., & Xu, H. (2022). The promise and potential of chatbots in academic libraries. *College & Research Libraries*, *83*(2), 193–208.

Mahlangu, L., Botha, A., & Buys, T. (2021). Designing an AI chatbot to support library services at a South African university. *Library Hi Tech*, *39*(1), 101–119.

Matteson, J. (2024). A systematic review of research on live chat service. *RUSQ: A Journal of Reference and User Experience*, 60(2).

Meyers-Martin, C. (2024). It's all about the chat and COVID-19: Virtual one-on-one research consultations surveyed. *Journal of Academic Librarianship*, *50*(4), 102904. DOI: 10.1016/j.acalib.2024.102904

Nahum-Shani, I., Smith, S. N., Spring, B. J., Collins, L. M., Witkiewitz, K., Tewari, A., & Murphy, S. A. (2016). Just-in-time adaptive interventions (JITAIs) in mobile health: Key components and design principles for ongoing health behavior support. *Annals of Behavioral Medicine*, •••, 1–17. PMID: 27663578

Nawaz, N., & Saldeen, M. A. (2020). ARTIFICIAL INTELLIGENCE CHATBOTS FOR LIBRARY REFERENCE SERVICES. *Journal of Management Information & Decision Sciences*, 23.

Nichols, M. (2020). Developing a virtual reference service with limited resources. *New Review of Academic Librarianship*, *26*(1-2), 158–172.

Ogbole, G. I., Adepoju, A. E., Ibrahim, A., Togunwa, T. O., & Nkeakam, F. A. (2023). MRI training in Africa. *Annals of Ibadan Postgraduate Medicine*, *21*(2), 75–80. PMID: 38298339

Ontario Council of University Libraries. (2024). *Scholars Portal Ask A Librarian service annual report 2024.*

Pike, G. (2021). Virtual reference services in academic libraries: Current issues and future directions. *Journal of Academic Librarianship*, *47*(4), 102398.

Qayyum, M. A., Mahmood, K., & Alqahtani, F. (2022). Students' perceptions of library virtual reference services: A case of an academic library. *Journal of Librarianship and Information Science*, *54*(2), 529–542.

Radford, M. L., & Gary, J. (2021). Ethics of AI and chatbots in reference services. *RSR. Reference Services Review*, *49*(3), 353–371.

Rodriguez, S., Szydlowski, N., & Cai, J. (2025, March). Library-led AI: Building a library chatbot as service and strategy (Workshop report). San José State University Library. https://www.ala.org/sites/default/files/2025-03/Library-LedAI.pdf

Shaw, M., & Matusiak, K. K. (2022). User perceptions of library chatbots: A case study. *Information Technology and Libraries*, *41*(4).

Springshare. (2023, October 10). Introducing the LibAnswers NPS® Module: Your new favorite feedback tool. Springshare Blog.

Springshare. (2025). LibAnswers product overview.

Tampa Bay Library Consortium. (2025). *Ask A Librarian fee structure and sustainability plan. Tampa Bay Library Consortium.* https://tblc.org/ask-a-librarian-fee-structure/

The Digital Librarian. (2025). AI in libraries: Opportunities and ethical considerations. Retrieved from https://thedigitallibrarian.org/ai-in-libraries

Twomey, B., Johnson, A., & Estes, C. (2024). *It takes a village: a distributed training model for AI-Based chatbots.*

University of Texas Libraries. (2024, November). *Building a Bot: An Exploration of AI to Assist Librarians.* TexLibris. https://texlibris.lib.utexas.edu/2024/11/building-a-bot-an-exploration-of-ai-to-assist-librarians/?utm_source=chatgpt.com

Urrea, G., & Yoo, E. (2023). The role of volunteer experience on performance on online volunteering platforms. *Production and Operations Management*, *32*(9), 3051–3071. DOI: 10.1111/poms.13879

Warren, S. J. (2021). A systematic review of virtual reference services in academic libraries. *College & Research Libraries*, *82*(4), 472–495.

Chapter 6
Content Provided to Mobile Users

ABSTRACT

This chapter explores the diverse spectrum of digital content optimized for mobile devices in libraries, including databases, multimedia, and mobile-ready platforms. It explains how recommender systems help users navigate information overload and how AI-enhanced personalization is reshaping discovery. The Semantic Web is presented as a powerful framework for deeper knowledge connections through linked data and ontology-driven search, alongside its ongoing challenges. The chapter also addresses how libraries balance global interoperability with local relevance, highlighting strategies for curating content that meets the varied needs of mobile users while supporting equitable access.

INTRODUCTION

The landscape of information consumption has undergone a profound transformation due to the widespread adoption of mobile devices. Libraries, as essential community hubs and repositories of knowledge, have continually adapted to these changes, recognizing the urgent need to deliver relevant and accessible content to users on their preferred platforms. The evolving strategies and technological advancements employed by libraries aim to cater to the unique needs of mobile users, focusing on the provision of digital resources, the emergence of recommender systems, the potential of the semantic web, and the delicate balance of acting locally while thinking globally.

Libraries, historically bastions of printed materials, have undergone a profound transformation with the rise of digital resources. These digital resources, accessible on mobile devices, include e-books, e-journals, and databases. They represent a paradigm shift in how users interact with library content and services. Library

DOI: 10.4018/978-1-7998-9175-8.ch006

websites offer a diverse range of digital resources optimized for mobile platforms, including e-books, audiobooks, and streaming media.

The digitization era has revolutionized library services. With the rise of smartphones and mobile devices, libraries have realized the importance of catering to the diverse content needs of mobile users.

The modern library extends beyond its physical walls, embracing the digital realm to serve a dispersed and mobile user base. The success of libraries in the mobile era hinges on their ability to curate, organize, and deliver a diverse array of content that is optimized for on-the-go consumption. This involves not only digitizing existing collections but also strategically acquiring and integrating new resources that are specifically designed for mobile environments.

DIGITAL RESOURCES AND DATABASES FOR MOBILE DEVICES

Libraries rely on digital resources and databases to provide mobile content. The cornerstone of mobile content provision in libraries is the accessibility of digital resources and databases. Patrons increasingly expect seamless access to scholarly articles, e-books, audiobooks, streaming media, and research tools directly from their smartphones and tablets. This demand has led libraries to invest in mobile-responsive websites and dedicated applications with intuitive interfaces and streamlined navigation systems.

Mobile-Optimized Resources

Libraries now provide mobile-optimized versions of their Online Public Access Catalogs (OPACs), enabling users to conduct searches, manage their accounts, and renew loans conveniently. Licensing agreements with database providers increasingly incorporate provisions for mobile access, ensuring that researchers, students, and general users can access extensive repositories of information irrespective of their geographical location. Platforms such as EBSCO, ProQuest, and Gale have crafted mobile-friendly interfaces and companion applications to acknowledge the transformation in user behavior.

In addition to traditional databases, libraries are actively exploring and integrating emerging digital content formats. This encompasses mobile-specific learning modules, podcasts such as "Research First Aid" designed for on-the-go researchers, and augmented reality experiences that enhance physical library visits or provide contextual information about collections. The primary focus is on delivering content that is not merely reduced in size for a small screen but reimagined for the mobile user experience, prioritizing convenience, efficiency, and immediate utility, (Liu & Lewis, 2023).

The integration of digital resources into mobile-friendly formats has democratized access to information. Libraries are digitizing special collections, historical documents, and local archives, making them accessible online to global audiences worldwide. For example, PressReader, a platform offering daily updated access to thousands of newspapers and magazines, has been integrated with library discovery systems such as EBSCO Discovery Service and Primo. This integration ensures that users can access the latest content on their mobile devices, bridging the gap left by traditional databases that are updated periodically. By providing mobile-optimized interfaces, libraries enable patrons to engage with content anytime and anywhere, fostering a culture of continuous learning and exploration.

Libraries are also investing in mobile-friendly databases, such as ProQuest and JSTOR, which are vital for academic research but are often overlooked in their mobile implementations. These platforms offer advanced search capabilities that adapt to mobile formats, enabling users to discover content as comprehensively as on traditional desktop interfaces.

Mobile-optimized databases have revolutionized the way libraries cater to diverse user needs. For instance, language students can access publications in over 60 languages through platforms such as PressReader, which are equipped with built-in translation features that facilitate comparisons across languages. This accessibility not only enhances the user experience but also caters to multilingual and multicultural audiences. As libraries expand their digital offerings, they position themselves as indispensable resources for mobile users seeking reliable and diverse content.

Delivering Library Content to Mobile Users

The mobile age has brought a paradigm shift in library content delivery. No longer confined to physical premises or desktop computers, users now access resources via smartphones and tablets. This evolution demands that libraries rethink resource discovery, enhance personalization, and maintain relevance in users' daily, digital experiences.

Responsive design is crucial for libraries to ensure that resources are accessible and user-friendly on smaller display screens. Many libraries have adopted platforms such as OverDrive and Hoopla, providing mobile-optimized access to a vast collection of digital titles. Additionally, the development of dedicated library apps has facilitated seamless connectivity, allowing users to check out e-books, manage their accounts, and search catalogs directly from their devices.

The rise of smartphones and tablets has revolutionized the way people access information. Recognizing this shift, libraries have embraced digital resources and databases that are specifically designed for mobile devices. These resources, including academic journals, e-books, and multimedia content, offer seamless access to vast repositories of knowledge. Platforms like JSTOR, EBSCO, and Semantic Scholar have become indispensable tools for researchers and students, providing direct access to peer-reviewed articles and scholarly materials on mobile devices. Over 60% of libraries are actively evaluating or planning to integrate artificial intelligence (AI)-powered tools to enhance patron access to these resources, highlighting the growing reliance on digital platforms.

In addition, third parties that libraries offer include such as OverDrive, Hoopla, and Libby, which provide intuitive mobile interfaces and seamless integration with library catalogs (Smith, Garcia, & Lee, 2022). These platforms empower users to borrow materials, create reading lists, and receive personalized recommendations, all from their mobile devices. Academic libraries extend these offerings further by providing access to scholarly databases, journal archives, and research tools that are mobile-responsive, such as JSTOR, ProQuest, and ScienceDirect (Johnson & Davis, 2024). The focus on mobile accessibility ensures that students, researchers, and lifelong learners can engage with high-quality content anytime and anywhere.

Libraries have embraced mobile-optimized interfaces to facilitate on-the-go access to eBooks, journals, and databases. Notably, many platforms, including those powered by interfaces like IEEE Xplore and Europeana, now incorporate recommender systems to automatically surface relevant literature. These systems utilize metadata, usage patterns, and collaborative signals to alleviate information overload, as highlighted by Beel et al. (2015), who noted the introduction of over 90 distinct recommendation methods in scholarly databases.

RECOMMENDER SYSTEMS: GUIDING USERS THROUGH INFORMATION OVERLOAD

As the volume of digital content continues to explode, users often face the challenge of information overload. Recommender systems (RecSys) offer a powerful solution by personalizing content discovery and suggesting relevant materials based on user preferences, borrowing history, and the behavior of similar users. In the context of mobile library services, recommender systems can significantly enhance the user experience by proactively offering tailored suggestions, transforming passive Browse into an engaging and efficient exploration of resources.

While commercial platforms like Amazon and Netflix have popular recommendation algorithms, libraries are uniquely positioned to utilize such technology in promoting less-known works or niche topics that align with user interests. For instance, implementations of tools like Bibliocommons—an integrated library system that incorporates patron-driven recommendations—have been successful in attracting users to resources that they may not otherwise explore.

Recommender systems for mobile library users are evolving beyond simple content similarity. Hybrid approaches—merging content-based and collaborative filtering—address cold-start problems and improve personalization. For example, in the Reading&Machine project, collaborative filtering outperformed content-only methods by up to 47%, particularly when augmented with external metadata sources like social book reviews.

Recommender Systems in Mobile Contexts

Recommendation systems have emerged as powerful tools for enhancing content delivery in libraries. As Liu (2007) noted in his article, "Utilizing Recommender Systems in Digital Libraries," the exponential growth of digital information has created a burden of information overload, making such systems a necessity. Originating from early projects like the Tapestry System at Xerox PARC, which allowed users to rate documents, these systems are designed to provide personalized pathways through vast collections, a key feature for mobile users who prioritize efficiency and personalization. In the context of digital libraries, recommender systems analyze user preferences, browsing history, and interactions to provide tailored recommendations, ensuring that patrons can easily discover relevant resources.

At their core, these systems employ several foundational techniques to generate suggestions. The most common are content-based filtering, which recommends items similar to those a user has previously liked, and collaborative filtering, which identifies like-minded users and suggests items that they have enjoyed. By analyzing user data, such as borrowing history and subject preferences, these methods help curate a more tailored approach to content delivery, moving beyond simple keyword searches.

Building on these foundations, modern implementations often favor more advanced hybrid models. Chen and Wang (2022) highlight the efficacy of hybrid systems that combine content-based and collaborative filtering methods, leveraging semantic relationships to recommend books, articles, and other resources. For instance, a library patron searching for literature on machine learning might receive recommendations for recent peer-reviewed articles, related e-books, and even multimedia tutorials, all tailored to their mobile device's interface. Similarly, Shariq Imran et al. (2023) explores pattern-based hybrid systems that use attributes such as genre, author, and user ratings to provide precise book recommendations. This level of personalization enhances user satisfaction and encourages deeper engagement with library resources.

However, the challenge of implementing these systems has always been balancing personalization with privacy. Liu (2007) emphasized this as a central issue, noting that libraries, as traditional sanctuaries for user confidentiality, must navigate the complexities of data collection. He pointed out that the very nature of recommender systems—which require collecting personal data to function effectively—is often at odds with the library's long-standing ethic of protecting patron privacy.

To address this critical issue, federated recommender systems represent a significant advancement. Varlamis et al. (2023) discusses how federated learning enables libraries to train global recommendation models without centralizing or compromising user data. In this model, the recommendation algorithm runs on a user's local device, maintaining privacy while still accessing a shared collection. Such innovations are particularly relevant for addressing the ethical considerations Liu raised. Interestingly, his research also noted that while users may be wary of tracking in commercial environments, they are often more willing to use recommendation features for academic or research purposes, recognizing the potential for discovery. Ultimately, the successful adoption of recommender systems in libraries will depend on their ability to resolve this tension, providing effective, personalized discovery without sacrificing the trust and privacy of the patrons they serve.

Libraries Embrace Advanced AI for Hyper-Personalized Recommendations

Libraries are rapidly moving beyond foundational recommender system techniques, integrating sophisticated artificial intelligence to provide users with hyper-personalized and contextually aware discovery experiences. While collaborative and content-based filtering remain core components, recent peer-reviewed studies highlight a significant leap towards more nuanced and effective systems powered by hybrid models, knowledge graphs, and the latest advancements in natural language processing.

The evolution of these systems is marked by a clear trend towards hybridization. Research from late 2024 demonstrates that combining collaborative filtering (CF) and content-based filtering (CBF) can significantly enhance recommendation efficiency. A study on a prototype digital library, which connects knowledge sources from multiple online publishers, found that a hybrid model heavily weighted towards collaborative data (80% CF, 20% CBF) yielded the highest precision and user satisfaction, showcasing a data-driven approach to system design.

Advanced architectures are also increasingly leveraging semantic technologies. The development of standardized knowledge graphs for recommender systems, such as the proposed "RecKG" in early 2025, aims to create a more interoperable and data-rich environment. By representing resources, user interests, and their relationships as a network of interconnected data, knowledge graph-based systems can overcome common challenges like data sparsity and the "cold start" problem for new users. This rich, contextual understanding allows for more serendipitous and accurate recommendations, connecting users with relevant materials in ways that simpler models cannot.

Furthermore, the "black box" nature of many recommender systems is being addressed through a growing emphasis on Explainable AI (XAI). A significant movement in 2025, particularly within academic and research libraries, is the push towards "human-centered explainable AI" (HCXAI). This approach advocates transparency in why a particular item is recommended. For instance, recent developments include the use of Large Language Models (LLMs) like GPT-2 to generate natural-language explanations for recommendations, articulating how a suggested book aligns with a user's past reading history or stated preferences. This not only builds user trust but also empowers them to refine their future recommendations. Library thought leaders are now urging the profession to become "explainability activists," demanding transparency from vendors and in the systems they build.

The integration of recommender systems into mobile library apps and websites is also becoming more sophisticated, incorporating these advanced techniques:

- *Personalized Homepages*: These now feature dynamic collections generated by transformer-based models, which are adept at understanding the sequential patterns in a user's borrowing history to predict future interests.
- *"You Might Also Like" Suggestions:* Powered by knowledge graphs, these suggestions can now highlight not just similar items, but also conceptually related resources, or works by authors who have influenced the one the user just viewed.
- *Reading Challenges and Pathways*: These are becoming more adaptive, using transformer models to dynamically adjust reading lists based on a user's progress and the nuances of their interactions with the library's collection.

The effectiveness of these advanced systems hinges on robust and ethical data practices. The use of federated learning, a technique that trains AI models across decentralized devices without centralizing user data, is being explored in conjunction with transformer-based models to enhance user privacy. As libraries continue to evolve into deeply personalized knowledge hubs, the fusion of advanced AI with a commitment to user privacy and transparency will be the hallmark of the next generation of digital library services. Major library system vendors are already beta-testing tools that combine their vast discovery indexes with LLMs to provide more reliable and context-aware search and recommendation experiences, signaling a broad industry shift towards these more intelligent systems.

SEMANTIC WEB: UNLOCKING DEEPER CONNECTIONS IN LIBRARY CONTENT

The Semantic Web, often referred to as the "Web of Data," or Web 3.0, represents a vision of the internet where data is structured and linked in a way that allows machines to understand its meaning, a paradigm shift in how information is organized and accessed online. By giving information "well-defined meaning," it enables computers to understand and process data more effectively, paving the way for enhanced information retrieval and knowledge discovery. This approach holds immense potential for libraries, enabling them to move beyond traditional cataloging and unlock deeper connections within their vast content, fostering richer user experiences and supporting the evolving landscape of learning and research.

As a framework that enhances the interoperability and meaning of data on the internet, the semantic web can be revolutionizing content delivery in libraries. By structuring data in a machine-readable format, the semantic web enables more intuitive and context-aware interactions between users and digital resources. Stănescu and Oprea (2025) emphasize the interdisciplinary nature of this technology, noting its applications in knowledge management, digital libraries, and e-learning. Libraries leveraging semantic web technologies can create knowledge graphs that connect related concepts, facilitating more accurate and contextually relevant content recommendations.

The Foundation of the Semantic Web:

The Semantic Web is built upon a layered architecture of technologies that work together to create a connected and meaningful data environment. Key components include:

- Resource Description Framework (RDF): A data model for representing information as triples (subject-predicate-object), facilitating the creation of linked data.
- Web Ontology Language (OWL): A language for defining ontologies, which are formal, explicit specifications of shared conceptualizations, enabling the definition of concepts, their properties, and relationships within a specific domain.
- SPARQL Protocol and RDF Query Language (SPARQL): A query language specifically designed for retrieving and manipulating data stored in RDF format.

These technologies, together with standards like XML and URIs, provide a framework for structuring and linking data in a way that is understandable to machines.

The Semantic Web presents Libraries with Profound Opportunities

For libraries, the Semantic Web holds immense potential to transform how information is organized, discovered, and delivered, particularly to mobile users seeking precise and contextually relevant results. By moving beyond mere keyword matching to a deeper understanding of content relationships, the Semantic Web can enable more intelligent and intuitive information retrieval.

Being as custodians of knowledge and information gatekeepers, libraries are continuously seeking innovative ways to enhance access and usability of their collections. With the exponential growth of digital resources and information overload, traditional methods of organization and search are becoming increasingly inadequate. The Semantic Web, by providing mechanisms to describe resources and their relationships in a machine-readable format, offers a powerful solution to this challenge. It extends the current web by making information more readily understandable and processable by computers, ultimately enabling a more intelligent and efficient information ecosystem.

Applying semantic technologies involves creating ontologies and metadata schemas that describe library resources with rich, machine-readable semantics. The purpose is to unlock deeper connections in library content; by embracing Semantic Web technologies, libraries can realize numerous benefits:

- *Enhanced Discoverability and Retrieval*: Semantic annotations, utilizing ontologies, enable more precise and context-aware searching, resulting in more relevant and accurate results for users. Instead of relying solely on keyword matching, users can search for concepts, relationships, and even explore related resources effortlessly. This allows users to search not just for titles or authors but for concepts, themes, and relationships between disparate pieces of information. For example, a user searching for "books on climate change policy" can retrieve not only specific books but also related journal articles, government reports, and even relevant news archives, all connected by their semantic connections.
- *Context-Aware Services:* Mobile devices, with their GPS capabilities and other sensors, can leverage semantic data to provide context-aware information. A user physically in a library could, for example, point their phone at a bookshelf and receive semantic overlays detailing reviews, related works, and author biographies.

- *Improved Interoperability:* Semantic Web technologies enable the seamless integration of information from diverse sources, eliminating data silos that separate information based on media types and metadata formats. This integration facilitates a unified and streamlined access to information across various systems. Consequently, a mobile library app can retrieve information from the library's catalog, external academic databases, and even open educational resources, providing a comprehensive and unified view of knowledge, (Moen & McClure, n.d.).
- Personalized Recommendations: Libraries can leverage semantic technologies to provide personalized recommendations based on user profiles, past searches, and areas of interest. This enhances the user experience and promotes deeper engagement with library resources.
- Support for Knowledge Discovery: By exposing the relationships between resources and their semantic meaning, libraries can facilitate knowledge discovery, enabling users to explore connections and gain insights from interconnected data.
- Increased Accessibility: Semantic Web technologies can improve accessibility for users with disabilities by providing rich and accurate descriptions of library resources.

While the full realization of the Semantic Web in libraries is an ongoing endeavor, efforts to adopt technologies like Resource Description Framework (RDF) and ontologies are laying the groundwork for a more intelligent and interconnected information landscape that will greatly benefit mobile users seeking nuanced and comprehensive information.

Ontology-Based Discovery

Semantic Web technologies—RDF, OWL, SPARQL—are essential for structuring library metadata so machines can interpret content meaningfully. In mobile settings, these semantics enable context-aware retrieval: for instance, recommender frameworks based on Bayesian networks query linked data ecosystems to detect relevant materials aligned with user intent.

A semantic web-based recommender system might use ontologies to map relationships between academic papers, authors, and research topics. A mobile user searching for articles on artificial intelligence could be directed to related works in machine learning, data science, or ethics, even if those terms were not explicitly included in the search query. This capability is exemplified in tools like Semantic Scholar, which uses natural language processing and machine learning to generate summaries and recommend papers based on user interests. Zhu and Hose (2025) highlight the role of knowledge graphs in link prediction and data federation, enabling libraries to integrate disparate data sources into a cohesive, mobile-accessible framework.

Embedding ontologies into mobile recommender systems enhances precision. App-store recommendation algorithms built on OWL ontologies showed promising results in clustering user preferences semantically. Hybrid frameworks combining RDF, OWL, and social metadata improve recommendation relevance in social and academic contexts.

The application of semantic web technologies also supports the creation of digital twins and smart applications in libraries. Zhu and Hose (2025) describe the use of ontologies in smartwatch apps to provide occupant-centered feedback, demonstrating how semantic technologies can enhance user interactions with digital content. By adopting these technologies, libraries can offer mobile users a richer, more interconnected experience, where information is not only accessible but also meaningfully contextualized.

Challenges and the Path Forward

While the Semantic Web offers significant opportunities, its implementation in libraries also presents challenges:

- Availability of Semantic Web Content: The availability of semantically annotated content in libraries is still evolving, requiring efforts to convert existing content and create new annotations.
- Ontology Development and Maintenance: Building and maintaining ontologies for library collections is a complex process, requiring expertise and resources.
- User Acceptance and Training: Librarians and users need to be trained on Semantic Web technologies and their applications to fully leverage their potential.
- Organizational Readiness: Libraries need to address organizational and strategic implications, including changes in workflows, policies, and resource allocation.

To overcome these challenges, a collaborative effort is needed between librarians, developers, and researchers. Libraries should embrace the use of emerging web technologies, update LIS curricula, and actively engage in the development and adoption of semantic technologies. By taking an incremental approach and focusing on user-centered design, libraries can gradually integrate Semantic Web technologies into their operations and transform their services for the future.

The Semantic Web holds the potential to revolutionize libraries, transforming them into intelligent and interconnected knowledge hubs. By unlocking deeper connections within library content, libraries can empower users with enhanced information retrieval, knowledge discovery, and personalized experiences. While challenges exist, libraries are uniquely positioned to contribute to the growth of the Semantic Web and remain vital resources in the ever-evolving information landscape.

ACTING LOCALLY, THINKING GLOBALLY: TAILORING CONTENT TO DIVERSE NEED

Information can now be accessed through computerized network systems. Acting locally and thinking globally is one of the phenomena described by many researchers, including Borgman (1997).

The philosophy of acting locally while thinking globally encapsulates the dual role of libraries as community anchors and participants in a global knowledge ecosystem. Locally, libraries tailor their digital offerings to meet the specific needs of their patrons, such as providing access to local archives or supporting community-specific educational goals. Globally, they align with international trends in technology and information access, ensuring that their services remain relevant in a connected world.

The principle of "acting locally, thinking globally" is particularly pertinent to content provision for mobile users in libraries. While libraries must offer access to a vast array of global knowledge, they also have a critical role in serving the specific needs and contexts of their local communities. Mobile technology can be a powerful tool for bridging this gap, allowing libraries to deliver globally relevant content with a local lens, and vice versa.

Achieving Global Interoperability and Collaboration

Libraries endeavor to implement systems that foster internal efficiency and deliver exceptional service while simultaneously considering global implications. They must ensure that local systems can facilitate data exchange with other systems globally, (Borgman, 1997). Numerous countries share cataloging resources and access to telecommunications infrastructure. Nevertheless, there are concerns regarding the challenges encountered by automated libraries in achieving "global interoperability" and the persistent issues of real-time data exchange between systems, (Borgman, 1997).

Borgman (1997), although outdated, highlights the need for continuous updates to common cataloging rules, data formats, and communication protocols. It is crucial for academic library administrators and librarians to demonstrate the vital roles of libraries in the global enterprise. Global exposure is essential for librarians to gain a deeper and broader understanding of international affairs, (Clause et al., 2015). We should actively engage with libraries and librarians worldwide to understand their operations, services, and patrons. By observing the activities in libraries globally, we can identify shortcomings and make necessary internal improvements. Internationalization prepares individuals, including students, faculty, and librarians, for success in an increasingly interconnected world and society, (Clause et al., 2015).

As libraries endeavor to adapt to the dynamic evolution of digital technologies, cross-cultural considerations assume a crucial role in the design of mobile-friendly library services. Libraries worldwide must consider cultural differences in user preferences, technological literacy, and information-seeking behavior to create content that resonates with a broad audience. (Farkas, 2008). By embracing cultural diversity in their design, libraries can create more engaging and accessible mobile experiences that resonate with patrons from diverse perspectives and backgrounds.

The British Library offers an application that enables access to its digital collections. Patrons can explore "over 100 of the library's great collection items, 250 high-definition images, over 40 videos providing expert commentary, textual interpretation for deeper understanding, and up-to-date information about the Library's current exhibitions," (Hartley, 2011). The British Library's mobile access seeks to enhance the accessibility of its vast collection to a global audience. This type of mobile service technology to a wide audience is essential for libraries to remain relevant; it enables impactful library experiences worldwide.

Collaborative efforts within libraries are essential for facilitating collaborative initiatives globally. These efforts play a crucial role in sharing best practices, resources, and enhancing the quality of services with libraries. OCLC (Online Computer Library Center) is a global library organization that provides resources and services to libraries worldwide. One of its primary services is its WorldCat database, which is one of the world's largest bibliographic catalogs. WorldCat enables libraries to exchange and share bibliographic records, streamlining interlibrary loan services. OCLC plays an important role in supporting library collaboration on a global scale. Initiatives, like OCLC, provide platforms for libraries to access resources globally and promote innovation in the library profession.

Mobile Strategies for Balancing Local and Global Access

The dual focus, "acting locally, thinking globally", entails several key strategies:

- *Local Content Digitization*: Libraries can digitize and make accessible local history archives, community narratives, and regional publications, enriching global content with unique local perspectives. Mobile platforms are ideal for delivering these localized collections, making them accessible to community members on the go.

- *Language and Cultural Relevance*: Mobile apps and websites can be tailored to support multiple languages and incorporate culturally relevant features, ensuring inclusivity for diverse user populations within a community. This might involve translating interface elements, highlighting resources in different languages, or featuring content that resonates with specific cultural groups.
- *Community Engagement via Mobile*: Libraries can use mobile platforms to facilitate local events, workshops, and community-sourced content. This fosters a sense of local ownership and participation, while potentially contributing to a broader, global knowledge commons. For instance, mobile apps can be used for civic engagement, connecting users with local government information or community initiatives.
- *Global Access to Local Knowledge*: By digitizing and optimizing local content for mobile devices, libraries make their unique collections discoverable to a global audience. A researcher across the world could access a rare local historical document through a mobile app, demonstrating the global reach of local efforts. This embodies the "thinking globally" aspect, as local actions contribute to a wider pool of accessible knowledge, (Hernon & Altman, 1994).

Striking this balance ensures that libraries remain deeply rooted in their communities while simultaneously connecting users to the vast resources of the global information sphere. Mobile technology plays a crucial role in facilitating this dynamic interplay. Decentralized "device-to-device" recommender systems, such as MobRec, enable mobile users to discover content through peer proximity, effectively merging local interactions with global recommendation models.

Libraries must reconcile local contexts—such as campus layouts, on-site collections, and user behaviors—with global digital networks. Semantic interoperability is critical for federating diverse metadata schemas through linked open data initiatives like SIMILE. Federated SPARQL queries also support mobile apps in aggregating personalized library services from institutional and public data endpoints.

The provision of content to mobile users in libraries is not merely an extension of traditional services but a fundamental reimagining of how libraries fulfill their mission in the digital age. By strategically leveraging digital resources and databases, implementing sophisticated recommender systems, exploring the transformative potential of the Semantic Web, and balancing local relevance with global accessibility, libraries are ensuring their continued vitality and relevance.

The ongoing evolution of mobile technology and user expectations demands a proactive and adaptive approach from library professionals. The emphasis must remain on creating intuitive, personalized, and deeply interconnected experiences that empower users to access, discover, and engage with information anytime, anywhere. As mobile devices become even more integrated into daily life, the innovative ways libraries deliver content will undoubtedly shape the future of information access and intellectual engagement for generations to come.

REFERENCES

Beel, J., Gipp, B., Lange, S., & Breitinger, C. (2015). Research-paper recommender systems: A literature survey. *International Journal on Digital Libraries.*

Borgman, C. L. (1997, June). *Multi-media, multi-cultural, and multi-lingual digital libraries: Or how do we exchange data in 400 languages?* D-Lib Magazine. https://www.dlib.org/dlib/june97/06borgman.html

Chen, T., & Wang, Y. C. (2022). Hybrid big data analytics and Industry 4.0 approach to projecting cycle time ranges. *International Journal of Advanced Manufacturing Technology, 120*(1), 279–295. DOI: 10.1007/s00170-022-08733-z

Clause, J., Forey, E., Lortie, C. J., Lambert, A. M., & Barot, S. (2015). Non-native earthworms promote plant invasion by ingesting seeds and modifying soil properties. *Acta Oecologica, 64*, 10–20. DOI: 10.1016/j.actao.2015.02.004

Farkas, M. G. (2008). *Embedded library, embedded librarian: Strategies for providing reference services in online courseware.* PDXScholar.

Hartley, M. T. (2011). Examining the relationships between resilience, mental health, and academic persistence in undergraduate college students. *Journal of American College Health, 59*(7), 596–604. DOI: 10.1080/07448481.2010.515632

Hernon, P., & Altman, E. (1994). *Patron satisfaction as a measure of service effectiveness: An analysis of public library services.* Washington, DC: ERIC Clearinghouse. (ERIC Document ED367336)

Johnson, K., & Davis, M. (2024). Bridging the digital divide: Mobile library services as equity tools in K-12 education. *School Library Research, 27*(1), 1–18.

Liu, Y. Q. (2007). Utilizing Recommender Systems in Digital Libraries, Current Study and Tech Issues. *Library and Information Service, 51*(12), 6–18.

Liu, Y. Q., & Lewis, S. (2023). Services to mobile users: The best practice from the top-visited public libraries in the US. *Information Technology and Libraries, 42*(1). Advance online publication. DOI: 10.6017/ital.v42i1.15143

Moen, W. E., & McClure, C. R. (n.d.). *Virtual reality and the academic library of the future.* ResearchGate. https://www.researchgate.net/publication/359601186_Virtual_reality_and_the_academic_library_of_the_future

Shariq Imran, A., Uddin, I., Fayyaz, M., & Sajjad, M. (2023). Pattern-based hybrid book recommendation system using semantic relationships. *Scientific Reports*, •••, 13.

Smith, P., Garcia, R., & Lee, S. (2022). Integrating digital resources through library apps: A user-centered perspective. *Digital Library Perspectives*, *38*(3), 157–175.

Stănescu, G., & Oprea, S. V. (2025). Recent Trends and Insights in Semantic Web and Ontology-Driven Knowledge Representation Across Disciplines Using Topic Modeling. *Electronics (Basel)*, *14*(7), 1313. DOI: 10.3390/electronics14071313

Varlamis, I., Sardianos, C., Chronis, C., Dimitrakopoulos, G., Himeur, Y., Alsalemi, A., Bensaali, F., & Amira, A. (2023). Using big data and federated learning for generating energy efficiency recommendations. *International Journal of Data Science and Analytics*, *16*(3), 353–369. DOI: 10.1007/s41060-022-00331-2

Zhu, Y., & Hose, K. (2025). A survey on knowledge-aware news recommender systems. *Semantic Web*, *15*(1), 18–25.

Chapter 7
Apps Libraries Use or Provide for Patrons

ABSTRACT

This chapter examines the expanding role of mobile apps as essential tools that bring library services directly to users' fingertips. It surveys widely used applications—such as Libby, EBSCO Mobile, AR tools, and visual search models—and analyzes their impact on user behavior and service adoption. Practical studies highlight how libraries choose, implement, and assess mobile apps to enhance circulation, learning, and community engagement. The chapter concludes with an exploration of Web 2.0, 3.0, and emerging decentralized paradigms, illustrating how new web generations foster collaboration, personalization, and innovation in mobile library environments.

INTRODUCTION

The rise of mobile apps necessitates that libraries offer access to their patrons. By compiling a list of popular apps and providing step-by-step instructions on how to use them, libraries can empower their patrons to maximize their library resources. Libraries have long been pillars of knowledge dissemination, but the rapid evolution of digital technology has reshaped how they engage with patrons. Today, libraries leverage mobile applications, artificial intelligence (AI) tools, and Web 2.0/3.0 technologies to deliver seamless, personalized, and accessible services. These tools not only enhance patron experiences but also empower libraries to remain relevant in an increasingly digital world.

As mobile devices become ubiquitous, patrons increasingly expect seamless, on-the-go access to library resources, personalized services, and innovative digital experiences. Libraries operate in a digital environment shaped by ubiquitous mobile devices, necessitating a fundamental transformation in service delivery to meet con-

DOI: 10.4018/978-1-7998-9175-8.ch007

temporary patrons' needs. The advent of smartphones and tablets has revolutionized how individuals access information and engage with services, compelling libraries to adapt by developing or implementing mobile applications and optimizing their digital offerings for seamless use on the go. This transformation ensures that library resources remain discoverable, accessible, and integral to daily life, irrespective of physical location. The emphasis on mobile accessibility reflects a broader commitment to innovation, convenience, and equitable access to information resources in the digital age.

This focus delves into the diverse range of app functionalities currently provided by libraries to enhance user experiences and foster community connections, drawing insights from recent peer-reviewed research. It examines current practices in library app provision, the integration of AI and Web2.0/3.0 technologies, the implementation of iPad lending programs, and the optimization of mobile services, all based on recent peer-reviewed literature and real-world examples.

THE EXPANDING ROLE OF LIBRARY APPS

Library apps are no longer simply digital card catalogs. Instead, they represent sophisticated platforms designed to streamline access to a wide spectrum of resources. A core function of many library apps is the facilitation of seamless material discovery and management. Patrons can browse the library's collection, search for specific titles, view availability, place holds, and renew borrowed items, all from the convenience of their smartphones or tablets, (Jones & Johnson, 2021). This digital self-service model empowers users and reduces the need for in-person visits or phone inquiries for routine tasks. Furthermore, some apps incorporate features such as personalized recommendations based on borrowing history and the ability to create and share reading lists, fostering a more engaging and individualized experience with the library's collection.

Beyond collection management, library apps are increasingly becoming conduits for accessing digital content. Many libraries provide access to e-books, audiobooks, and digital magazines directly through their dedicated apps, often integrating with popular third-party platforms (Smith, Garcia, & Lee, 2022). This integration simplifies the user experience, allowing patrons to discover, borrow, and consume digital materials within a unified interface. Moreover, some libraries are expanding their digital offerings to include streaming video services, online learning platforms, and digitized archival materials, making a wealth of information and entertainment readily accessible through their mobile applications.

Recognizing the importance of community engagement, many library apps incorporate features designed to connect patrons with library events, programs, and staff.

These features can include a calendar of upcoming events, the ability to register for programs, receive notifications about library news, and even contact librarians via chat or email for research assistance (Brown & Davis, 2023). By providing these interactive channels, libraries are fostering a sense of community and ensuring that patrons remain informed and connected to the resources and expertise available to them. Some innovative apps also include features such as digital library card access, location finders for library branches, and even integration with local community resources, further solidifying the library's role as a central point of connection within the community.

The development and implementation of effective library apps require careful consideration of user needs and technological capabilities. Libraries must prioritize user-friendly interfaces, intuitive navigation, and accessibility for individuals with diverse needs. Furthermore, ongoing evaluation and updates are crucial to ensure that the app remains relevant, secure, and continues to meet the evolving needs of the community. As technology continues to advance, library apps will undoubtedly play an increasingly vital role in how libraries serve their patrons, bridging the gap between physical spaces and digital access, and fostering vibrant and engaged communities in places like Oxford, Connecticut, and beyond.

Bringing Libraries to Patrons' Fingertips

Searching can be done on a computer, but it's more likely to be done on a mobile device. People across the world use mobile media to talk to one another, order food, and attend meetings and appointments. "Mobile media—referring to devices, services, and content accessible on the go—has in a decade rapidly become a part of urban culture, and the habit of using a mobile device in public is only increasing," (Brandtzaeg, 2020). After the COVID-19 pandemic hit, mobile media usage surged to facilitate connections as shelter-in-place orders were imposed nationwide. Even as people resume their outdoor activities, mobile media will likely remain essential for certain services. The pandemic has profoundly reshaped the world, prompting organizations like libraries to explore innovative ways to engage with their patrons.

With time, the availability of mobile-friendly services has become critical to a library's long-term viability. The ability to search a library's catalog, for example, has aided a traditional library's modern relevance. "The digital library on mobile devices has been a milestone in library industry development, leading to huge changes in knowledge carrying, spreading, acquiring, processing, and sharing of cloud computing and big data in online and offline forms." 10 Users want data at their fingertips, whether it is an online order or a library catalog. "Social sharing functions such as reading, borrowing, sharing, comment tracking, automatic retrieval, and new book recommendations on mobile devices such as mobile phones have become the popular trend of digital library development," (Liu et al., 2020).

When evaluating mobile library websites and apps that libraries use to offer services, the literature appears to imply that a mobile version of a library website is more prevalent than a dedicated (or "native") app. This is because of the development cost, the different mobile platforms (Apple, Android, etc.), and the maintenance required. 12 Instead of creating an app, it is often easier to optimize the library's website for mobile devices through responsive design techniques.

Clearly, contemporary mobile users are no longer content with rudimentary mobile websites that offer only a limited subset of the information and functionalities accessible on desktop websites. While the compact screen size of mobile devices may complicate certain tasks, mobile users anticipate increasingly performing a broader range of activities on their mobile devices, (LeMire et al., 2017).

Studies have revealed the significance of understanding user needs when designing an app. A study found that "perceived situation efficiency and perceived mobile library quality positively affect intention to use mobile libraries," indicating that both quality and situation efficiency are crucial for satisfying users' needs in the mobile era, (Zhang et al, 2016).

While the quality of the library app or mobile website is undoubtedly important, knowing the services it will provide is equally vital. Users are unlikely to use an app if it doesn't offer the information they require. Studies have identified that underscores the desire of mobile users for quick and efficient information retrieval, (Liu & Briggs, 2015). Literature from the fields of information science and library science shows that in recent years, typical library services have been adapted to be accessible via a mobile device rather than a service being developed specifically for a mobile platform.

Transforming Library Services Through Mobile and Intelligent Applications

Artificial intelligence is reshaping library services by introducing intelligent tools that enhance user engagement and automate routine tasks. AI-powered chatbots and virtual assistants, such as the Xiaotu AI library developed by Tsinghua University, allow patrons to interact via mobile apps or social networks, providing real-time reference services. Xiaotu's self-learning function updates its knowledge base with user feedback, ensuring accuracy and relevance. Similarly, the Aisha chatbot, built using the ChatGPT API, assists patrons at a Malaysian university library with queries about resources and services, (Lappalainen & Narayanan, 2023).

Artificial intelligence is rapidly transforming the ways we work, learn, and serve our communities. Library staff are eager to understand and apply AI tools like ChatGPT in practical and meaningful ways. While they are aware of the ethical considerations that come with these emerging technologies and are careful not to over-promise what AI can do, librarians are discovering how AI can enhance people's daily work, streamline operations, and improve patron engagement,

Their discoveries delve into the potential of ChatGPT, generative AI, and large-language models to revolutionize library services. They aim to understand how these tools function, explore their capabilities, and develop essential skills in crafting effective prompts to achieve specific goals. By doing so, they develop practical strategies for utilizing AI to support their professional work and development, ranging from enhancing communications to enriching research and streamlining project planning.

Furthermore, they explore AI's transformative potential and emphasize the importance of responsible use. Initial practical frameworks are provided to address key considerations such as privacy, bias, and information quality. This knowledge equips them with the practical skills to handle copyright considerations, protect patron privacy, and ensure equitable AI access across their community. They learn specific approaches for implementing AI tools in ways that align with their libraries' values and mission, ultimately strengthening community trust.

The discoveries reveal concrete ways to enhance both their daily work and library services through AI tools. They develop practical strategies for integrating AI into their workflows, exploring real-world applications, and utilizing AI as a personal assistant. From assisting with writing projects and replying to challenging emails to providing immediate help with spreadsheet issues and technical questions and even brainstorming innovative programming ideas and enriching reference services, AI offers numerous benefits. These practical approaches empower them to support their work, enhance their library's services, and help their teams save time and work more creatively.

AI tools support advanced functions like text analysis and data mining. Libraries use natural language processing (NLP) to generate metadata, abstracts, and summaries, freeing librarians for strategic tasks, (Jyoti & Kumar, 2024). For example, the Urban Libraries Council advocates for AI to improve performance through prompt engineering, where librarians use keywords to refine AI outputs. Research by Cox, Pinfield, and Rutter (2018) indicate that AI enhances reference services, information retrieval, and cataloging, with university libraries in Pakistan employing AI for patron recognition via QR codes and voice search, (Ali et al., 2021).

By integrating AI, libraries not only enhance service delivery but also position themselves as leaders in technological innovation.

POPULAR SERVICE APPLICATIONS OFFERED ACROSS VARIOUS LIBRARIES

Mobile technology has revolutionized information access and service delivery, prompting libraries to adopt mobile apps as essential tools for community engagement. The mobile app landscape at libraries varies, reflecting diverse missions, needs, and capabilities. This session explores the functionalities and offerings of mobile apps across libraries, illustrating their role in democratizing access, fostering literacy, and building community connections.

Practical Studies on Mobile Apps Used in Libraries

At the foundational level, most library apps provide core functionalities that mirror and enhance traditional library services. These often include comprehensive catalog search capabilities, allowing patrons to discover physical and digital materials with sophisticated filters and intuitive interfaces. Beyond mere discovery, these apps empower users to manage their accounts effortlessly, facilitating tasks such as viewing borrowed items, renewing loans, and placing holds on desired titles, thereby increasing convenience and reducing the administrative burden on library staff. The seamless integration of digital content, from e-books and audiobooks to streaming media and digital magazines, is also a staple, often leveraging partnerships with platforms like OverDrive, Libby, or Hoopla, presenting a unified gateway to vast digital collections.

Beyond these essential services, many libraries extend their mobile app offerings to include enhanced features designed to deepen patron engagement and streamline access to specialized resources. Event calendars and registration modules are common, enabling users to browse upcoming workshops, story times, and cultural programs, and often register directly within the app. Some apps incorporate interactive maps for navigating larger library facilities, virtual tours, or even room booking functionalities for study spaces or meeting rooms. Notifications for due dates, hold availability, and library news further personalize the user experience, ensuring patrons remain informed and connected.

The evolution of library apps also reflects a commitment to lifelong learning and community enrichment. Many applications provide access to online learning platforms, language learning tools, or specialized databases that might otherwise be underutilized. Children's sections within apps often feature educational games, interactive storybooks, or curated content designed to promote early literacy. Furthermore, some libraries develop apps that highlight local history collections, genealogical resources, or community archives, offering a unique digital window into the heritage of the region they serve. These specialized offerings underscore the library's role not just as a content provider, but as a cultural custodian and educational partner.

Innovative libraries are also exploring apps that support more dynamic, interactive experiences. This can include augmented reality (AR) features for guided tours or interactive displays within the library, or gamified elements to encourage reading and learning, (Lee & Kim, 2023). Apps associated with makerspaces might offer equipment booking, tutorial videos, or even project inspiration. The overarching goal remains consistent: to provide a versatile, accessible, and intuitive digital interface that meets the diverse needs and expectations of a modern, digitally-savvy populace. The success of these varied mobile app initiatives across libraries underscores their critical role in ensuring that libraries remain vibrant, relevant, and indispensable community assets in the digital age.

Libraries are increasingly utilizing mobile applications to enhance patron services and streamline access to their extensive resources. The specific applications and services accessible to patrons will differ depending on the library. Further details are provided in Section II, categorized by types of libraries.

The proliferation of mobile applications in libraries, as highlighted in the "Mobile Apps Offered Across Various Libraries" Canvas document, necessitates a deeper examination of their practical impact and effectiveness. Beyond theoretical discussions of features and functionalities, empirical studies provide crucial insights into how these digital tools are actually utilized by patrons, their perceived value, and the challenges and successes encountered during implementation.

Mobile Apps Used to Facilitate Adoption and Implementation of Library Services

Studies on mobile apps used in libraries underscore that the availability of comprehensive, well-designed mobile offerings is a significant factor in user adoption and overall positive appraisal of library services.

A study by Davies and Miller (2020) investigated the usage patterns of a major public library's mobile app, finding that while basic functions like catalog searching and account management were highly utilized, advanced features such as event registration or online learning platform access saw varied engagement depending on proactive library promotion. Their findings emphasized the importance of intuitive design and clear communication of available features to drive higher engagement. Another study by Kim and Park (2021) specifically explored the impact of app design on user satisfaction, concluding that apps with clean interfaces, minimal navigation steps, and personalized content significantly correlated with positive user experiences.

Studies by Professor Liu and his colleagues (Liu, Lu, & Liu, 2014; Liu & Briggs, 2015; Liu & Lewis, 2023) have consistently shown that mobile services, including dedicated apps, are crucial for meeting user expectations for on-demand access to library resources. Their research on university and top-visited public libraries in the US highlights that key mobile services, such as catalog searching, account management, and access to e-resources, are highly valued and contribute significantly to user satisfaction and the perception of the library as a modern, accessible institution.

In Scotland, a study on mobile apps was conducted. While mobile apps and mobile webpages aren't precisely the same (as proven by previous literature), the study "Public Library Mobile Apps in Scotland: Views from the Local Authorities and the Public" shed some insight on mobile apps. It seems that many libraries do not have or are not interested in developing a mobile app (this coincides with the findings from LeMire, et al.'s study in the United States). However, when the public was surveyed, they did want more from their libraries. In order to increase the use of library services, apps should be considered (Kerr and Pennington, 2018). "By using an app instead of a mobile-enabled website, all the functionalities of smart technology can be incorporated to the library's advantage. Improved communication with patrons increases exposure to communities who otherwise would not use library services. (Kerr & Pennington, 2018)."

A similar study was done in Malawi, where mobile services to libraries were analyzed. As Malawi is more of a developing country than Scotland or the United States, the number of mobile services was fewer. The country recognized the potential of using mobile devices to access library services. Computer shortages are often a problem within libraries in developing countries, so by creating mobile services, the users could still access the library through their own mobile devices.

Studies have also found that it is important to know what your users will be looking for and use that information when creating your app design. "Perceived situation efficiency and perceived mobile library quality positively affect intention to use mobile library, demonstrating that both quality and situation efficiency are necessary to satisfy library users' needs in mobile era (Zhang, Shen, Zhu, and Yang, 2016)." The quality of the library app or mobile website is obviously important but knowing what kind of services will be provided is just as important. Users are not going to turn to an app if it is not going to provide the information they need. This study proved that mobile users want to quickly obtain the information they want in the most effective way possible.

The Impact of Apps on Specific Library Services and User Behaviors

Another critical domain for practical studies is the impact of apps on specific library services and user behaviors. Particularly, how do apps influence physical visits, borrowing patterns, or participation in library programs? A compelling case study by Foster (2016) examined the effect of an event-centric library app on program attendance. By analyzing registration data and comparing it to previous manual registration methods, they demonstrated a measurable increase in sign-ups for workshops and cultural events following the app's introduction, particularly among younger demographics. This study provided strong evidence that app-based registration modules actively streamline access and encourage participation. Similarly, research on the integration of e-content via apps often assesses the shift in borrowing habits, noting a consistent trend towards increased digital consumption, particularly during periods when physical access to the library is limited (Li et al., 2025).

Practical studies also frequently address the effectiveness of apps in promoting lifelong learning and literacy initiatives. Libraries leverage their apps to deliver educational content, host virtual story times, and offer access to specialized learning tools. A qualitative study by Thompson et al. (2002) interviewed parents about their use of a library's children's app, revealing that features like interactive storybooks and curated educational games were highly valued for supplementing home learning and fostering early literacy skills. Furthermore, the ability to access language learning platforms or professional development courses directly through the app has been shown in various studies to democratize access to these resources, especially for individuals facing geographical or time constraints (Patel & Singh, 2022).

Innovative and emerging app functionalities are fertile ground for practical investigation. The concept of augmented reality (AR) or gamification within library apps, as mentioned in the Canvas document, is being explored through pilot programs and user acceptance tests. Lee and Kim (2023), as cited previously, contribute to this by exploring user perceptions of AR features for library navigation and interactive learning. Practical studies in this domain often involve small-scale deployments and user feedback loops to refine the technology and understand its potential for broader application. These studies are crucial for guiding future development and ensuring that cutting-edge features genuinely enhance the library experience.

Practical studies on mobile apps in libraries move beyond conceptual discussions to provide empirical evidence of their efficacy. By examining user behaviors, service impacts, and the effectiveness of specific features, these studies offer invaluable insights for libraries seeking to optimize their digital offerings, ensure robust patron engagement, and solidify their role as essential community assets in an increasingly mobile world.

Libby and the EBSCO Mobile App

Two such apps, Libby and the EBSCO Mobile App, are two that I find personally interesting. The EBSCO Mobile App & Web Interface allows libraries to offer unlimited access to academic content wherever and whenever patrons need it. This reading app is designed for mobile researchers and includes academic journals, ebooks, articles, and other related content that information-seeking patrons can search through, browse, and borrow as needed. The app includes sharing and citation tools, alongside the ability to save articles to folders for later reading (Löff et al., 2022).

Libby is a similar reading app, albeit geared more towards the general public. It offers users the ability to search, browse, and borrow ebooks, audiobooks, magazines, and other digital items from local libraries for free (OverDrive, Inc., 2023). This app has grown exponentially in their global reach, with "over 90% of public libraries in North America….[and] 78 countries worldwide" (OverDrive, Inc., 2023) having access to it. Both apps' content are able to be accessed via a phone, tablet, web browser, or e-reader device (ex. kindle).

Below, I've included an interesting data table that Löff et al. (2022) discussed surrounding information-seeking behavior amongst researchers. This data was used to enhance EBSCO's mobile app, making it more intuitive and helpful. As shown, the mobile app was used most for on-the-go content, for organizational purposes, and for long-term research needs. This showcases that mobile apps such as these can enhance patron's information seeking experiences.

Figure 1. Attitude using mobile app on-the-go (OverDrive, Inc., 2023)

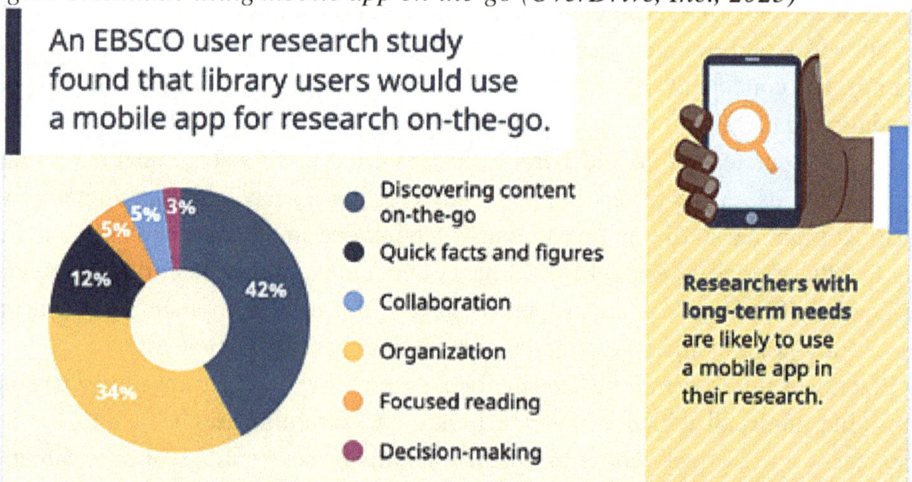

Augment Reality Practice

A live direct or indirect view of a physical, real-world environment whose elements are augmented (or supplemented) by computer-generated sensory input such as sound, video, graphics, or GPS data is described as being an example of augmented reality (AR). In the last few years, there has been a sharp surge in the application of augmented reality (AR) in the entertainment and business sectors, bringing to life objects, printed materials, and even physical locations by enhancing what we see around us with new digital resources. In essence, the application of augmented reality (AR) adds a virtual overlay to our current scenario in order to improve it by including rich media.s

The UK retailers Tesco and Marks & Spencer have incorporated AR into their customer magazines to give readers the opportunity to view videos of recipes being turned into meals by simply holding their smartphone or tablet over a particular image. A number of brands are using AR to add these virtual layers to physical objects. Recently, we've started to notice educational institutions investigating the opportunities that augmented reality (AR) presents for teaching and improving student learning, in addition to public engagement and marketing.

Visual Search Models

The world has become much more beautiful during the last ten years. Facebook updates used to be mostly text-based, but they have since shifted more toward images. Instagram and Pinterest both saw phenomenal development as visual social media platforms. Expect to see more of this type of visual search in the future years as the technology spreads to smaller-sized businesses.

Given that the human brain processes visual information for more than half of the time, it should come as no surprise that people prefer to search in this manner. How could visual search function, though, given that a search box is normally used for text input?

The saying "a picture is worth a thousand words" was coined by Fred R. Barnard. For visual search, there aren't many different methodologies. Using artificial intelligence is another technique for characterizing and classifying photos into structured data (language, numbers, categories) that can be looked up using a conventional search. Utilizing actual photographs as the query is one approach. In this article, I'll describe how visual search works and how, in our opinion, a site may currently use it most profitably.

Have you ever utilized Google Lens or Pinterest Lens ("shop the look") apps to identify anything in a photo? Bing Visual Search from Microsoft also has its own distinct flavor. using your phone's and your fingertips' amazing visual search technology!

Visual Search Functionalities

At their core, digital images are nothing more than vast grids of pixels—tiny points of red, green, and blue light (RGB) arranged in different patterns to form recognizable pictures. When viewed closely on a television or computer screen, these illuminated pixels reveal how millions of such units combine to create the images we see. Visual search technology builds on this principle by training neural networks to interpret these pixel grids, enabling them to recognize patterns, identify similarities, and distinguish specific objects such as faces or everyday items.

Beyond recognizing shapes and patterns, visual search can also be enhanced by incorporating text. Through a process called vectorization, text, alternative image descriptions (alt text), and associated images are converted into mathematical representations. This allows models to learn the relationships between language and visuals, improving search accuracy. For example, Google has successfully applied this technique in services like Google Photos and its broader search engine, making it possible to retrieve images based on text queries with remarkable precision.

Neural networks have become essential to this process, providing advanced machine learning capabilities for interpreting, classifying, and describing visual data. Modern AI models are now capable of identifying objects within images, categorizing them into groups, and generating descriptive results for users.

Today, visual search functionalities are widely integrated across platforms such as Google Search, Bing, DuckDuckGo, and Google Images. These services rely on vast databases containing millions of photographs and contextual metadata gathered from websites, blogs, and online articles. By combining this extensive image corpus with AI-driven recognition, search engines can deliver results for similar items, products, or related concepts with increasing sophistication.

A practical example of this in libraries is the use of visual search to support discovery in digital collections. For instance, the University of Illinois Library's digital collections platform allows users to perform image-based queries, enabling them to locate similar photographs, maps, or illustrations without relying solely on text-based metadata. This approach is particularly valuable when users cannot easily describe an item in words—such as historical photographs or architectural drawings—making visual search a powerful complement to traditional cataloging and keyword searching.

Visual Searches Implement

Visual search technologies are increasingly being used to support creativity, accessibility, and engagement. They offer powerful tools for visual storytelling, writing prompts, and multimedia projects, while also making creative expression more accessible to neurodiverse and differently abled learners.

Research indicates that when search results are displayed visually, users are about 50% more likely to be persuaded to complete a purchase, underscoring the importance of visual search for industries such as e-commerce and home furnishings, where product appearance plays a central role. In response, entire companies have emerged to develop AI-driven "computer vision" technologies tailored to retail and shopping applications. Major corporations, such as Amazon, have also integrated visual search capabilities into their platforms and extended these services through AWS, further expanding its commercial relevance.

The applications of visual search, however, extend well beyond online retail. Augmented reality (AR), which is projected to generate billions in commerce-related sales in the coming years, relies heavily on visual search to create immersive shopping experiences. Platforms like Snapchat have already embraced this trend, developing AR-optimized visual search to appeal to Millennials and Gen Z consumers who increasingly value interactive and visually driven digital experiences.

Libraries have also begun exploring visual search as a tool for discovery and engagement. For example, the New York Public Library's Digital Collections platform incorporates image-recognition features that allow users to browse and retrieve visually similar materials, such as historical photographs, manuscripts, and illustrations. This functionality enables patrons to find resources even when they lack precise keywords, bridging gaps in traditional metadata and opening new pathways for research and learning. Such applications demonstrate that visual search is not only shaping commerce but also enhancing cultural heritage access and library services.

Application of Visual Search in Real Life

The future of search is rapidly evolving beyond the traditional boundaries of text, audio, or even visual queries. Emerging research points toward brain-powered search technologies that integrate all modalities into seamless discovery experiences. Yet, for the present, visual search has established itself as a powerful complement to text- and voice-based tools, particularly in contexts where visual recognition is faster or more intuitive than language-driven queries.

One practical way visual search is implemented today is through tools such as the Google Vision API. This technology enables developers to integrate visual search capabilities into databases by automatically extracting features like color, shape, or other metadata from images as they are indexed. For instance, in e-commerce, the API can automatically generate filters and facets based on image properties, allowing users to refine searches efficiently whenever new products are added to a catalog.

Libraries, too, are exploring the potential of visual search to enhance discovery. A notable example is Europeana, the digital cultural heritage platform that aggregates millions of items from libraries, museums, and archives across Europe. Its visual search functionality allows users to explore collections by uploading or selecting images, which the system then matches with visually similar items. This approach is particularly valuable for uncovering related artworks, historical photographs, or artifacts when users may not know the precise keywords to describe what they are seeking. By integrating visual recognition with traditional metadata search, Europeana demonstrates how libraries can broaden access and improve user engagement through innovative search modalities.

Mobile users, in particular, benefit from these advancements. With increased access to digital services and digital libraries, mobile applications offer a range of advantages, including broader audience reach, greater accessibility, 24/7 availability, offline functionality, personalization, interactivity, and encouragement of reading habits (Löff et al., 2022).

NEW WEB GENERATION FOSTERING COLLABORATION AND DECENTRALIZATION

The digital landscape has undergone significant transformations, primarily influenced by the concepts of Web 2.0 and Web 3.0. Each iteration of the web has fundamentally reshaped the way users interact, share information, and access resources.

Web 2.0 and Library 2.0

Concurrently, libraries, traditionally repositories of knowledge, are undergoing their own transformation through Library 2.0 and Library 3.0. This discourse elucidates the salient characteristics of these interrelated phenomena, juxtaposing them with pertinent real-world instances and contemporary practices.

Web 2.0 technologies, characterized by user-generated content and interactivity, have transformed library services by fostering collaboration. Web 2.0, often dubbed the "social web," represents a significant shift from the static web of the past. With the rise of social networking sites, blogs, and user-generated content, the internet transformed into a participatory platform where individuals could both consume and create content. Platforms like Facebook, Twitter, and YouTube exemplify this shift, facilitating dialogue and engagement among users.

A notable example of Web 2.0's impact is the creation of Wikipedia, a freely accessible online encyclopedia that relies on public contributions. According to a study published in *Nature*, Wikipedia's accuracy is comparable to that of traditional encyclopedias, showcasing the efficacy of crowd-sourced information (Giles, 2005). This paradigm shift toward collaborative creation has had implications beyond general knowledge sharing; it has stimulated the democratization of content, allowing diverse voices to shape narratives that previously remained the domain of established authorities.

A practical example is the integration of Web 2.0 features into mobile apps like PressReader, which provides access to thousands of digital newspapers and magazines. This platform's faceted search and remote access capabilities reflect Web 2.0's emphasis on user participation and dynamic content.

Simultaneously, libraries have been on a journey of evolution, transitioning from Library 1.0, characterized by passive information dissemination, to the more interactive Library 2.0, which embraced the principles of Web 2.0. Library 2.0 fosters user engagement, collaboration, and community building—critical components that enhance the value of library services. The concept of Library 2.0, coined by Michael Casey in 2006, emphasizes participatory services like social media integration and user tagging. For instance, libraries use platforms like Twitter to communicate with patrons, as seen with the Osfoora 2 app, which allows librarians to manage multiple accounts for outreach (Masters in Library Science, 2014). Web 2.0 tools also enable patrons to contribute reviews and tags, enhancing resource discovery and creating a dynamic information ecosystem.

One practical application of Library 2.0 principles is the integration of social media platforms in library outreach. Libraries such as the New York Public Library have effectively utilized Twitter and Instagram to engage users, promote events, and share librarians' recommendations. A study by Kim & Abbas (2010) published in *Library Review* details how libraries adopting social media reported increased community interaction and user satisfaction, demonstrating the effectiveness of user-centric service models

Web 3.0 and Library3.0

Emerging from the foundations of Web 2.0, Web 3.0 strives for a more intelligent and interconnected web experience. Often encapsulated by the term "Semantic Web," this iteration emphasizes data interoperability and machine learning, enabling devices to understand and process information contextually rather than merely associatively. Web 3.0, the next evolution, introduces decentralization and semantic web capabilities, leveraging AI and blockchain for greater utility. Libraries are exploring Web 3.0 to create decentralized apps (dApps) that enhance data security and interoperability. While still emerging, Web 3.0's potential to connect information across platforms aligns with library goals of universal access. Research by Vasishta, Dhingra, and Vasishta (2025) suggests that Web 3.0 could revolutionize resource sharing by enabling libraries to operate on peer-to-peer networks, reducing reliance on centralized systems.

For example, blockchain technology, a core component of Web 3.0, facilitates decentralized applications that prioritize user privacy and data sovereignty. Cryptocurrencies like Bitcoin and Ethereum exemplify this principle, allowing users to transact without centralized oversight. The emergence of decentralized finance (DeFi) applications further illustrates the potential of Web 3.0, offering financial services that rely on smart contracts rather than traditional banking systems.

Recent studies have begun to explore the implications of Web 3.0 technologies in various sectors. A review by Furlan, Rizzo, and Alesiani (2022) in the *Journal of Information Technology* highlights the potential for enhanced data security and personalized user experiences, with applications ranging from data management to digital identity verification

Transitioning into Library 3.0 entails adopting the more structured and data-driven approaches facilitated by Web 3.0 technology. This involves leveraging artificial intelligence to tailor user experiences and improve resource discovery. For instance, libraries are beginning to implement intelligent cataloging systems using AI that adapt to individual user preferences and reading histories. This shift not only enhances user experience but promises to revolutionize how libraries curate and deliver their collections.

The University of North Texas Libraries have integrated augmented reality (AR) into their user services, allowing patrons to experience immersive interactions with library artifacts and gain instant access to additional digital content through their smartphones. This innovative approach embodies the Library 3.0 ethos: experiential, user-focused, and adaptive.

Similarly, in a study published in *Library Hi Tech*, Williams, Marks, and Kwan (2021) explored the use of blockchain technology in libraries to enhance transparency in digital loans and diversify licensing agreements. This application not only fosters user trust but aligns library practices with the ethos of data ownership that Web 3.0 promotes.

As libraries embrace Web 3.0, they will further enhance these services, creating more interconnected and intelligent platforms for patrons.

Libraries have long been the cornerstone of knowledge dissemination, but the rapid advancement of digital technology has transformed their engagement with patrons. Today, libraries utilize mobile applications, artificial intelligence (AI) tools, and Web 2.0/3.0 technologies to deliver seamless, personalized, and accessible services. These tools not only enhance patron experiences but also empower libraries to maintain relevance in an increasingly digital environment.

As we transition from Web 2.0 to Web 3.0, and similarly from Library 2.0 to Library 3.0, the fundamental principles guiding these transformations are evident: user engagement, data empowerment, and a commitment to adapting to evolving societal needs. Libraries must persistently adapt, leveraging emerging technologies and integrating user-centric practices to remain pertinent in an increasingly digitized world.

New Wave of Web Generations for Mobile Services in Libraries

As library services expand beyond buildings and desktop portals, a new wave of "web generations" is reshaping how collections, conversations, and communities operate on mobile devices. Three movements are especially consequential: (1) federated social technologies that enable institution-run, interoperable social spaces; (2) decentralized storage and preservation that harden public information against link rot and single-point failure; and (3) open interoperability layers that let cultural objects flow across mobile viewers while supporting collaborative annotation in teaching and research. Together, these shifts reimagine libraries as hubs in a distributed, read-write web—privacy-respecting, portable, and built for participation.

Federated social spaces for community engagement. The W3C's ActivityPub standard provides a protocol for decentralized social networking in which independently run servers (including those operated by cultural institutions) federate updates and conversations, (W3C, 2018). Libraries and archives can host or join these networks to engage patrons where they are—on mobile apps for Mastodon and other ActivityPub platforms—without ceding control to a single corporate platform. A concrete example is the American Archive of Public Broadcasting (a collaboration of the Library of Congress and GBH), which maintains a Mastodon presence to share audiovisual heritage and interact with the public; patrons can follow and reply from any compatible instance using mobile clients, (AAPB, n.5d.). Federated approaches let libraries set local moderation policies, align with institutional values, and still benefit from a larger, cross-instance public sphere.

Decentralized storage to preserve public information. Mobile discovery is only as strong as the persistence of what it points to. Decentralized storage networks now complement traditional repositories by distributing redundancy across independent nodes. The Internet Archive's End of Term Web Archive 2024/2025 adds an extra preservation layer by uploading data to the Filecoin network for long-term storage, continuing earlier collaborations with the Filecoin Foundation, (Hanamura, 2021). For mobile users who encounter government documents and datasets through apps, catalogs, or social feeds, this backing reduces the risk that cited resources vanish—mitigating link rot and strengthening the reliability of mobile learning and civic engagement.

Open, interoperable media for mobile access. On the access side, the International Image Interoperability Framework (IIIF) defines open APIs that allow any compliant mobile viewer to request the exact region, size, or sequence needed for an object—whether a single image, a multi-page manuscript, or time-based media packaged with structural metadata (IIIF, n.da; n.db). Europeana and partner libraries promote IIIF for scalable delivery and curation; guidance emphasizes how IIIF's tiled imagery and manifests enable responsive, zoomable access in web and mobile viewers, (EuropeanaPRO, 2022). The practical effect for patrons is fluid, app-agnostic experiences: a medieval map or photographic album can be deep-zoomed on a phone, embedded in a course site, or remixed into exhibits without rehosting assets.

Collaborative annotation across devices. Pedagogically, social annotation platforms let students and researchers read "together" on phones or tablets, overlaying highlights, questions, and citations onto articles, OERs, and primary sources. Hypothesis—widely adopted across higher education—integrates with LMSs and supports browser- and app-based use; case studies report hundreds of institutions using annotation to boost engagement and critical reading in mobile contexts. Libraries can partner with teaching faculty to scaffold source evaluation, citation practice, and disciplinary discourse directly within digital texts that students access on mobile devices.

Implications for library strategy. These developments collectively tilt library technology toward institutional autonomy plus interoperability. Federated social services align with library ethics around privacy and community governance. Decentralized preservation reduces single-vendor dependency and strengthens the integrity of links circulating in mobile environments. IIIF and social annotation operationalize the library's role in participatory and portable scholarship. Strategically, libraries can: pilot ActivityPub engagement from institution-managed servers; add decentralized replicas for high-value web archives; publish IIIF manifests for digitized collections; and embed social annotation in mobile-first instruction. Each step enhances resilience and collaboration without sacrificing openness.

The emerging web is less a destination than a fabric—stitched from open standards, federated services, and distributed storage that meet patrons on their phones while preserving institutional control and public memory. Libraries that embrace this fabric can deliver richer mobile experiences, cultivate local communities that connect globally, and ensure that the public record remains accessible—today and over the long term.

REFERENCES

AAPB (n.d.). AAPB (@amarchivepub) [Mastodon profile].

Ali, S., Habes, M., Youssef, E., & Alodwan, M. (2021). A cross-sectional analysis of digital library acceptance, & dependency during [*International Journal of Computing and Digital System.*]. *COVID*, ●●●, 19.

Brandtzaeg, P. B. (2020). How mobile media impacts urban life: Blending social cohesion and social distancing. *Interaction*, *27*(6), 52–56.

Brown, A., & Davis, K. (2023). Mobile engagement: Libraries connecting with communities through app technology. *Journal of Library Outreach and Engagement*, *3*(1), 45–62.

Cox, A. M., Pinfield, S., & Rutter, S. (2018). The intelligent library: Thought leaders' views on the likely impact of artificial intelligence on academic libraries. *Library Hi Tech.*

Davies, R., & Miller, S. (2020). User engagement and satisfaction with public library mobile applications: A quantitative study. *Journal of Library and Information Science*, *52*(3), 291–305.

Documentation, 74(5), 1092–1107.

Europeana, P. R. O. (2022). Recommendations on the implementation of IIIF (ARMA/GS project, July 2022). https://pro.europeana.eu/files/Europeana_Professional/Projects/Recommendations_on_implementation_of_IIIF_by_ARMA_GS_project_July_2022.pdf

Foster, M. J., & Evans, M. R. (2016). Libraries creating sustainable services during community crisis: Documenting Ferguson. *Library Management*, *37*(6/7), 352–362.

Furlan, E., Rizzo, S., & Alesiani, F. (2022). Unlocking the Value of Data with Web 3.0 Technologies. *Journal of Information Technology*, *37*(2), 120–135. DOI: 10.1057/s41265-022-00284-1

Giles, J. (2005). Special Report Internet encyclopaedias go head to head. *nature, 438*(15), 900-901.

Hanamura, W. (2021, April 1). Filecoin Foundation grants 50,000 FIL to the Internet Archive. Internet Archive Blogs. https://blog.archive.org/2021/04/01/filecoin-foundation-grants-50000-fil-to-the-internet-archive/

IIIF. (n.d.a). Presentation API 3.0. https://iiif.io/api/presentation/3.0/

IIIF. (n.d.b). API specifications (Image, Presentation, Auth, and more). https://iiif.io/api/

Jones, M., & Johnson, L. (2021). The evolution of library mobile applications: Features and user expectations. *Information Technology and Libraries*, *40*(2), 1–18.

Jyoti, J., & Kumar, J. (2024). Mapping the landscape of virtual word of mouth (v-WOM) in the tourism sector: A bibliometric review.

Kerr, A., & Pennington, D. R. (2018). Public library mobile apps in Scotland: Views from the local authorities and the public. *Library Hi Tech*, *36*(2), 237–251.

Kim, Y., & Park, J. H. (2021). Factors influencing the intention of knowledge sharing in public Libraries: Based on self-determination theory. *Journal of the Korean BIBLIA Society for library and Information Science, 32*(1), 247-265.

Kim, Y. M., & Abbas, J. (2010). Adoption of Library 2.0 functionalities by academic libraries and users: A knowledge management perspective. *Journal of Academic Librarianship*, *36*(3), 211–218.

Lappalainen, Y., & Narayanan, N. (2023). Aisha: A custom AI library chatbot using the ChatGPT API. *Journal of Web Librarianship*, *17*(3), 37–58.

Lee, J., & Kim, S. (2023). Enhancing library user engagement through gamification and augmented reality in mobile applications. *Library Hi Tech*, *41*(2), 273–289.

LeMire, S., Gilbert, S., Graves, S., & Faultry-Okonkwo, T. (2017). *Selfie as Guide: Using Mobile Devices to Promote Active Learning and Student Engagement*. Mobile Technology and Academic Libraries: Innovative Services for Research and Learning, 55-71.

Li, I. W., Jackson, D., & Koshy, P. (2025). Student's reported satisfaction at university: The role of personal characteristics and secondary school background. *Higher Education*, *89*(6), 1513–1531.

Liu, L., Su, X., Akram, U., & Abrar, M. (2020). The user acceptance behavior to mobile digital libraries. [IJEIS]. *International Journal of Enterprise Information Systems*, *16*(2), 38–53.

Liu, Y. Q., & Briggs, S. (2015). A Library in the Palm of Your Hand: Mobile Services in Top 100 University Libraries. *Information Technology and Libraries*, *34*(2), 133–148.

Liu, Y. Q., & Lewis, S. (2023). Services to mobile users: The best practice from the top-visited public libraries in the US. *Information Technology and Libraries*, *42*(1).

Löff, J., Hoffmann, R. B., Pieper, R., Griebler, D., & Fernandes, L. G. (2022). Dsparlib: A c++ template library for distributed stream parallelism. *International Journal of Parallel Programming*, *50*(5), 454–485.

OverDrive, Inc. (2023). Libby: The library reading app. OverDrive, Inc. https://www.overdrive.com/apps/libby

Patel, A. K., & Singh, K. (2022). A Bibliometric Study of Global Publications on the Effect of Social Media in Libraries from 1991 to 2021. *Management of Modern Libraries in New Normal*, 216-226.

Smith, P., Garcia, R., & Lee, S. (2022). Integrating digital resources through library apps: A user-centered perspective. *Digital Library Perspectives*, *38*(3), 157–175.

Thompson, B., Cook, C., & Thompson, R. L. (2002). Reliability and structure of LibQUAL+ scores: Measuring perceived library service quality. *portal. Portal (Baltimore, Md.)*, *2*(1), 3–12.

Vasishta, P., Dhingra, N., & Vasishta, S. (2025). Application of artificial intelligence in libraries: A bibliometric analysis and visualisation of research activities. *Library Hi Tech*, *43*(2/3), 693–710.

W3C. (2018). ActivityPub (W3C Recommendation). https://www.w3.org/TR/activitypub/

Williams, S., Marks, L., & Kwan, Y. (2021). The Potential Role of Blockchain in Libraries: A Systematic Review. *. *Library Hi Tech*, *39*(3), 520–535. DOI: 10.1108/LHT-06-2021-0114

Zhang, M., Shen, X., Zhu, M., & Yang, J. (2016). Which platform should I choose? Factors influencing consumers' channel transfer intention from web-based to mobile library service. *Library Hi Tech*.

Section 2
Library Services to Mobile Users

Chapter 8
Mobile Services in Academic Libraries

ABSTRACT

This chapter examines the rapid expansion of mobile services in university and college libraries and the diverse tools now supporting student success. It surveys mobile-optimized websites, catalog access, messaging services, e-books, social media, QR codes, printing tools, and tablet-lending programs, highlighting both established and emerging service models. The chapter explores evolving mobile content strategies, including digital textbooks, streaming media, OERs, and personalized recommendations shaped by analytics. It also analyzes remote services for distance learners and the technologies that enable interactive engagement at scale. Virtual reference modalities and ADA-compliant design principles are reviewed to underscore accessibility, equity, and the changing expectations of mobile-first academic communities.

INTRODUCTION

Academic libraries are committed to maintaining their position at the forefront of adaptation to significant shifts and trends in higher education, as well as in the fields of library and information science. Digital shift services for academic libraries in the new era offer a great deal to the communities they serve. What is the state of mobility in libraries? Academic librarians are required to meet the demands of post-secondary students, faculty, and staff in intricate, frequently networked institutions ranging from community colleges to independent, private, and public universities with a focus on basic and applied research.

Mobile devices, such as smartphones, tablets, e-book readers, handheld gaming tools and portable music players are practically omnipresent in today's society. According to Walsh (2012), "Mobile data traffic in 2011 was eight times the size

DOI: 10.4018/978-1-7998-9175-8.ch008

of the global internet in 2000 and, according to forecasts, mobile devices will soon outnumber human beings". Studies have revealed that the use of mobile devices is widespread and continues to increase. As of 2013, 56% of Americans owned a smartphone, (Smith 2015). This number is even higher among people ages 18 to 29. However, Peters and Bell (2013) point out that mobile phones at least can be found among people of all ages, nationalities and socioeconomic classes. He writes, "We truly are in the midst of a global mobile revolution." The ACRL Research Planning and Review Committee found that 55% of undergraduates have smartphones, 62% have iPods, and 21% have some kind of tablet, (Elmore & Stephens, 2012). Over 67% of these students use their devices academically. Elmore and Stephens (2012) write, "Academic libraries cannot afford to ignore this growing trend. For many students a mobile phone is no longer just a telephonic device but a handheld information retrieval tool."

It is clear from these studies that academic libraries can expect their patrons to be accessing their sheer number of mobile devices on the market and the myriad ways libraries could offer mobile services can be daunting. Additionally, offering mobile services requires investing time, money, and personnel. Bomhold (2014) examined 73 academic libraries at RU/VH-rated institutions to assess their mobile library services. The results showed that even while most academic libraries at research universities offer some form of mobile service, the options available are extremely diverse and unreliable. Many eminent university libraries still don't have a mobile presence.

Liu and Briggs (2015) conducted a comprehensive study on the services provided by academic libraries in the United States to their communities. Their study, titled "A Library in the Palm of Your Hand: Mobile Services in Top 100 University Libraries," was published in Information Technology and Libraries. The study examines the current status of mobile services in the United States' top 100 universities' libraries as a model, specifically focusing on the services offered, their intended use, and the challenges faced in providing mobile services.

Two key questions arise from this study: what is the current state of mobile services among academic libraries in the country's top-ranked universities, and what insights can be gained from the experiences of these libraries regarding best practices for mobile services at the university level? Universities are often more progressive than small-town libraries. For instance, a study of 100 university libraries revealed that many of the top universities in the United States are already utilizing mobile-friendly services, (Liu & Briggs, 2015). Typically, the services offered to mobile users include mobile websites, mobile applications, mobile OPACs, mobile access to databases, text messaging services, QR codes, augmented reality, and e-books. While public libraries may not offer all of these services (augmented reality is not typically available on public library websites or mobile applications), many of the others are, particularly the mobile OPAC, or library catalog.

GROWTH OF MOBILE SERVICES IN ACADEMIC LIBRARIES

Academic libraries have adopted responsive web design for device-agnostic access. Top university libraries offer mobile sites, text messaging, e-books, and chat, with a trend toward AI-enhanced tools for routine queries, (Liu & Briggs, 2015). Virtual reference services (VRS) gained traction post-COVID, with high user satisfaction for chat-based support in Hong Kong academic libraries, (Chow & Croxton, 2022).

There is a lot of data regarding the prevalence of mobile services in academic libraries. A 2010 study found that 35% of the English-speaking members of the Association of Research Libraries had a mobile website for either the university, the library, or both, (Canuel & Crichton, 2017). A study of Chinese academic libraries revealed that only 12.8% surveyed had a section of their web pages devoted to mobile library service, (Li, 2024). Canuel and Crichton (2017) found that 13.7% of Association of Universities and Colleges of Canada members had some mobile services, including websites and apps. They found that that 44% of academic libraries offered some type of mobile service. 39% had a mobile website, and 36% had a mobile version of the library's catalog. Half of the libraries which did not offer mobile services were in the planning process for creating a mobile website, catalog, and text notifications. Additionally, 40% planned on implementing SMS reference services, and 54% wanted the ability to access library databases on mobile devices. However, it is widely assumed that mobile services will expand rapidly in the future, (Canuel & Crichton 2017).

Ashford and Zeigen (2012), a survey of academic libraries in the Pacific Northwest, found that 50% had a mobile version of the library's website and/or catalog, 40% used QR codes, 38% had a text messaging service, and 18% replied "other" with mobile interfaces for databases being a popular offering. However, 31% of survey respondents still did not have any mobile services. Osika and Kaufman (2012) surveyed community and junior colleges nationwide to determine what mobile services were being offered. 73% offered mobile catalog access, 62% offered vendor database apps, two were creating a mobile app for the library, and 14.7% had a mobile library website.

Academic libraries are undergoing a significant transformation, driven by the rapid proliferation of mobile technologies. Yan Quan Liu, a prominent scholar in library and information science, has extensively studied the integration of mobile services in academic libraries, offering valuable insights into their development, implementation, and impact. Empirical research, such as the on- and off-line surveys conducted in academic libraries, is crucial for identifying preferred service types and satisfaction factors, thereby providing strategies for quality improvement

Liu and Briggs (2015) highlight that mobile services in academic libraries have evolved from basic SMS notifications to sophisticated mobile applications that offer access to catalogs, databases, e-books, and interactive services. The study provides a detailed examination of how libraries are adapting to meet the needs of mobile-first users. The study found that students increasingly expect seamless access to library resources through their smartphones and tablets, prompting libraries to invest in mobile-friendly websites, apps, and responsive design.

One of Liu's key contributions is his comparative analysis of mobile services across top academic institutions. In *A Library in the Palm of Your Hand*, Liu and Briggs (2015) surveyed mobile services in the top 100 university libraries, revealing a growing trend toward mobile integration. The study categorized services into four main areas: access to library resources, communication and reference services, library instruction, and administrative functions. It found that while most libraries offered mobile access to catalogs and databases, fewer provided mobile reference services or instructional content, indicating areas for further development.

Liu also emphasized the importance of user-centered design in mobile library services. His findings suggest that successful mobile implementations are those that prioritize usability, accessibility, and relevance to student needs. Libraries that engaged in continuous user feedback and iterative design processes were more likely to see higher adoption and satisfaction rates.

Despite the progress, Liu identified several challenges facing academic libraries. These include limited budgets, lack of technical expertise, and the need for ongoing staff training. Moreover, the rapid pace of technological change requires libraries to be agile and forward-thinking in their mobile strategies.

Liu's research provides a comprehensive overview of the current status and future potential of mobile services in academic libraries. His work underscores the necessity for libraries to embrace mobile technologies not just as a convenience, but as a core component of their service model. As mobile usage continues to rise, academic libraries must innovate to remain relevant and responsive to the evolving needs of their users.

Study results presented compelling evidence that mobile services are already ubiquitous among the country's top universities. The most recognized ones are mobile sites, mobile apps, mobile OPACs, mobile access to databases, text messaging services, QR codes, augmented reality, and e-books. These service forms confirm those commonly named in the literature as library mobile services.

TYPES OF MOBILE SERVICES PUBLIC LIBRARIES PROVIDE

The integration of mobile technologies into academic libraries represents a paradigm shift in how information is accessed, delivered, and consumed. As mobile devices become ubiquitous among students, faculty, and researchers, academic libraries have responded by developing a suite of mobile services designed to enhance accessibility, convenience, and user engagement. These services are not merely add-ons but are increasingly central to library strategy and operations.

As mobile devices become the primary access point for information among students and faculty, libraries have responded by developing a wide range of mobile services. Yan Quan Liu and his co-authors, along with subsequent studies, has identified a core set of mobile services that have become standard across many academic institutions. These services are designed to meet users where they are—on their smartphones and tablets—and to provide seamless access to library resources and support.

Liu and Briggs (2015) conducted a survey and analysis of the top 100 university libraries in the United States. Providing a foundational understanding of how academic libraries have embraced mobile services, their study identified eight core mobile services: mobile websites, mobile apps, mobile OPACs (Online Public Access Catalogs), mobile access to databases, text messaging, QR codes, augmented reality, e-books and other local based services.

Via the website visits and survey, they found all libraries surveyed offered at least one mobile service, with the majority providing five or more at the top 100 universities are offering multiple mobile services. Mobile websites, mobile access to the library's catalog, mobile access to the library's databases, e-books, and text messaging services being the most common. QR codes and especially augmented reality are not as common.

Table 1. Number of mobile services offered

Number of mobile services offered	Number of libraries	Percentage
1 mobile service	1	1%
2 mobile services	0	0%
3 mobile services	7	7%
4 mobile services	13	13%
5 mobile services	24	24%
6 mobile services	46	46%
7 mobile services	8	8%
8 mobile services	0	0%

The results showed that the libraries do not appear to be offering mobile services in isolation; they have taken several of their most popular services (such as websites, reference, and search functions) and mobilized all of them. In fact, the average number of mobile services offered among the eight services we examined is 5.31.

- 100% of libraries offered at least one mobile service.
- The average number of mobile services offered was 5.31.
- 46% of libraries offered six services, and 8% offered seven.
- No library offered all eight services, and only one offered just one.

Such a data pattern demonstrates not only that mobile services are very widespread at these universities' libraries, but also that the vast majority of these libraries offer multiple mobile services. This widespread adoption signifies a strategic shift towards meeting the evolving needs of academic communities in the digital age.

Today, mobile platforms are essential conduits for accessing a rich range of content, extending far beyond the traditional catalo. This pattern demonstrates that libraries are not implementing mobile services in isolation. Instead, they are mobilizing multiple core services—such as catalogs, databases, and communication tools—to create a comprehensive mobile experience.

Below is a detailed exploration of the eight fundamental types of mobile services commonly offered by academic libraries.

Mobile-Optimized Websites

A mobile-optimized or responsive website is the foundation of any library's mobile strategy. These websites automatically adjust their layout and functionality based on the user's device, ensuring a smooth browsing experience.

The development of mobile-optimized websites was one of the most foundational services from Liu and Briggs (2015). These sites are designed with responsive layouts that adapt to various screen sizes, ensuring users can access catalogs, databases, and library hours seamlessly from any device. Liu's study found that nearly all top 100 university libraries had implemented mobile-friendly websites, recognizing their importance in user engagement.

The findings showed that all libraries with mobile sites include library hours on their site, making this the most common feature. The next two most common features are library contact information and a search function for the catalog, which both received 96.67%. Searching within mobile-friendly databases, such as EBSCOhost Mobile, JSTOR and PubMed, is the next most popular feature, although it trailed a little behind library hours, contact information, and catalog searching at 70%. Book renewal received 56.67%, and access to patron accounts received 53.33%.

Interlibrary loans are the least common feature by far, offered by only 26.67% of respondents. See Chart below, (Liu & Briggs, 2015)

Figure 1. Components of libraries' mobile sites

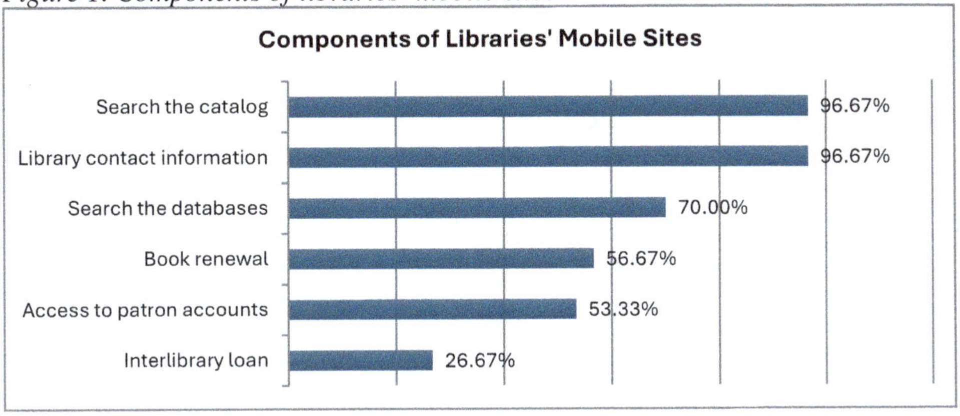

A quick rundown of the main features includes:

- Access to the library catalog (OPAC)
- Links to databases and e-resources
- Hours of operation
- Contact information
- News and announcements

A 2016 investigation into mobile access provision by elite academic libraries—titled Mobile Web Adoption in Top Ranked University Libraries: A Preliminary Study—revealed that a substantial majority (88%) of the top 50 university libraries (as ranked by the Web Ranking of World Universities) offered some form of mobile access, whether through a mobile-optimized website or a dedicated application. Specifically, 80 percent (40 libraries) supported access via the mobile web, whereas only 34 percent (17 libraries) provided an app. These findings underscore that, as early as 2016, leading academic libraries had already embraced mobile access, although the level of adoption varied significantly between mobile web and apps, (Torres-Pérez, Méndez-Rodríguez, & Orduna-Malea, 2016).

Singh and Narayan (2021) explore the growing potential of Progressive Web Applications (PWAs) as an innovative, mobile-first solution to enhance academic library services. Emphasizing Ranganathan's principle that libraries are dynamic, they argue that PWAs provide a cost-effective, installation-free method for delivering information ubiquitously via handheld devices, offering mobility that aligns with contemporary user behavior. By leveraging modern web technologies—such as service workers and manifests—PWAs enable academic libraries to offer responsive, app-like experiences that users can access anytime and anywhere, without the dependence on physical infrastructure or staff, (Singh & Narayan, 2021). This approach facilitates more seamless interaction with library resources, advancing the reach and impact of academic library services in the digital era.

Mobile Access to Catalogs and Databases

Libraries offer mobile access to their Online Public Access Catalogs (OPACs) and subscription databases. This allows users to search for resources, place holds, and access full-text articles from their devices. Mobile access to library catalogs and subscription databases allow users to search for books, articles, and multimedia resources from anywhere. Many libraries integrate federated search or discovery tools like EBSCO Discovery Service or Primo, which are optimized for mobile use.

Features often include:

- Keyword and advanced search
- Filtering and sorting results
- Access to full-text articles
- Citation tools

These services allow users to:

- Search for books, articles, and multimedia
- Access full-text content
- Manage holds and renewals

This service is especially valuable for students conducting research on the go or accessing materials during lectures and study sessions.

Liu and Briggs (2015) revealed that mobile OPACs were offered by 87.8% of libraries (website visits) and 88.24% (survey), while mobile access to databases was slightly more variable: 92.9% (website visits) versus 70.59% (survey). This discrepancy may reflect differences in how libraries define or promote mobile database access.

While not the primary focus of the four studies, mobile access to catalogs and databases is often integrated into mobile apps and websites. Saragossi, Costello, and Kasten (2024) note that mobile apps increasingly include access to core library functions like catalog search and database access, which are essential for research on the go.

The CHFSR Library at Assam Agricultural University developed a mobile app to enhance OPAC services, reaching a broader user base and saving time for patrons, (Thurlow, Duke, & Davies, 2006). Peer-reviewed studies, such as Beg, Russell, and Hornig (2022), emphasize that mobile apps improve data security through encryption, addressing privacy concerns while enhancing user trust. By investing in mobile infrastructure and staff training, libraries can create robust, user-centric services that align with modern expectations.

Customized Mobile Implementations

Custom mobile apps are increasingly common, offering a centralized platform for accessing services such as account management, book renewals, event calendars, and even augmented reality tours of library spaces. These apps often include push notifications for due dates and announcements.

Dedicated mobile apps offer a centralized platform for accessing a wide range of library services. These apps are often custom-built or adapted from third-party vendors and may include:

- Account management (renewals, holds, fines)
- Event calendars and news
- Study room reservations
- Augmented reality tours
- Push notifications for due dates and events

Apps provide a more integrated and personalized experience than mobile websites, and they often support offline access to certain features.

Liu and Briggs (2015) examined the presence of eight specific mobile services. For mobile apps:

- 20.2% of libraries were found to offer mobile apps based on website visits.
- 38.24% of libraries reported offering mobile apps in the survey responses.

This discrepancy was noted and discussed in the study. The authors suggested that mobile apps might not always be prominently advertised on library websites, which could explain why they were underreported in the website analysis compared to the survey.

In their findings, the researchers discovered that while website visits and survey results indicated that library apps are not widely used, both tools revealed a widespread use of apps for various purposes. The most commonly mentioned app is BrowZine, which is used for accessing e-journals. Several respondents also mentioned in-house apps developed for library services, such as an app for reserving study rooms, accessing university archives, and sending catalog records to mobile devices. Libraries also use vendor-provided or third-party apps, including those for accessing PubMed, ScienceDirect, Naxos Music Library, AccessMyLibrary (for Gale resources), a mobile medical dictionary, and the American Chemical Society. To support student research, libraries loan iPads preloaded with popular apps like EndNote, Notability, GoodReader, Pages, Numbers, and Keynote, among others. The apps libraries either use or provide access to include iResearch (for storing articles locally), Boopsie (for building a library mobile app), ebrary (for accessing e-books), and Safari (for accessing books and videos online). These results suggest that the use of apps is robust and diverse among these libraries. Additionally, it appears more common for libraries to use and/or provide apps created by third parties rather than develop in-house apps, possibly due to the expertise and expense involved in creating and maintaining an app.

Text Messaging and Chat Services

SMS-based services allow users to receive notifications about holds, due dates, and overdue items. Additionally, many libraries offer real-time chat or instant messaging (IM) services for reference inquiries, which are accessible via mobile devices. Liu noted that these services were among the most popular due to their immediacy and convenience

Text messaging and mobile chat services are essential for real-time communication between users and library staff. These services typically include:

- SMS alerts for due dates, holds, and overdue items
- Live chat for reference and research help
- Integration with platforms like WhatsApp or LibChat
- Real-time reference support via SMS or chat

Liu and Brigg's (2015) study revealed that Text messaging services were widely adopted, with 80.8% (website visits) and 73.53% (survey) of libraries offering them. Their research highlighted the popularity of these services due to their immediacy and ease of use. They are particularly effective for quick questions and support outside of traditional service hours.

SMS and chat services are often embedded within mobile apps or websites. Singh and Madhusudhan (2023) mention that real-time communication tools like chatbots and WhatsApp integration are becoming more common in mobile library services.

E-Books and Mobile-Optimized Content

Access to e-books and mobile-compatible academic content is a major component of mobile services. Libraries subscribe to platforms like ProQuest, JSTOR, and

EBSCOhost, which offer mobile-friendly interfaces or dedicated apps for reading and downloading materials.

Academic libraries increasingly provide access to e-books and digital content that is optimized for mobile consumption. This includes:

- E-book platforms like ProQuest Ebook Central, JSTOR, and SpringerLink
- Mobile-friendly PDF and EPUB formats
- Mobile-friendly reading interfaces
- Downloadable content for offline reading

These services are crucial for remote learning and flexible study; they support flexible learning and are especially beneficial for distance learners and users with accessibility needs. Singh and Madhusudhan (2023) discuss how mobile apps are increasingly used to access e-book platforms and digital content repositories. They note that libraries are focusing on cross-platform compatibility and offline access, which are crucial for mobile users.

Social Media and Communication Tools

Academic libraries use social media platforms to engage with users, promote events, offer services, build community, and share updates and news. These tools are particularly effective for reaching students who spend significant time on social media apps. Common platforms academic libraries offer include:

- Twitter for announcements and updates
- Instagram for visual storytelling and event promotion
- Facebook for community engagement and live events

Engaging with students in informal ways, social media has become a standard communication tool for academic libraries to foster a sense of connection beyond the physical library space. While not a central focus of the studies, Singh and Madhusudhan (2023) acknowledge that mobile apps often integrate with social media platforms to enhance communication and outreach. This integration helps libraries maintain visibility and engagement with users in digital spaces they frequent.

QR Codes and Location-Based Services

QR codes strategically placed throughout library spaces serve as digital guides, tutorials, or catalog entries. Some libraries even leverage geolocation to provide contextually relevant assistance, such as guiding users to specific resources or services within the building. Academic libraries have experimented with QR codes to offer instant access to digital resources. Others have explored geolocation services to enhance navigation within large library buildings.

Using QR codes to bridge the physical and digital library environments, based on Liu and Brigg's (2015) study, QR codes were offered by 61.6% of libraries (website visits) and 55.88% (survey).

They are often placed:

- On bookshelves to link to subject guides
- In study areas to provide access to tutorials
- On posters to promote events or services
- Link to subject guides or tutorials
- Provide quick access to digital content
- Enhance signage and navigation

Led by Liu, Guo et al. (2025) expanded upon Liu's 2015 study, which revealed that some libraries also experiment with geolocation services that provide context-aware assistance, such as guiding users to specific resources or services within the building. Their study explored augmented reality (AR) and virtual navigation in academic libraries and found that 86% of top U.S. university libraries had implemented VR/AR technologies.

- AR-enhanced navigation
- Virtual exhibitions
- Immersive learning environments

Libraries are increasingly leveraging QR codes to bridge the gap between their physical and digital collections. These two-dimensional codes enhance accessibility by instantly linking patrons to e-books, audiobooks, research papers, and multimedia content. They also foster deeper engagement through interactive learning experiences, gamified activities, and personalized recommendations. On the operations side, QR codes streamline processes such as self-service checkouts, inventory management, and event registration—ultimately boosting efficiency and helping libraries reach broader audiences while promoting lifelong learning, (Ibrahim & Sanyal, 2025).

Mobile Printing and Reservation Systems

Mobile printing services allow users to send documents from their devices to library printers. Additionally, mobile interfaces for reserving study rooms, equipment, or event spaces are becoming standard. These mobile interfaces are used for:

- Reserving study rooms or equipment
- Booking appointments with librarians
- Registering for workshops or events

These services streamline administrative tasks and reduce the need for in-person interactions, enhancing user convenience. From Liu and Brigg (2015) revealed that:

- 41.18% of libraries offered mobile book renewal
- 20.59% offered mobile interlibrary loan
- 32.35% offered mobile-friendly library instruction

These services reflect a growing trend toward mobilizing administrative and instructional functions, further enhancing user convenience. Singh and Madhusudhan (2023) mention that mobile apps are increasingly supporting administrative functions such as:

- Study room reservations
- Mobile printing
- Interlibrary loan requests

These features enhance user convenience and reduce the need for in-person interactions. Studies conducted through surveys conducted in academic libraries have shown that mobile service adoption is widespread and multifaceted. The average number of services offered is continuously increasing, indicating a strong institutional commitment to mobile accessibility.

Tablet Lending Services

Academic libraries also use iPads to support instruction and reference services. Boise State University's Albertsons Library provides iPads with access to its mobile website, which includes animated floor maps and study room reservations. Kim (2013) highlights that iPad programs enhance patron engagement by offering portable access to library resources, particularly for students conducting research on the go. These programs also promote digital literacy, as librarians guide patrons in using apps for citation management and database searches.

Providence College's Phillips Memorial Library began lending iPads to students, faculty, and staff in 2012. The library curated a selection of subject-specific and task-based apps, aiming to enhance both academic and personal productivity. Despite efforts to promote the use of educational apps, assessment data revealed that patrons primarily used the iPads for internet access, social media, and media consumption. This finding prompted library staff to rethink their approach, leading to the deployment of iPad kiosks within the physical collection to better connect patrons with relevant digital resources, (Providence College, 2017).

The University at Buffalo Health Sciences Library implemented an embedded librarian service using iPads. Librarians brought iPads directly to users, offering point-of-need reference and instructional services. While the project demonstrated the value of mobile technology in delivering library services outside traditional spaces, it also highlighted the importance of aligning device use with specific educational goals and user expectations, (University at Buffalo Health Sciences Library, 2013).

The study "iPad and Digital Sheet Music Delivery," conducted at the University of Manitoba Libraries (UML), delves into how music libraries can harness the power of iPads and digital sheet music to foster student growth and promote the adoption of digital scores. The study involved distributing iPads equipped with Apple Pencils, protective cases, charging stations, and pedals for page turning. These iPads were configured with a single Apple ID to streamline the app installation and data transfer process. Basic apps such as Adobe PDF Reader and nkoda Reader were pre-installed on the iPads. By providing access to iPads and digital sheet music, libraries empower students to gain hands-on experience working in a digital environment. They also enable students to utilize applications like nkoda Reader for marking on uploaded scores and potentially explore the use of wireless page turners for performing, (Penner, 2024).

Services Commonly Considered or in Development

The planning to expand mobile services is continually evolving based on the evolving needs and capabilities of technology. A chart below summarizes the findings from Liu and Briggs (2015), which provides strong evidence and older data suggesting a high existing adoption and continued interest in mobile services.

Figure 2. Percentage of the libraries seeking to add specific mobile services

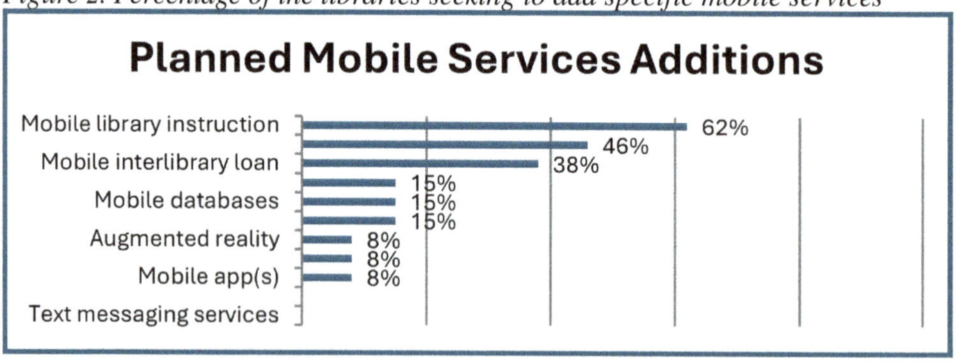

Liu and Briggs (2015) revealed that mobile-friendly library instruction is the most popular selection, with 61.54% of respondents choosing this option. Mobile websites are the second most common choice, with 46.15% of respondents selecting this option. Mobile interlibrary loan is chosen by 38.46% of respondents. Less common services planned include adding mobile access to the library's OPAC, mobile access to the library's databases, and mobile book renewal, each of which were chosen by 15.38% of respondents. 7.69% of respondents plan to add mobile apps, e-books, and augmented reality, respectively. No one indicated plans to add text messaging services or QR codes. These results indicate that libraries anticipate continued growth in demand for traditional library services in a mobile-friendly format. While mobile-friendly library instruction was only offered by 32.35% of respondents, 61.54% have plans to offer this service in the future. Mobile interlibrary loan is currently offered by 20.59% of respondents, so the fact that 38.46% would like to add it represents a significant change.

Where older studies (e.g., from 2010 or 2016) indicated that a significant portion of academic libraries were in the planning stages for mobile services. It is likely that many of these plans have now been implemented or are progressing. Recent surveys and reports indicate that academic libraries are actively exploring and implementing new mobile services to cater to the evolving needs of their users.

Some key trends and planned initiatives can be identified by observing the similar pattern of planning development.

- Mobile-friendly websites or apps: Providing access to library information and resources optimized for mobile devices.
- Mobile access to the library catalog and databases: Enabling users to search and access library resources from their phones or tablets.
- Text messaging services (SMS): Facilitating communication with patrons for services like renewal reminders or notifications.
- QR codes and Augmented Reality: Exploring innovative ways to provide information and interact with library spaces.
- Access to e-books: Allowing users to read and browse digital books on their mobile devices.
- Mobile-based information literacy instruction: Leveraging mobile devices for teaching information literacy skills

Growing Expectation and Demand

- Several articles highlight the "explosive growth of mobile devices" and the increasing user expectation for mobile access to library services.

- The 2022 EDUCAUSE Review article "Moving to Mobile: Space as a Service in the Academic Library" notes that students have "almost entirely transitioned to mobile computing," with 91% of students owning a laptop and 95% owning a smartphone, (EDUCAUSE, 2022). This indicates a strong user-driven need for mobile library services.
- A 2024 Emerald Insight article mentions that "modern library users have positive attitudes towards mobile technology and are starting to demand libraries to provide services that are accessible via mobile technologies," (Mustapha, et al., 2024).

Although research conducted in the field has indicated that not all library services for mobile devices are fully utilized, libraries are proactively seeking solutions to address this issue. Mobile services are a must! (Liu & Lewis, 2023) Academic libraries are actively pursuing the development of websites that are responsive to a wide range of devices, including desktops, laptops, tablets, and smartphones. This commitment to adapting to the mobile environment is driven by the growing expectations and requirements of their patrons.

CONTENT OFFERED BY MOBILE SITES IN ACADEMIC LIBRARIES

Digital content ecosystems of academic libraries to their mobile users are growing. Academic libraries have progressively embraced mobile-friendly interfaces, revolutionizing how students, faculty, and researchers interact with their services. This transformation, which occurred in the digital age, goes beyond simply migrating content from print to digital formats. It involves a comprehensive redesign of services, user engagement strategies, and content curation approaches. Academic libraries now offer a wide range of digital content, including eBooks, multimedia, databases, and interactive tools, directly to users through smartphones, tablets, and laptops. Current research, particularly the influential work of Dr. Yan Quan Liu and his research teams, has provided valuable insights into how academic libraries curate, deliver, and personalize digital content for users on mobile devices.

Mobile Content Strategy: A Shift Toward Accessibility and Ubiquity

The ubiquity of mobile devices has prompted academic libraries to rethink how they provide information access. Dr. Yan Quan Liu, a leading voice in mobile library services, emphasizes that libraries must embrace "mobility" not just as a technological function, but as a paradigm for equitable information delivery, (Liu & Briggs, 2015). His studies show that more than 90% of surveyed academic libraries in the U.S. provide mobile-accessible catalogs, and approximately 75% offer access to licensed databases through mobile interfaces.

Liu's findings reveal that academic libraries are increasingly investing in responsive web design and app-based platforms to support mobile access. Libraries at institutions such as the University of Illinois at Urbana-Champaign and the University of Michigan provide mobile-optimized interfaces for their digital collections. These platforms allow users to search, browse, and download content ranging from scholarly articles to streaming educational videos, all from a smartphone or tablet.

Digital Textbooks: Enhancing Learning on the Go

One of the most prevalent types of content provided to users on digital devices is the eBook. Academic libraries have partnered with vendors like ProQuest, Springer, JSTOR, and EBSCO to provide access to extensive eBook collections that support curricula across disciplines. According to the 2023 Ithaka S+R Faculty Survey, over 78% of faculty consider eBooks essential or very important for course-related reading, especially for graduate-level instruction, (Ithaka S+R, 2023).

Libraries like those at the University of California system have adopted demand-driven acquisition (DDA) models, enabling students to trigger purchases or temporary access based on real-time usage. This not only ensures content relevance but also aligns acquisitions with actual learning needs.

Moreover, digital textbooks integrated into Learning Management Systems (LMS) allow students to annotate, highlight, and search within texts. These features improve study efficiency and are accessible across platforms, reinforcing continuity between in-class and remote learning environments.

Streaming Media and Multimedia Content: Visual Learning Redefined

Academic libraries now host robust streaming platforms that extend access to educational videos, films, and documentaries, increasingly important for visual learners and hybrid classrooms. Kanopy, Films on Demand, and Alexander Street Press are among the leading providers. These services allow students to view licensed content without geographical or temporal constraints.

Statistics from the Association of College and Research Libraries (ACRL) show a 35% increase in academic libraries' investment in streaming media between 2019 and 2022. The University of California, Los Angeles (UCLA) Library, for example, offers thousands of titles via Kanopy and Swank, which are integrated with the course syllabi and accessible directly through mobile apps.

Multimedia content also includes recorded lectures, language learning modules, and virtual lab simulations. These are increasingly mobile-optimized, ensuring students in remote or underserved locations are not disadvantaged by lack of physical access to campus facilities.

Access to scholarly databases remains a cornerstone of digital library services. Libraries subscribe to thousands of academic journals and research databases like Scopus, PubMed, IEEE Xplore, and Academic Search Premier, all of which now offer mobile-friendly interfaces. Mobile-optimized discovery layers, such as Ex Libris Primo and EBSCO Discovery Service, integrate seamlessly with library catalogs, allowing users to conduct advanced searches, save results, and cite sources—all from their phones or tablets.

In a 2022 EDUCAUSE report, over 60% of students indicated that mobile access to academic content enhanced their ability to complete coursework on time. Libraries have responded with "one-click" authentication systems (e.g., OpenAthens, Shibboleth) that simplify login processes, particularly on mobile devices, (EDUCAUSE, 2022).

Dr. Liu's work corroborates these developments, noting that ease of access directly impacts usage rates. His survey across 130 academic institutions revealed that mobile-responsive library platforms saw an average increase of 25% in search sessions compared to non-responsive counterparts.

Personalized Content and Recommender Systems: Enhancing User Experience

The evolution of academic libraries into digital service hubs has also introduced intelligent recommender systems to deliver personalized content. These systems use user behavior, search history, and course enrollment data to suggest relevant articles, eBooks, or videos. Libraries at institutions like North Carolina State University and Stanford have implemented machine learning-based recommendation tools within their discovery platforms.

These tools resemble commercial algorithms found in platforms like Netflix or Amazon, yet they are tailored to academic needs. They reduce information overload and enhance user satisfaction by surfacing high-value resources.

For example, the "bX Recommender" by Ex Libris suggests related scholarly articles during the search process. A study published in the Journal of Academic Librarianship found that libraries using such tools observed a 30% increase in user engagement with digital resources (Reid et al., 2020).

Open Educational Resources (OERs) and Institutional Repositories

The rise of Open Educational Resources has further democratized access to academic content. Libraries are increasingly curating and hosting OERs that users can freely download, adapt, and share. Platforms such as MERLOT, OER Commons, and institutional repositories provide seamless mobile access to lecture notes, textbooks, and instructional videos.

Libraries at institutions like Rice University and SUNY system schools have taken leadership roles in OER dissemination. These resources are optimized for mobile use and are often embedded in course modules. The affordability and adaptability of OERs have made them an essential tool in expanding educational access.

Additionally, institutional repositories hosted through platforms like DSpace or Digital Commons allow users to access theses, dissertations, and faculty publications on their personal devices. These repositories are open access, providing a vital avenue for scholarly communication beyond paywalls.

Analytics and User Behavior: Driving Data-Informed Content Delivery

Understanding how users interact with digital content is critical for academic libraries. Usage analytics allow libraries to refine digital offerings based on patterns of engagement. Tools like Google Analytics, Springshare's LibInsight, and vendor-provided dashboards enable real-time monitoring of which resources are accessed most often, what devices are used, and where users encounter friction.

Heat maps and session tracking improved the mobile layout of their discovery system, resulting in a 40% increase in successful searches on mobile devices.

Such analytics not only guide content acquisition decisions but also inform interface design, accessibility improvements, and resource marketing strategies.

Academic libraries are no longer peripheral to the mobile learning experience—they are central, shaping how knowledge is accessed, interpreted, and applied in a digitally mediated world. Through eBooks, databases, multimedia, personalized tools, and open content, libraries now function as mobile information ecosystems that empower learners globally. Dr. Yan Quan Liu's pioneering work underscores the necessity of continual adaptation, user-centered design, and strategic partnerships in realizing this vision. As academic libraries refine and expand their digital offerings, the commitment to accessibility, inclusivity, and scholarly excellence remains unwavering.

The digital device is not just a tool; it is a portal to the academic library's evolving mission—a mission rooted in connecting people with knowledge, wherever they are.

REMOTE RESOURCE PROVISION FOR DISTANCE LEARNERS

Distance education has reshaped higher education, making online learning a dominant modality. Academic libraries have transformed to meet the needs of this dispersed learning community. This focus explores the role of academic libraries in providing online course resources and services to distance learners, drawing from peer-reviewed studies. It examines best practices, provides examples, and highlights effective strategies and emerging trends in this dynamic field.

The rapid expansion of online education, catalyzed by the COVID-19 pandemic, has led to an involuntary shift to online and blended learning, now considered the "new normal" in higher education. (Jabarali, Sathya Kumar, & Barak, 2024) This abrupt transition posed challenges, especially for institutions with less established online infrastructures. Despite these challenges, global investment in online education is projected to reach $350 billion by 2025, underscoring the strategic imperative for educational institutions, including libraries, to fully embrace and optimize online learning environments.

Academic libraries have shifted from managing physical collections to focusing on digital resources to meet changing user needs. This reflects the broader shift towards digital information consumption. Libraries play a crucial role in fostering innovation by providing information access and facilitating new knowledge creation. The internet transformed libraries into virtual entities, enabling remote access to resources and services globally, (Ashikuzzaman, Islam, & Islam, 2025). This paradigm shift emphasizes seamless access to digital content over physical material ownership. The ACRL Standards for Distance and Online Learning Library Services ensure equal quality of services, resources, and collections for all distance and online members of higher education institutions, promoting equity in access and support.

While professional standards clearly articulate the right to equivalent services for distance learners, practical observations reveal a persistent gap where distance learners may be unaware of services or lack adequate remote access. This disparity underscores a significant challenge: the mere provision of digital resources does not automatically guarantee equitable access or effective utilization. The underlying issue is not solely technological availability but also encompasses awareness, usability, and the varying levels of technological preparedness among students.10 Achieving true equity necessitates libraries actively bridging these gaps through robust outreach, comprehensive user education, and accessible design, ensuring that the digital divide does not become an information divide.

Distinct Needs and Information-Seeking Landscapes for Distance Learners

Effective library support for distance learners hinges on a nuanced understanding of their unique challenges and information-seeking behaviors, which often differ significantly from those of traditional on-campus students. Addressing these distinct needs is paramount for providing relevant and impactful services.

Online students face unique challenges, including time management difficulties, communication gaps, technical difficulties, isolation, adapting to new pedagogical approaches, distractions in the home environment, and psychological impacts due to rapid shifts to online instruction, (Theiss, 2022). The rapid and involuntary shift to online instruction has been shown to have a deep psychological impact on students, exacerbated by technology glitches and a perceived lack of connection with faculty. Research also indicates that "library anxiety" can significantly affect distance or online learners' willingness and ability to utilize academic library resources and services.

The information-seeking behaviors and resource preferences of distance learners reveal distinct patterns. They are prominent users of digital scholarly environments, often accounting for the highest number of sessions and pages viewed in digital journal and e-book collections. Their behavior is frequently efficiency-driven, prioritizing the easiest and most convenient methods for information seeking and valuing the time-saving characteristics of electronic resources. Students often exhibit a "satisficing" tendency, preferring keyword searching over browsing and frequently ceasing their information search once they have found a sufficient number of sources to fulfill assignment requirements, (Nicholas et al. 2009).

Electronic resources are overwhelmingly preferred (72% of respondents), with search engines, email, and e-books being the most popular, while e-databases, e-theses, dissertations, and physical CD/DVDs are less frequently used. The main purposes for using e-resources are writing assignments, preparing for examinations, and updating general knowledge. Despite the availability of specialized library resources, most distance students (80.7% in one study) primarily use general search engines like Google to find information for their academic needs. Contacting a librarian is often the least utilized method for finding information (only 7.6% of participants), and a significant proportion of students (55.2%) have never participated in formal library instruction sessions. This suggests a critical gap in awareness or a reluctance to seek professional assistance. Compounding this, many faculty members mistakenly assume that distance students are already proficient in library use, which can lead to a lack of referrals to library resources or the omission of library instruction within their courses, (Oliveira, n.d.).

Researchers noticed something interesting: distance learners are super tech-savvy and love using online resources for convenience. But guess what? They often skip the fancy library tools and just use regular search engines. And they hardly ever talk to librarians or get formal help with finding information. So, even though they have easy access to digital stuff, they don't always know how to use academic resources well. They just take the easiest way out, which can end up with less thorough and high-quality work. This means libraries need to step up their game and teach students how to find and use information effectively, not just give them access.

The challenges faced by distance learners extend far beyond mere access to academic resources, encompassing technical barriers, feelings of isolation, and difficulties with self-regulation. This reveals that library support cannot be effective if it operates in a silo, focusing solely on information provision. Instead, effective library services must be conceived as part of a holistic support system that integrates robust technical assistance, proactive community-building initiatives, and deep collaboration with faculty to ensure clear pedagogical instructions and timely feedback. Addressing the psychological impact and fostering a sense of belonging is as crucial as providing access to databases for overall student success and well-being in the online environment.

Based on the studies by Theiss (2022), National University (2025), and other researchers the key challenges faced by distance learners, along with corresponding library support strategies, can be summarized in the table below:

Table 2. Key challenges of distance learners and corresponding library support strategies

Distance Learner Challenge	**Corresponding Library Support Strategy/Service**
Ineffective Time Management	Providing time management resources/apps
Technical Difficulties	Robust technical support & digital literacy training
Isolation & Lack of Community	Facilitating virtual study groups/discussion boards
Library Anxiety	Embedded librarians & proactive outreach
Lack of Awareness of Library Services	Clear instructions/tutorials for online services
Difficulty Adapting to New Learning Styles	Comprehensive online information literacy instruction
Unclear Instructions/Expectations	Collaboration with faculty for clear rubrics and resource integration

Core Services and Resources: Bridging the Distance Gap

Online services have grown and adapted significantly, especially during the transition to fully virtual operations. Institutions have expanded online instruction and virtual reference services, pivoting from in-person to online one-on-one information literacy sessions and research consultations. Online communication became the primary method of interaction between libraries and patrons, leading to a substantial increase in online reference questions and chat usage. Remote answering of reference desk calls became common, and some libraries implemented "proactive chat" features. Librarians are developing more online tutorials and utilizing screencasting for asynchronous classes. There's a growing demand for embeddable content, such as modules, PDF handouts, and short videos, designed for direct integration into LMS. Libraries have also strategically acquired more e-books, streaming videos, and other digital content to support online instruction and research.

Ensuring equitable access to resources and services is fundamental. The ACRL Standards for Distance and Online Learning Library Services mandate reliable, rapid, and secure access to online resources for distance learning communities. Libraries must develop clear policies for delivering both hard copy and electronic materials and regularly update collection development and acquisitions policies to reflect distance community needs. Direct marketing and promotion of library collections and services to distance and online learning communities are crucial for awareness and utilization. Libraries support open access and OER initiatives for equitable and affordable access to educational materials. They provide resources and services in multiple formats to ensure accessibility and choice, adhering to guidelines like the ADA and Rehabilitation Act, and utilizing UDL in instructional design. This ensures learners with varying technological access and abilities receive equal experiences.

While the initial pandemic response involved rapid, reactive modifications and librarians' self-education, the sustained demand for online services and ACRL's formal standards signal a deeper, strategic evolution. Developing embeddable content for LMS and implementing "proactive chat" features demonstrate a deliberate move towards designing services that meet users' needs within their native online learning environments. This indicates a growing sophistication in understanding and anticipating online user behavior, shifting from mere digital presence to integrated functionality, (Association of College and Research Libraries, 2023).

Despite the significant expansion and sophistication of online library services, studies reveal a persistent and concerning issue: distance learners are often unaware of the available services or how to effectively utilize them.7 This directly contradicts the goal of equitable access articulated in the ACRL standards. Therefore, the explicit inclusion of marketing and promotion benchmarks in the ACRL standards is not merely a suggestion for outreach, but a critical, strategic component of ensuring that robust digital services are utilized. Without effective and targeted awareness campaigns, even the most advanced and well-designed library services will remain underutilized, undermining the library's value proposition and impact on distance learner success.

TECHNOLOGIES FACILITY REMOTE ACCESS AND ENGAGEMENT

The effective provision of academic library services to distance learners hinges on a robust and continually evolving technological infrastructure. These technologies enable seamless distance access and engagement, integrating services into learning environments and enhancing the discoverability and personalization of resources.

Remote Access Foundations in Library Services

Essential technologies for remote access are foundational to supporting distance learners. High-speed internet access and pervasive mobile technologies have become as fundamental as electricity, enabling libraries to deliver services directly to users' mobile devices, embodying the principle of libraries going to the users. Libraries facilitate remote access to their comprehensive digital collections, including e-books, e-journals, and research databases, from any location with an internet connection, thereby removing the physical limitations of traditional libraries.

Common authentication methods include requiring active library membership credentials (username/password). Many libraries utilize proxy servers (e.g., OCLC's EZproxy) as intermediaries; when a user logs in remotely, the proxy server routes their connection through the library's network, making it appear as if the user is accessing resources from within the library's whitelisted IP range. This provides seamless access to licensed content. Some institutions provide Virtual Private Networks (VPNs), which create a secure, encrypted connection between the user's device and the library's network, allowing safe access to resources from remote locations. Single Sign-On (SSO) systems, such as OpenAthens and Shibboleth, built on standards like SAML (Security Assertion Markup Language), simplify access by allowing users to log in with their existing institutional credentials (e.g., university ID) to access multiple library resources, (Lisedunetwork, 2024). SSO offers centralized control and enhances user convenience.

Strategic integration within Learning Management Systems (LMS) is a critical component of modern library services. Integrating library services directly into LMS platforms is a strategic imperative for open and distance learning institutions, aiming to democratize access to educational resources and significantly enhance the overall learning experience. This integration typically includes easy accessibility of e-books and journals, seamless connections to the library's digital catalog and databases, and effective embedding of library resources directly within course materials. Library tutorials, guides, and tools for citation and referencing are increasingly available within the LMS environment. (Aliyu, 2025) Faculty have shown increased openness to incorporating library instruction directly into their LMS courses, (Germain, n.d.).

Studies show several models for embedding library remote services exist:

- Macro-level (a single, global library presence for all online courses)
- Micro-level (customized library presence at the program or course level)
- Nano approach (tailoring services to the information architecture of individual courses), (Farkas, 2008).
- Hybrid approaches, combining elements of macro and micro models, are often adopted for practicality, (Farkas, 2008).
- The "embedded librarian" concept involves librarians actively participating within the online classroom environment to provide course-specific reference assistance and information literacy instruction, (Kumar, 2025).

Library-delivered, web-scale discovery systems (e.g., Primo, Summon, EBSCO EDS) have become successfully embedded in many academic libraries. These systems consolidate all resource indexes (catalogs, institutional repositories, databases) into a single search portal, effectively emulating the simplicity of general search engines like Google, (Akeroyd, n.d.). These systems have emerged as a significant starting point for scholarly information searches, challenging both general search engines (like Google Scholar) and discrete bibliographic databases. Users, particularly students, have consistently expressed satisfaction with these discovery systems through surveys and research, (Akeroyd, n.d.). A comparative study found that while Google Scholar excels in perceived ease of use, system quality, and user satisfaction, academic library discovery systems are stronger in perceived comprehensiveness, subjective norm (recommendation by credible sources), user loyalty, and intended future use, (Asher & Duke, 2019).

Innovative Technologies and Interactive Engagement

Artificial Intelligence (AI) is poised to transform academic libraries into dynamic, cutting-edge solutions, enhancing search efficiency, facilitating scholarly publications, and delivering highly personalized resource recommendations. Practical applications include AI chatbots and virtual helpers offering 24/7 support for common questions and research guidance. Machine learning tools aid in providing personalized suggestions and organizing content, helping users find relevant resources tailored to their needs, (Ogbebor, 2024). Natural Language Processing (NLP) can enhance search functions, allowing users to query resources using natural language and receive more helpful results. AI holds the potential to improve academic outcomes and enhance student engagement through personalized learning experiences and intelligent tutoring systems that provide tailored guidance and feedback. However, significant challenges accompany AI adoption, including ensuring data privacy, addressing algorithmic bias, developing robust ethical guidelines, managing integration complexity, high implementation costs, the need for continuous upgrades, and ensuring adequate librarian training, (Alharbi, 2025).

The Internet of Things (IoT) offers additional possibilities. IoT devices provide convenient and secure access to library resources using smartphones or other IoT-enabled devices, potentially eliminating the need for physical library cards. They can be deployed to continuously monitor environmental conditions (e.g., temperature, humidity, air pollution) within library spaces, ensuring a comfortable and healthy atmosphere for users. IoT also enables personalized messaging to effectively engage with users, keeping them informed and connected. The integration of IoT and AI-equipped robots can revolutionize collection management, assisting in sorting, organizing, and retrieving books, thereby optimizing staff efficiency, (Roy, 2025).

Virtual Reality (VR) and Augmented Reality (AR) are also expanding the academic library's mission by enhancing access to information resources and providing new analytic tools that promote "embodied understanding"—a deeper, more engaging interaction with information, (Moen & McClure, n.d.). A significant proportion (86%) of the top one hundred US university libraries have already implemented VR and AR technologies, with applications in VR/AR studios, makerspaces, immersive learning services, and virtual exhibitions. These immersive technologies have the potential to transform libraries into highly interactive and engaging service environments, like the profound impact of the internet, (Phillips, 2024). Key challenges for VR/AR implementation include ensuring accessibility for disabled students, allocating sufficient staff time and technical resources, and fostering a critical discourse around their ethical and pedagogical integration. (Moen & McClure, n.d.) The widespread adoption of remote access tools (proxies, VPNs, SSO) and the strategic push for deep LMS integration collectively indicate that academic libraries are striving to create a user experience where accessing resources for distance learners is as frictionless and intuitive as possible. The comparative analysis with Google Scholar further underscores that convenience, ease of use, and intuitive design are paramount for user adoption and satisfaction in the digital realm. The reported low perception of usability despite high technical integration in some cases serves as a critical reminder that mere technical deployment is insufficient; the user experience must be actively designed, continuously refined, and effectively communicated to truly achieve seamless access and engagement.

Table 3. Key technologies in academic library distance learning support

Technology Category	Specific Technologies/ Tools	Primary Function/Benefit for Distance Learners
Remote Access & Authentication	EZproxy, OpenAthens, VPNs, Shibboleth	Secure off-campus access to licensed resources (Lisedunetwork, 2024)
Learning Management System (LMS) Integration	Embedded Librarian, Library Modules	Seamless integration of library resources into course environments (Kumar, 2025)
Discovery Systems	Primo, Summon, EBSCO EDS	Single search point for all library content (Akeroyd, n.d.)
Emerging Technologies	AI Chatbots, VR/AR, IoT	Personalized support and enhanced search, Immersive learning experiences and efficient operations (Ogbebor, 2024)

Strategic Frameworks for Effective Support

Studies have shown that to effectively serve distance learners, academic libraries must adopt strategic frameworks and adhere to established best practices that ensure quality, accessibility, and pedagogical integration. These practices are crucial for maximizing the impact of library services on student success in online environments.

Adherence to professional standards is foundational. The ACRL Standards for Distance and Online Learning Library Services serve as a comprehensive framework, outlining benchmarks across key areas: vision and strategic planning, diversity, equity, inclusion, and accessibility (DEIA), personnel, collections and access, research and instruction, assessment, and financial support, (Association of College and Research Libraries, 2023). These standards ensure that services for distance learners are equivalent to those offered on campus. Libraries must also meet all national and regional guidelines, laws, and regulations, including the Americans with Disabilities Act (ADA), Sections 504 and 508 of the Rehabilitation Act, and Web Content Accessibility Guidelines (WCAG), to ensure that all resources and services are accessible to learners with diverse needs.6 Furthermore, the Code of Best Practices in Fair Use for Academic and Research Libraries provides crucial guidance for librarians on ethically and legally digitizing copyrighted materials for student use, making special collections available online, and archiving websites, (Association of Research Libraries, n.d.).

The "embedded librarian" model has emerged as a highly effective best practice. Librarians are increasingly being embedded directly into courses within the university Learning Management System (LMS). This model provides students with immediate research help and information literacy instruction at their point of need, (Kumar, 2025). Embedding the library within the CMS streamlines access for online learners, making them significantly more likely to utilize available library resources and services. (Farkas, 2008) Key practices for successful implementation include proactively contacting faculty at the beginning of each semester, discussing the desired level of librarian activity in the course, adding a clear "librarian link" to the LMS navigation menu (often with the librarian's name), tracking assignment due dates to send timely reminders and offer proactive assistance, and sending introductory emails to students explaining the embedded librarian's role, (Kumar, 2025).

The pivotal role of Open Educational Resources (OER) cannot be overstated. Libraries actively promote and support OER, defined as teaching, learning, and research materials that are either in the public domain or licensed to allow free and perpetual engagement in the "5R activities" (retaining, remixing, revising, reusing, and redistributing), (Vogus, 2019). For students, OER significantly lower textbook costs, provide immediate and permanent access, and offer access from any device. For faculty, OER provide immediate access for students, allow content modification to fit course needs, and do not go out of print.30 Studies demonstrate that OER adoption does more than just save money; it improves end-of-course grades and decreases DFW (D, F, and Withdrawal) rates for all students, with even greater positive impacts observed for Pell recipient students, part-time students, and populations historically underserved by higher education. Academic libraries play a crucial role in facilitating OER adoption by creating guides to open repositories, actively promoting OER, reviewing their quality, and even assisting in their creation. They collaborate extensively with faculty and educational technology managers, offer workshops, and provide expertise on copyright and Creative Commons licensing. Libraries are working to integrate OER programs into foundational library operations, such as collection development and reserves, to enhance visibility and demonstrate their investment in student academic success, (Dill & Cullen, 2020).

The evolution of library services for distance learners, particularly evident in the widespread adoption and refinement of the embedded librarian model and the comprehensive support for OER, signifies a profound shift in the librarian's role. Libraries are no longer merely providers of resources but have become integral pedagogical partners, actively shaping learning outcomes. This transition reflects a deeper engagement with the teaching and learning process, where librarians contribute directly to curriculum design and instructional delivery, moving beyond traditional support functions to become essential collaborators in academic success. This collaborative approach recognizes that effective information literacy instruction and resource utilization are best achieved when integrated directly into the learning environment, rather than being treated as separate, supplementary activities.

The evolving role of librarians in this digital-first, distance-learning environment requires a blend of traditional expertise with advanced digital and pedagogical competencies. Librarians are increasingly expected to be proficient in web design, database administration, and network administration, enabling them to develop online services and act as liaisons to instructional designers and IT staff, (Farkas, 2008) They must also possess strong instructional design skills to create engaging and accessible learning materials. Beyond technical skills, librarians are becoming crucial facilitators of information literacy, guiding students through complex digital landscapes and helping them develop critical thinking skills essential for navigating vast online information.1 This expanded role means librarians are now digital navigators, pedagogical partners, and advocates for equitable access, continuously adapting to new technologies and learning modalities to support the diverse needs of remote scholars, (Ashikuzzaman, Islam, & Islam, 2025).

VIRTUAL REFERENCE AND CHAT SERVICES: SUPPORT BEYOND WALLS

With increased access to digital content comes the need for virtual support. Chat-based reference services such as LibChat, Ask-A-Librarian, and Text-a-Librarian provide real-time assistance to users navigating complex research databases or citation tools. Many of these are now embedded directly into mobile apps and websites, providing instant help without requiring users to switch platforms.

The rise of virtual reference services (VRS) in academic contexts shifts toward digital engagement that allows patrons to connect with librarian expertise remotely—through modalities like live chat, SMS/text messaging, email, and occasionally social media. These approaches emerged as essential alternatives to in-person reference, especially for digital-native students and distance learners who demand rapid, flexible service delivery.

Dr. Liu's research in collaboration with Sarah Briggs, investigated virtual and mobile reference in top-tier academic libraries in the U.S. In one foundational study, Liu and Briggs (2015) surveyed the top 100 university libraries and found widespread adoption of mobile-compatible reference tools, including chat, SMS texting, mobile apps, and mobile-optimized websites. Their findings highlighted that nearly all leading institutions integrated virtual reference channels to enhance user accessibility and convenience.

Core Modalities of Virtual Reference Services

Live Chat / Instant Messaging (IM). Liu and Briggs (2015) report that 100% of surveyed top academic libraries offered live chat or IM reference services—making it the most ubiquitous channel for virtual help. These services are typically embedded within library websites, catalogs, and research guides, allowing users to instantly initiate consultation without leaving their current web activity. Platforms such as

LibChat or proprietary IM systems enable librarians to manage multiple simultaneous inquiries efficiently and maintain quality.

AI has the potential to elevate thinking, learning, and writing—but only when used thoughtfully. As Scott and Ward (2025) point out, AI-driven tools can streamline tasks like metadata extraction or reference support, yet they cannot replicate the human insight and ethical reasoning librarians provide. Ensuring AI remains a facilitator—not a crutch—requires careful cultivation of AI literacy skills. Lo and Stark (2020) underscore that without sufficient training and ethical frameworks, library professionals may struggle to implement AI in ways that genuinely support critical thinking and purposeful learning.

SMS / Text Messaging Reference

While slightly less universal, text-based reference remains prevalent: Liu and colleagues documented SMS reference offerings in approximately 81% of institutions surveyed. The key advantage of SMS is its minimal data requirements and accessibility—users can text questions without needing internet connectivity or specialized apps, making it ideal for mobile or remote usage situations.

Email Reference

Though not strictly "mobile," email reference still plays a vital role—particularly for complex, graduate-level research questions that benefit from longer, well-supported responses. Users often begin with a chat or SMS inquiry and follow up by email for more detailed help.

Social Media Platforms

Reference via social media remains less common. In large academic settings, only a minority of libraries offer reference through platforms like Facebook or Twitter. When adopted, it often serves more as outreach rather than an official consultation channel, due to privacy concerns and platform usability limitations.

Implementation & Integration Trends

Liu and his collaborators examined how institutions embed virtual reference within users' academic workflows. For example, many libraries integrate chat/IM widgets into learning management systems (e.g. Canvas), discovery tools, and mobile library portals—making reference services context-sensitive and immediately available when and where users need them.

Moreover, asynchronous support tools like FAQ knowledge bases and instructional video guides are often accessible from within the digital platforms themselves. These services ensure that the digital library experience remains user-centric and inclusive, irrespective of user location or device.

These integrations align with user expectations for seamless, on-demand support. The ability to initiate a chat pop-up while searching the catalog or completing an assignment enhances service visibility and reduces friction in help-seeking behavior. Liu and Briggs (2015) emphasizes that mobile-optimized interfaces and well-integrated reference services significantly boost user engagement and satisfaction.

Recognized Examples

Liu's studies revealed consistent patterns: every surveyed library supported live chat, and most offered SMS, showing institutional commitment to virtual reference (Liu & Briggs, 2015) These libraries often also provide email reference, embedded chat widgets in course pages, and mobile-optimized access points for easy user contact. In addition to assessing whether libraries had a mobile site, the study's survey asked libraries that already have a mobile site what is included on the site. Among these surveyed libraries looked offering virtual and/or mobile reference services. 100% offer chat/IM, 81.25% offered text/SMS messaging, and 21.88% offer reference services via a social media account. These reference services in these top universities are summarized in Figure below:

Figure 3. Popular mobile reference services

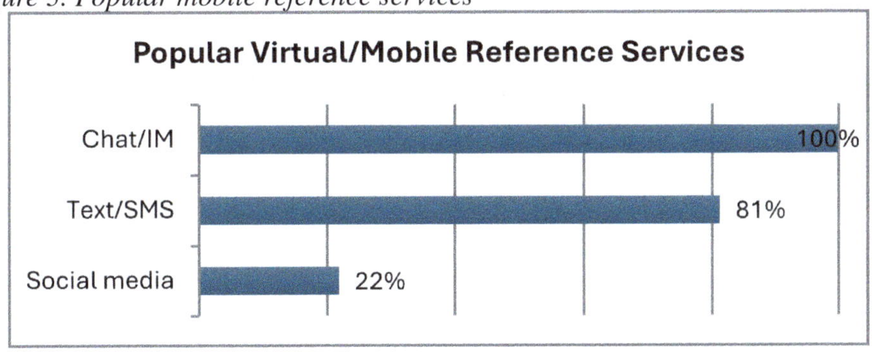

(Liu & Briggs, 2015)

Chat/IM is the most popular way to get virtual/mobile reference services. Text/SMS is also pretty popular. Their results showed that social media isn't as good as texting or chat/IM. Most social media sites have a mobile version that libraries can use for free, but it might not be the best way to get reference help.

Tasks and Forward-Looking Insights

Despite widespread offer of virtual reference modalities, several challenges persist:

- Staffing and scheduling: Providing consistent chat/SMS service during off-hours or academic breaks often requires cross-departmental coordination.
- Quality control: Training librarians to maintain consistent tone, speed, and responsiveness across synchronous (chat/SMS) and asynchronous (email) formats is essential.
- Evaluation and feedback: Many libraries now analyze chat transcripts, response times, and post-interaction surveys to refine services and staffing models.

Liu's more recent work underscores the potential of AI-driven chatbots to manage routine inquiries, while reserving human expertise for complex research support. Yet, he also cautions that automated systems must be carefully implemented to avoid diminishing service quality.

Academic libraries have embraced virtual reference services in response to evolving user habits and technological capabilities. Live chat/IM stands out as a universal entry point, with SMS/text messaging used extensively, and email remaining critical for in-depth consultation. Virtual reference, especially when embedded into academic platforms and mobile workflows, represents a strategic commitment to accessibility and user-centered support—insights that Liu's research has rigorously documented.

ADA COMPLIANCE IN ACADEMIC LIBRARY MOBILE SERVICES

Offering resources and services optimized for smartphones and tablets is no longer a luxury but a necessity for reaching today's students and researchers. However, this drive towards mobile accessibility must be intrinsically linked with the principles of the Americans with Disabilities Act (ADA), ensuring that these advancements serve all users, regardless of their abilities. The proliferation of mobile devices has fundamentally reshaped how users interact with information, and academic libraries have been at the forefront of adapting their services to this mobile-first paradigm. Numerous studies in the literature have extensively explored the critical intersection of the Americans with Disabilities Act (ADA) and mobile services within academic libraries. These studies have examined the evolution of this intersection, analyzed key standards and schemas, and offered practical examples to create truly inclusive digital environments.

The Evolving Landscape of Accessibility in Academic Libraries

The ADA, enacted in 1990, prohibits discrimination based on disability. While initially focused on physical spaces, its scope has rightfully expanded to encompass digital realms, particularly with the increasing reliance on online resources and services. Academic libraries, as institutions committed to education and equal access to information, bear a significant responsibility in ensuring their websites, applications, and digital content are accessible to individuals with disabilities.

The initial efforts towards digital accessibility often centered on desktop-based websites. However, the surge in mobile usage has necessitated a parallel focus on the accessibility of mobile websites and applications. Simply having a mobile-friendly website is insufficient; it must be designed and developed with accessibility in mind from the outset. Liu (2023) emphasizes the critical importance of ADA digital accessibility specifically within academic library websites. Their research findings reveal a significant gap between the prevalence of mobile-optimized websites and their actual accessibility, with nearly 80% of public university libraries exhibiting accessibility errors such as low color contrast and missing alternative text for images. This stark reality underscores the urgent need for a more concerted effort to integrate accessibility considerations into mobile service development. As Liu and Briggs (2015) noted in their study of mobile services in top university libraries, the focus was primarily on the adoption and types of mobile services offered. The subsequent years have highlighted the necessity of ensuring the inclusivity of these very services.

Foundational Standards: WCAG and Section 508

Two primary frameworks guide digital accessibility efforts: the Web Content Accessibility Guidelines (WCAG) and Section 508 of the Rehabilitation Act. Understanding and implementing these standards is fundamental to achieving ADA compliance in mobile library services.

Web Content Accessibility Guidelines (WCAG)

Developed by the World Wide Web Consortium (W3C), WCAG provides internationally recognized, technology-agnostic guidelines for making web content more accessible to people with disabilities. WCAG is organized around four core principles, often remembered by the acronym POUR:

- Perceivable: Information and user interface components must be presentable to users in ways they can perceive. This includes providing text alternatives for non-text content (such as images, audio, and video), offering captions and other alternatives for multimedia, ensuring content can be presented in different ways without losing meaning (e.g., resizable text), and making it easier for users to see and hear content (e.g., sufficient color contrast).
- Operable: User interface components and navigation must be operable. This involves making all functionality available from a keyboard, providing users enough time to read and use content, avoiding content that could cause seizures, and helping users navigate, find content, and determine where they are.
- Understandable: Information and the operation of the user interface must be understandable. This includes making text content readable and understandable, making the appearance and operation of web pages predictable, and helping users avoid and correct mistakes.
- Robust: Content must be robust enough that it can be interpreted reliably by a wide variety of user agents, including assistive technologies. This principle emphasizes the importance of using standard web technologies and ensuring proper coding.

WCAG exists at three levels of conformance: A (basic accessibility), AA (addressing most common barriers), and AAA (highest level of accessibility). While striving for AAA conformance is ideal, achieving Level AA is widely considered the benchmark for accessibility and is often the target level for legal compliance.

Section 508 of the Rehabilitation Act

Section 508 mandates that federal agencies and the information and communication technology (ICT) they develop, procure, maintain, or use must be accessible to people with disabilities. While directly applicable to federal entities, Section 508 has significant implications for academic libraries, particularly those receiving federal funding or collaborating on federal projects. The current Section 508 standards closely align with WCAG 2.0 Level AA, effectively harmonizing accessibility requirements across public and federally funded institutions.

Applying Standards to Mobile Library Services: Practical Examples

Translating these abstract guidelines into tangible practices within the context of mobile library services is paramount. Here are some illustrative examples demonstrating how WCAG principles and Section 508 standards are applicable:

Perceivable

- Alternative Text for Images: When displaying book covers, library logos, or informational graphics on a mobile website or app, providing descriptive alternative text ("alt text") is essential. For example, instead of an alt text of "book cover," a more informative alternative would be "Cover of 'The Great Gatsby' by F. Scott Fitzgerald, featuring a stylized art deco design." Liu's (2023) findings highlight the prevalence of missing alt text, underscoring the need for greater awareness and training in this area.
- Color Contrast: Ensuring sufficient color contrast between text and background is vital for users with low vision or color blindness. Tools are available to check color contrast ratios against WCAG guidelines. For instance, a light gray text on a white background would likely fail the contrast requirements, making it difficult for many users to read.
- Resizable Text: Mobile websites and apps should be designed to allow users to easily resize text without loss of content or functionality. This can be achieved through relative font sizing (using percentages or ems) rather than fixed pixel values.
- Captions and Transcripts for Multimedia: Any video or audio content embedded in mobile library resources, such as online tutorials or author interviews, must include accurate captions for users who are deaf or hard of hearing, and transcripts for those who prefer to read the content.

Operable

- Keyboard Navigation: While mobile devices are primarily touch-based, users with motor disabilities may rely on external keyboards or assistive technologies that emulate keyboard input. Mobile library websites and apps should be navigable and operable using only a keyboard interface. This means ensuring that all interactive elements, such as buttons, links, and form fields, can be accessed and activated using keyboard commands.
- Time Limits: If mobile services include timed activities (e.g., completing a form or accessing a temporary resource), users should be given sufficient time to complete the task or the ability to extend the time limit. Avoid rapidly moving or blinking content that could trigger seizures.
- Clear Navigation: Mobile interfaces should have clear and consistent navigation structures. Menus should be easily understandable and accessible, allowing users to quickly find the information or service they need. Breadcrumbs or clear headings can help users orient themselves within the mobile site or app.

Understandable

- Readable Content: Use clear and concise language, avoiding jargon or overly complex sentence structures. Organize content logically with headings, subheadings, and bullet points to improve readability.
- Predictable Operation: Ensure that interactive elements behave in a consistent and predictable manner. For example, links should be clearly identifiable, and form submissions should provide clear feedback to the user.
- Input Assistance: For online forms (e.g., interlibrary loan requests or registration forms), provide clear labels and instructions. Offer error messages that are informative and guide users on how to correct mistakes.

Robust

- Semantic HTML: Using semantic HTML elements (e.g., `<nav>` for navigation, `<article>` for articles, `<button>` for buttons) helps assistive technologies understand the structure and purpose of web content, leading to a more accessible experience.
- ARIA Attributes: Accessible Rich Internet Applications (ARIA) attributes can be used to provide additional semantic information to assistive technologies about dynamic content and interactive elements that are not inherently accessible with standard HTML alone. For instance, ARIA can define the role, state, and properties of custom widgets used in a mobile interface.

The Schemas' Role and Ongoing Accessibility Practices

While not directly mandated by ADA, the implementation of structured data markup, often using schemas like schema.org, can indirectly contribute to accessibility. Schema.org provides a collection of schemas that webmasters can use to markup their HTML content in ways that help search engines and other applications understand the information on their pages.

Academic libraries can enhance the discoverability of accessible resources and services by implementing relevant schemas. Marking up information about library events with schema.org's `Event` type, including details about accessibility accommodations such as wheelchair accessibility and sign language interpretation, makes it easier for users with disabilities to find and participate in these events. Additionally, using the `Library` or `EducationalOrganization` schema can help highlight the library's commitment to accessibility within search engine results.

Achieving and maintaining ADA compliance in mobile library services is an ongoing process, not a one-time task. The following practices are essential:

- Accessibility Audits: Regularly conduct accessibility audits of mobile websites and apps, using both automated testing tools and manual review by individuals with expertise in accessibility. Incorporating user testing with individuals with disabilities can provide invaluable feedback. Bielefield, Liu, and Waimon's (2021) study on private post-secondary library websites highlights the importance of ongoing evaluation.
- Accessibility Training: Provide comprehensive training to library staff involved in web development, content creation, and service design on accessibility principles and best practices. Raising awareness and fostering a culture of accessibility is crucial.
- Incorporating Accessibility into the Development Lifecycle: Integrate accessibility considerations into every stage of the mobile service development process, from initial planning and design to development, testing, and deployment. "Accessibility by design" is far more effective and cost-efficient than retrofitting accessibility features later.
- Clear Accessibility Statements: Publish clear and easily accessible accessibility statements on the library's website, outlining the library's commitment to accessibility, the steps taken to achieve it, and contact information for users to report accessibility barriers or request accommodations.
- Procurement Policies: When selecting third-party mobile applications or digital resources, prioritize vendors that demonstrate a commitment to accessibility and can provide Voluntary Product Accessibility Templates (VPATs) outlining their conformance with accessibility standards.

Ensuring ADA compliance in academic library mobile services is not merely a legal obligation; it is a fundamental ethical imperative. By embracing the principles of universal design and diligently implementing WCAG and Section 508 standards, libraries can create mobile environments that are truly inclusive and provide equitable access to information for all users. As Liu's (2023) research underscores, while progress has been made in adopting mobile technologies, a significant gap remains in ensuring their accessibility. By prioritizing accessibility in the development and maintenance of mobile services, academic libraries can reaffirm their commitment to inclusivity and empower all members of their communities to fully engage with the wealth of resources and opportunities they offer. The continued focus on research and practical application in this area, building upon the foundational work highlighted by Dr. Liu and their colleagues (Boateng & Liu, 2014; Guo et al., 2020; Liu & Briggs, 2015; Zhao et al., 2017), will be crucial in advancing library services for all mobile users.

REFERENCES

Akeroyd, J. (n.d.). Role of library discovery systems for online learners. Learned Publishing. https://library.sulross.edu/wp-content/uploads/2023/02/Learned-Publishing-2017-Akeroyd-Discovery-systems-Are-they-now-the-library.pdf

Alharbi, A. (2025). Emerging trends in academic library distance learning: AI and personalized resources. Education Sciences, 15(3), 343. MDPI. https://www.mdpi.com/2227-7102/15/3/343

Asher, A. D., & Duke, L. M. (2019). Comparison of Google Scholar and academic library discovery systems for scholarly information seeking. ScholarCommons. https://scholarcommons.sc.edu/cgi/viewcontent.cgi?article=1366&context=libsci_facpub

Ashford, R., & Zeigen, L. (2012). State of the Mobile Landscape: Mobile Literacy and What It Means for Libraries.

Ashikuzzaman, M., Islam, M. S., & Islam, M. N. (2025). *Open Educational Resources (OER)*. Enhancing Open Learning Provision in Bangladesh.

Association of College and Research Libraries. (2023). Standards for distance and online learning library services. https://www.ala.org/acrl/standards/standardsdistancelearning

Association of College and Research Libraries. (2023). Standards for distance and online learning library services. https://www.ala.org/acrl/standards/standardsdistancelearning

Association of Research Libraries. (n.d.). The code of best practices in fair use for academic and research libraries. https://www.arl.org/code-of-best-practices-in-fair-use-for-academic-and-research-libraries/

Beg, R., Russell, A. J., & Hornig, G. (2022). Evolution, structure, and topology of self-generated turbulent reconnection layers. *The Astrophysical Journal, 940*(1), 94. DOI: 10.3847/1538-4357/ac8eb6

Bielefield, A., Liu, Y. Q., & Waimon, V. (2021). Private post-secondary library websites and the ADA: Compliancy and COVID-19. *Universal Access in the Information Society*, •••, 1–16.

Boateng, F., & Liu, Y. Q. (2014). Web 2.0 applications' usage and trends in top US academic libraries. *Library Hi Tech, 32*(1), 120–138. DOI: 10.1108/LHT-07-2013-0093

Bomhold, C. (2014). Mobile services at academic libraries: Meeting the users' needs? *Library Hi Tech*, *32*(2), 336–345. DOI: 10.1108/LHT-10-2013-0138

Canuel, R., & Crichton, C. (Eds.). (2017). *Mobile technology and academic libraries: Innovative services for research and learning*. Association of College and Research Libraries.

Chow, A. S., & Croxton, R. A. (2022). Virtual reference services in academic libraries: A case study in Hong Kong. *Journal of Academic Librarianship*, *48*(1), 102–115.

Dill, E., & Cullen, M. A. (2020). Introduction: Open educational resources and the academic library. *Library Trends*, *69*(2), 335–342. DOI: 10.1353/lib.2020.0036

EDUCAUSE. (2022). Students and technology report: Reimagining the student experience. https://library.educause.edu/resources/2022/6/educause-student-technology-report-2022

Elmore, L., & Stephens, D. (2012). The application of QR Codes in UK academic libraries. *New Review of Academic Librarianship*, *18*(1), 26–42. DOI: 10.1080/13614533.2012.654679

Germain, C. A. (n.d.). Current services academic libraries provide to distance learners. College & Research Libraries. https://crl.acrl.org/index.php/crl/article/view/25864/34597

Guo, R., Pang, Y., Xu, Y., Liu, Z., Chen, Y., & Guo, Y. (2025). Application of artificial intelligence technologies in library services at the top 100 US universities. *The Electronic Library*, *43*(4), 619–648. Advance online publication. DOI: 10.1108/EL-12-2024-0386

Ibrahim, A., & Sanyal, M. (2025). Benefits of QR codes in libraries: Enhancing accessibility, engagement, and efficiency. *Organization Development Journal*. Retrieved February 26, 2025, from ResearchGate.

Ithaka, S. +R. (2023). Faculty Survey 2023: Information practices and needs of higher education faculty. https://sr.ithaka.org/publications/ithaka-sr-us-faculty-survey-2023/

Jabarali, A. K., Sathya Kumar, H., & Barak, Y. S. (2024). Analyzing the COVID-19 pandemic's effects on college students: E-learning experience, challenges and resource awareness. *Journal of Library & Information Services in Distance Learning*, *18*(1-2), 18–37. DOI: 10.1080/1533290X.2024.2332806

Kim, B. (2013). The present and future of the library mobile experience. *Library Technology Reports*, *49*(6), 15–28.

Kumar, S. (2025). *Embedded librarian online courses: Academic library examples*. College & Research Libraries News. https://crln.acrl.org/index.php/crlnews/article/view/24813/32645

Li, Y. (2024). Factors influencing the behavioral intention to use mobile library services: An empirical study of university students in China. AU-GSB e-Journal, 17(2), 45–57. https://auojs.au.edu/index.php/Scholar/article/view/7474

Lisedunetwork. (2024, September 25). Remote access to library resources. https://www.lisedunetwork.com/remote-access-to-library-resources/

Liu, Y. Q., & Briggs, S. (2015). A Library in the Palm of Your Hand: Mobile Services in Top 100 University Libraries. *Information Technology and Libraries*, *34*(2), 133–148. DOI: 10.6017/ital.v34i2.5650

Liu, Y. Q., & Lewis, S. (2023). Services to mobile users: The best practice from the top-visited public libraries in the US. *Information Technology and Libraries*, *42*(1). Advance online publication. DOI: 10.6017/ital.v42i1.15143

Lo, P. C., & Stark, A. J. (2020). *Examining the relationship between social inclusion and mobile libraries in the age of Internet connectivity: A qualitative study of mobile librarians* around the globe. *Journal of Librarianship and Information Science*. Advance online publication.

Moen, W. E., & McClure, C. R. (n.d.). Virtual reality and the academic library of the future. ResearchGate. https://www.researchgate.net/publication/359601186_Virtual_reality_and_the_academic_library_of_the_future

Mustapha, A., Ibrahim, F., Garba, K. D., & Muhammed, M. (2024, December). *Transforming academic libraries: The impact of mobile technology*. ResearchGate. https://www.researchgate.net/publication/390121750_Transforming_Academic_Libraries_The_Impact_of_Mobile_Technology

National University. (2025, July 15). What are the challenges of online learning for students? National University Blog. https://www.nu.edu/blog/challenges-of-distance-learning-for-students/

Nicholas, D., Rowlands, I., Clark, D., & Williams, P. (2009). Student information-seeking behaviour in a digital academic environment. *The Journal of Documentation*, *65*(1), 106–130. DOI: 10.1108/00220410910926149

Ogbebor, O. A. (2024, December 19). Artificial intelligence transforms academic libraries into dynamic, cutting-edge solutions. International Journal of Library and Information Science. Academic Journals. https://academicjournals.org/journal/IJLIS/article-full-text-pdf/52BB02772891

Oliveira, S. M. (n.d.). Information seeking behavior of online students: Academic library perspectives [Preprint]. Preprint Press JHU. https://preprint.press.jhu.edu/portal/sites/default/files/oliveira.pdf

Osika, B., & Kaufman, C. (2012). 'MOBILIZING'COMMUNITY COLLEGE LIBRARIES: Trends and Solutions. *Searcher (Medford, N.J.), 20*(9).

Penner, K. (2024). iPad and Digital Sheet Music Delivery: Supporting Student Development and Encouraging Digital Score Use Through Library-Circulated Hardware. *Fontes Artis Musicae, 71*(3), 179–189. DOI: 10.1353/fam.2024.a939059

Peters, T. A., & Bell, L. A. (Eds.). (2013). *The Handheld Library: Mobile Technology and the Librarian: Mobile Technology and the Librarian*. Libraries Unlimited. ABC-CLIO. DOI: 10.5040/9798400661679

Phillips, A. L. (2024, November 1). Embracing the metaverse: A survey of virtual reality and augmented reality practices at the United States' top one hundred university libraries. College & Research Libraries. https://crl.acrl.org/index.php/crl/article/view/25664/34458

Providence College. (2017). *A novel application: Using mobile technology to connect physical and virtual reference collections*. Digital Commons @ Providence. https://digitalcommons.providence.edu/facstaff_pubs/49/

Roy, L. (2025). Using data analytics to identify information gaps and create inclusive library programs. Journal of Web Librarianship, 19(2), 123–138. Taylor & Francis Online. https://www.tandfonline.com/doi/full/10.1080/1941126X.2025.2497729

Saragossi, J., Costello, L., & Kasten, K. (2024). Mobile applications in academic libraries. Library Resources & Technical Services, 62(4), 198–... DOI: 10.5860/lrts.62n4.198

Scott, S. M., & Ward, L. (2025). The use of AI in academic libraries. In *Examining AI Disruption in Educational Settings* (pp. 59–100). IGI Global. DOI: 10.4018/979-8-3693-7873-1.ch003

Singh, A. K., & Narayan, R. (2021). Progressive Web Apps: A Smart Way to Build Mobile-Web Apps for Academic Libraries. Applying Mobile Technologies in Transformation of Library Services, 91-117.

Singh, B. P., & Madhusudhan, M. (2023). Mobile apps–based applications in libraries and information centers: A systematic review of the literature and future research agendas. *International Journal of Librarianship, 8*(3), 83–102. DOI: 10.23974/ijol.2023.vol8.3.294

Smith, G. (2015). The future of library mobile websites: A usability study. *Library Hi Tech*, *33*(4), 574–587.

Theiss, D. (2022). Distance and online learners and library anxiety: An exploration into the causes, impact, and recommendations for practice. *Journal of Library & Information Services in Distance Learning*, *16*(2), 152–167. DOI: 10.1080/1533290X.2022.2112799

Thurlow, I., Duke, A., & Davies, J. (2006). Applying semantic web technology in a digital library. In *Demos and Posters of the 3rd European Semantic Web Conference, available at: www. eswc.*

Torres-Pérez, P., Méndez-Rodríguez, E., & Orduna-Malea, E. (2016). Mobile web adoption in top ranked university libraries: A preliminary study. *Journal of Academic Librarianship*, *42*(4), 329–339. DOI: 10.1016/j.acalib.2016.05.011

University at Buffalo Health Sciences Library. (2013). Providing mobile reference and instruction services using iPads. *Journal of the Medical Library Association: JMLA*, *101*(3). https://pmc.ncbi.nlm.nih.gov/articles/PMC4733882/

Vogus, B. (2019). Open educational resources (OER) and textbooks: Opportunities for academic libraries. *Public Services Quarterly*, *15*(3), 242–247. DOI: 10.1080/15228959.2019.1629858

Walsh, A. (2012). *Using mobile technology to deliver library services a handbook.* Facet. DOI: 10.29085/9781856048996

Zhao, F., Liu, Y. Q., & Wang, Y. (2017). Research on the Current Situation and Countermeasures of the Mobile Information Service in Chinese Academic. *Library and Information Service, 61*(21), 49-55. https://www.lis.ac.cn/CN/10.13266/j.issn.0252-3116.2017.21.006

Chapter 9
Mobile Services in Public Libraries

ABSTRACT

This chapter explores how public libraries expand community reach through mobile platforms that reflect principles of nomadicity, digital equity, and responsive design. It reviews research on post-pandemic service trends, usability practices, and the integration of websites and dedicated apps. Major mobile service features—circulation tools, mobile printing, tablet lending, learning programs, and community engagement initiatives—are examined alongside patterns of adoption. The chapter provides a deep analysis of mobile and virtual reference services, including AI-driven models, market growth, and quality assurance considerations. ADA compliance frameworks, real-world examples, and universal design strategies illustrate accessibility in practice. The chapter concludes with mobile services tailored to diverse populations, emphasizing multilingual support, inclusive programming, and adaptive assistance for patrons with specialized needs.

INTRODUCTION

In the contemporary landscape, public libraries are evolving beyond their traditional roles as repositories of physical books, increasingly becoming dynamic community hubs offering a wealth of digital resources and services. The advent of mobile technology has transformed how information is accessed, disseminated, and consumed worldwide. Integral to this transformation is the strategic deployment of mobile applications, or apps, which serve as virtual extensions of the library's physical space, providing patrons with unprecedented access to information, services, and engagement opportunities.

Mobile services in public libraries have undergone significant transformation in the digital age, evolving from traditional bookmobile operations to sophisticated

DOI: 10.4018/978-1-7998-9175-8.ch009

technological platforms that serve diverse community needs. The significance of mobile services has grown exponentially as libraries adapt to changing demographic needs and technological capabilities. According to Liu and Lewis (2023), libraries are increasingly providing mobile services to bridge digital divides and ensure equitable access to information resources across all population segments. This evolution requires careful consideration of accessibility standards, staff training protocols, and specialized services for diverse patron groups.

Drawing upon insights from recent peer-reviewed research, public libraries have explored the diverse range of application functionalities they offer to enhance user experiences and foster community connections. Traditionally perceived as physical repositories of knowledge, libraries have been compelled to innovate and expand their service delivery to include mobile technologies, aiming to meet users' expectations for anytime, anywhere access. Mobile library services exemplify this transformation through various critical dimensions, representing a strategic evolution from fixed-location service points to dynamic, portable, and user-centered platforms that facilitate broad accessibility and inclusivity.

SERVICE EXPANSION AND COMMUNITY REACH

Public libraries across the United States have dramatically expanded their mobile service offerings to address community needs and bridge accessibility gaps. The pervasive nature of mobile technology has profoundly reshaped how individuals access information and services, and public libraries have actively embraced this shift to better serve their communities. Mobile services in public libraries encompass a wide array of offerings designed to deliver library resources and assistance directly to users' mobile devices, ensuring access anytime and anywhere. This evolution from traditional brick-and-mortar services to ubiquitous mobile accessibility represents a significant advancement in library outreach and engagement.

Conceptual Foundation: Nomadicity and Responsive Design

The concepts of nomadicity and ubiquitous information access are central to this transformation, (Saravani & Haddow, 2017). Nomadicity reflects the user's mobility and the ability to remain connected, a phenomenon that libraries increasingly acknowledge as integral to modern service frameworks. This means that libraries recognize the importance of providing services that cater to users who are on the move, whether they are accessing resources from their homes, on public transportation, or while visiting various locations.

To achieve this, libraries are leveraging advanced technologies such as responsive web design. Responsive web design is crucial because it allows web pages to automatically adjust to the screen size and resolution of the device being used, ensuring that users have an optimal viewing experience on smartphones, tablets, laptops, and desktop computers. This feature is instrumental in creating seamless user interfaces that enhance user interaction with library catalogs, databases, and digital resources.

The development of mobile services in libraries has progressed significantly since the early 2000s, moving beyond basic mobile-friendly websites to sophisticated applications and integrated digital experiences. Early efforts often focused on providing simple access to library catalogs and basic account management functions. As mobile technology advanced and user expectations grew, so too did the complexity and scope of mobile library services. Libraries are changing to better fit into a world where people use their phones and tablets a lot. "Mobile services" in this context has been referred to the provision of library resources and support that are accessible and optimized for use on portable devices such as smartphones and tablets. This includes mobile-friendly websites, dedicated library apps, and SMS-based services, all designed to offer seamless interaction for users on the go.

Research Insights on Mobile Services in Public Libraries

Whether libraries have adopted mobile devices to provide mobile service, in their 2014 study, The Library Mobile Services in USA: A Case Study of New York State Libraries, Liu and his research team revealed the facts and issues relating to the change by surveying public and academic libraries in New York State, USA. The investigation resulted: More than 90% libraries provide services for mobile users; Over half libraries provide six or more service categories to various mobile devices; While remotely access to library catalog and digital content is the most frequently activity supported by the mobile-optimized library websites, multiple social networking activities through third-party platforms are undertaken in many public and university/college libraries; Almost half of the libraries have strategic plans to help guide mobile services in their libraries. The data hopefully provide librarians, technologists, and information professionals with an increased awareness of the current mobile services in the US to endorse libraries' mobile services.

The impact of digital resources and services on urban residents utilizing New England public libraries has also been investigated. Liu and Wnuk (2009) explored this impact through a survey report, providing insights into how mobile access to digital resources shapes user engagement and satisfaction within urban communities. Such studies are vital in understanding the tangible benefits that mobile services bring to library users, including increased convenience, expanded access to information, and enhanced engagement with library offerings.

Guo, Liu, and Bielefield (2018) examined how urban public libraries utilized services for mobile users. Their study wanted to see what services and content were being offered. One of the goals of their analysis was to show other public libraries what was being done in an urban setting to potentially help them plan and create mobile serviceS.

The accessibility of these mobile services to all members of the community, including those with disabilities, has been a critical area of focus. Liu, Bielefield, and McKay (2019) conducted a study examining whether urban public library websites are accessible to Americans with disabilities, underscoring the importance of inclusive design in mobile service development. This highlights a crucial aspect of mobile service provision: ensuring equitable access for all users, regardless of their abilities, by adhering to accessibility standards and guidelines.

Technology Expansion, Digital Equity, and Service Trends

Research conducted by Liu and Lewis (2023) specifically investigated services to mobile users, identifying best practices from top-visited public libraries in the US. This research offers valuable insights into the strategies employed by leading institutions to effectively cater to their mobile patrons, focusing on optimizing user experience and maximizing accessibility of resources. Research by Liu and Lewis (2023) analyzing 151 top-visited public libraries found that mobile services have become integral to library operations, with every examined library maintaining at least one mobile platform for service delivery.

The geographic distribution of mobile services reflects strategic community planning. Libraries serving populations of varying sizes have adapted their mobile service models accordingly. Data indicates that mobile applications, mobile reference services, mobile library catalogs, and mobile printing are among the most frequently offered services by public libraries, (Liu & Lewis, 2023).

In their 2023 study, Liu and Lewis found that 83% of library websites are optimized for mobile use. Public libraries adapt to the prevalence of mobile devices by offering mobile-friendly services. The 2023 Public Library Technology Survey, conducted by the Public Library Association, gathered responses from 1,511 public libraries across the United States. This survey provides a comprehensive view of current technology trends and challenges.

Their survey reveals a notable expansion in the technological capabilities and digital services offered by public libraries across the United States. Over 95% of libraries now provide access to e-books, digital literacy training, and vendor-supplied mobile apps such as Libby and Hoopla. Maker technologies, including 3D printers and digital media production tools, are increasingly present, especially in urban libraries. However, rural and town libraries show slower adoption rates.

Technological lending for off-site use, including hotspots (provided by 46.9% of libraries) and laptops (24.8%), has experienced substantial growth since 2020. Additionally, early learning devices and recreational technology have garnered significant circulation. Nevertheless, approximately 28.4% of libraries continue to operate with broadband that falls below FCC standards. Infrastructure limitations, staffing shortages, and budgetary constraints are frequently cited as impediments to technological enhancement, (Liu & Lewis, 2023).

Despite these challenges, libraries are proactively addressing community needs through digital equity initiatives, technology instruction, and partnerships with organizations such as AT&T. This reflects their evolving role as essential hubs for digital access and inclusion. This is a significant increase from previous years. Every library surveyed in a 2023 study had at least one mobile-optimized website, mobile catalog, or mobile app, (Public Library Association, 2024).

The Library Research Service's 2024 report, Public Libraries and Technology: Beyond the Public Library Survey, highlights the growing emphasis public libraries place on serving mobile users through a range of digital and technology-driven services. The findings indicate that libraries are increasingly investing in mobile-responsive websites, library-branded mobile apps, and remote access to digital collections to meet users where they are—on their smartphones and tablets. Many libraries now offer mobile-friendly catalogs, eBook lending platforms, and remote authentication systems that enable users to access materials without needing to visit a physical branch. Additionally, the study reveals a rise in virtual reference services, including live chat and SMS-based assistance, designed specifically to support users seeking help via mobile devices. These developments demonstrate libraries' commitment to expanding equitable access to resources and ensuring that patrons can engage with library services anytime, anywhere.

Usability, Functionality and Post-Pandemic Service Models

With time, the availability of mobile-friendly services has become critical to a library's long-term viability. The ability to search a library's catalog, for example, has aided a traditional library's modern relevance. "The digital library on mobile devices has been a milestone in library industry development, leading to huge changes in knowledge carrying, spreading, acquiring, processing and sharing of cloud computing and big data in online and offline forms," (Liu et al., 2020). Users want data at their fingertips, whether it is an online order or a library catalog. "Social sharing functions such as reading, borrowing, sharing, comment tracking, automatic retrieval and new book recommendations on mobile devices such as mobile phones have become the popular trend of digital library development," (Liu et al., 2020).

When assessing mobile library websites and applications for specific library information, the literature suggests that a mobile version of a library website is more prevalent than a dedicated (or "native") application. This preference is attributed to the development costs, the diverse mobile platforms (e.g., Apple, Android), and the ongoing maintenance required, (LeMire et al., 2017). Consequently, it is more feasible to optimize the library's website for mobile devices rather than developing an application.

The challenge for mobile devices lies in ensuring they provide the same functionality as desktop viewers. "Today's mobile users are no longer content with basic mobile websites that offer only a limited subset of the information and features available on desktop websites. Although the small screen size of mobile devices can make certain tasks more tedious or cumbersome, mobile users anticipate performing increasingly more tasks on their devices," (LeMire et al., 2017). LeMire, et al. (2017) conducted a study examining the advancements in mobile services since 2010. While the study primarily focused on academic libraries and conducted surveys, the findings are relevant to the study of public libraries as well.

Following the widespread spread and substantial impact of the COVID-19 virus on various public services sectors globally, it is imperative to identify areas where libraries can enhance their services. As more individuals rely on mobile services, libraries must adapt to meet these evolving needs. Dr. Yan Quan Liu's research teams conducted several studies from 2015 to 2023 to examine mobile services provided by public libraries across the United States. These studies analyzed the influence of the pandemic on these services and the libraries' future plans. Libraries, particularly in suburban areas, had to evolve to adapt to the changing environment. Public libraries have experienced substantial growth and now offer a broader spectrum of patron services. Many libraries continued to utilize the service methods established during the COVID-19 pandemic, including curbside and contactless services, which eliminate the necessity for direct interaction with librarians. These services have gained popularity and transcend geographical and temporal limitations, thereby reducing the risk of disease transmission through contact with others.

MOBILE PLATFORMS: WEBSITES AND DEDICATED APPS

Mobile services in public libraries emerged as a direct response to the widespread adoption of smartphones and tablets, which transformed user expectations for on-demand, location-independent access to information. Early efforts centered on creating mobile-friendly websites and optimizing online catalogs for smaller screens, but these services have since evolved into multifaceted digital ecosystems. Today, public libraries increasingly provide native mobile apps, remote printing capabilities, device lending programs, real-time notifications, and mobile-enabled outreach initiatives designed to meet the needs of diverse user groups.

Research led by Dr. Yan Quan Liu has documented this progression. Guo, Liu, and Bielefield (2018) found that all public library websites in their study were accessible via mobile devices, with many institutions introducing mobile services specifically in response to the COVID-19 pandemic. Their survey revealed that all libraries have an online public access catalog (OPAC) accessible through mobile devices, 63% of participating libraries had added new mobile offerings within the previous year, highlighting the pandemic's role in accelerating digital service adoption. Building on this work, Liu (2023) conducted a nationwide survey of 151 public libraries, confirming that mobile apps, mobile reference, catalog access, and mobile printing are among the most frequently provided mobile services, reflecting the sector's ongoing commitment to expanding mobile access.

Responsive Websites & Channels

With all public library websites now accessible via mobile devices, as highlighted by Guo, Liu, and Bielefield (2018), each website provides essential contact information and details about the libraries' locations and operating hours. Notably, many libraries have developed websites that are designed to accommodate various screen sizes, ensuring a user-friendly experience on mobile devices. The most important details in this text are the differences between the websites of the investigated libraries. Web-mediated services have become increasingly prevalent since the onset of the COVID-19 pandemic, with curbside pickup for books and other materials becoming a common feature. Library events are now almost exclusively offered virtually.

Furthermore, many contemporary systems are adopting Progressive Web Apps (PWAs) to provide a mobile-like experience through a user's web browser. These self-contained PWAs offer functionalities such as catalog browsing, account management, and notifications, effectively bridging the gap between websites and native mobile applications.

Progressive Web Apps (PWAs) are applications that you build by using web technologies, and that can be installed and can run on all devices, from one codebase. PWAs provide native-like experiences to your users on supporting devices. They adapt to the capabilities supported by each device, and they can also run in web browsers, like websites.

- Responsive Design:

Both responsive web apps and PWAs are designed to adapt to different screen sizes and devices, ensuring a consistent user experience.

- PWAs as Enhanced Web Apps:

PWAs are essentially web applications that leverage web technologies to provide a native app-like experience. They can be installed on devices and offer features like offline access, push notifications, and faster loading times, similar to native apps.

- Progressive Enhancement:

PWAs utilize progressive enhancement, meaning they work on all browsers but offer more features and a richer experience on modern browsers.

- Cross-Platform Compatibility:

PWAs, like responsive web apps, are built with a single codebase that works across multiple platforms and devices.

- Installation and Offline Functionality:

PWAs can be installed on devices, allowing for offline access to content and features, and they can also utilize device features like cameras and microphones.

Customized Mobile Implementations

Beyond general service provision, specific advanced applications of mobile technology within public libraries have been explored Mobile applications have become a cornerstone of library services, enabling patrons to access resources and manage their accounts from anywhere. Libraries develop apps that allow users to access resources and manage their accounts from their smartphones or tablets. Popular library apps include Libby, Hoopla, and CloudLibrary. These apps offer intuitive interfaces that allow users to search catalogs, place holds, renew materials, and access digital collections on their smartphones or tablets.

The Ex Libris Library Mobile App, integrated with systems like Alma and Primo, provides a streamlined experience with features such as digital patron ID cards and push notifications to replace costly SMS alerts. This app allows libraries to deliver hyper-personalized content, ensuring patrons receive timely updates about events or account activities, (Breeding, 2023). Similarly, the Vega Mobile app, used by public libraries with Sierra or Polaris systems, supports self-checkout via RFID or barcode scanning and curbside pickup, enhancing convenience while reducing staff workload.

A study examining public library mobile apps in Scotland revealed that many local authorities either lack such apps or have no plans to develop them, often due to cost constraints and competing organizational priorities, (Kerr & Pennington, 2018). While library authorities tended to be satisfied with their current provision of basic mobile services, they rarely explored expanded functionality, and none had engaged in public consultation regarding app development. In contrast, members of the public—including those living in areas without a library app—expressed interest in mobile access to library resources, though they voiced frustration with the complexity of accessing certain external services. The findings highlight that both the quality of the app or mobile website and the clarity of the services offered are critical to user engagement. Moreover, larger institutions such as university libraries are more likely than smaller, rural libraries to offer a wide range of mobile services, including mobile-friendly websites, dedicated apps, online public access catalogs (OPACs), mobile database access, text messaging services, QR codes, augmented reality features, and e-books, pointing to a significant disparity in mobile service provision.

Guo, Liu, and Bielefield (2018) uncovered a significant prevalence of dedicated mobile applications within libraries. Notably, 52% of these libraries possess at least one app specifically tailored for their institution or system. These applications empower patrons with seamless access to their accounts, enabling them to monitor their ongoing checkouts, conduct comprehensive searches of the library catalog, initiate holds, and request desired items. It is noteworthy that this study excluded other library applications, such as Hoopla and Libby/Overdrive, as well as those that provide information on upcoming events and other library-related details, including locations and operating hours.

Library mobile apps offer a range of services, including event listings, webinars, recent arrivals or recommendations, and ways to contact the library. However, gathering information from these apps can be challenging, as some require users to be affiliated with the library and log in using a library card. Despite these limitations, it's worth noting that mobile apps have become the most popular method of connecting with patrons. However, maintaining and ensuring that these apps work seamlessly on all mobile platforms and devices can be a complex and ongoing task.

According to Liu's research, the majority of public library mobile applications—approximately 67%—are developed by commercial vendors, while only 33% are created in-house by library staff, (Liu, 2023). Among the commercial solutions, Solus UK Ltd. stands out as one of the most widely adopted providers, supplying scalable and adaptable mobile platforms that can be customized to meet specific user and institutional needs. In an analysis of 78 public library apps, nearly one-quarter (24%) were powered by Solus software, reflecting its strong market presence. Libraries that use Solus often cite its user-friendly, intuitive interface and flexible architecture, which allow for streamlined navigation, efficient service delivery, and the integration of emerging features without requiring significant redevelopment.

The use of recommender systems in digital libraries, a concept that can be seamlessly integrated into mobile platforms, was examined by Liu (2007). Similarly, Speidel and Liu (2007) questioned the efficacy of recommender systems for libraries, prompting a deeper consideration of their practical application in mobile environments. More recently, Liu and Wang (2023) published a book on research into library's smart services, derived from recommender systems and knowledge discoveries, further illustrating the potential for advanced, personalized mobile library experiences. The potential of smart recommendation services, based on knowledge discovery, has also been a subject of research by Wang, Dong, and Liu (2018), which can enhance the mobile user's experience by offering tailored content and suggestions. These systems, when integrated into mobile apps, can offer personalized book recommendations, suggest relevant events, or highlight new acquisitions based on a user's borrowing history and interests, thereby enriching their mobile library experience.

Content Delivered via Mobile Websites and Apps

Catalog Search and Resources Access

Mobile OPACs provide instant access to bibliographic information, enabling patrons to locate materials by author, title, or subject from their smartphones for books, movies, music, and other materials in the library catalog. Access digital magazines through apps like Flipster. Online resources: Access online databases, research guides, and other digital learning tools.

E-Books and Audiobooks

Borrow and download digital books and audiobooks for offline use. E-book and audiobook lending has become a cornerstone of mobile services. Platforms like Libby and OverDrive allow patrons to borrow digital materials with a single tap, with usage surging during the pandemic. The Los Angeles County Public Library reported 1,184,289 e-book circulations in the fiscal year ending June 30, 2018, a figure that has likely grown with increased mobile adoption. These platforms offer content tailored to diverse interests, from children's books to academic journals, ensuring broad appeal.

Streaming Media

Stream movies, TV shows, and music through platforms like Hoopla and Kanopy.

Event Registration

Register for library programs and events. Another area of study related to mobile services is the analysis of public library events. Kukiolczynski and Liu (2021) delved into the "who, what, where, when, and why" of public library events, an area increasingly promoted and managed through mobile platforms. This suggests that mobile services are not just about accessing static content but also about dynamic engagement with library programming and community activities, allowing users to discover, register for, and even attend virtual events directly from their mobile devices.

TYPES OF MOBILE SERVICES AND THEIR FEATURES

Public libraries have developed a diverse portfolio of mobile services to meet evolving patron needs. These services can be broadly categorized into resource access, circulation management, and community engagement tools for their patrons.

A comprehensive investigation led by Dr. Yan Quan Liu revealed that 95% of U.S. urban libraries provide at least one type of mobile service—such as a mobile website, catalog, or app—which underscores the widespread adoption of mobile-enabled access, (Guo, Liu, & Bielefield, 2018). Using data sourced from the Institute of Museum and Library Services, researchers selected 151 of the most-visited public libraries in states with a substantial number of libraries and at least 3 million annual in-person visits. They compiled details on each library's state, population served, total circulation, program offerings, number of visits, and other variables into an Excel spreadsheet for further analysis. The study employed a mixed-methods approach—examining mobile website and app visits, conducting content analysis, and distributing email surveys—to gather and interpret data, (Guo, Liu, & Bielefield, 2018).

Circulation Management

Mobile services streamline circulation tasks, enhancing patron convenience. Traditional circulation functionalities have been adapted for mobile use, allowing users to perform tasks like book renewals and holds directly through mobile applications or responsive websites. By managing their library accounts, patrons eliminate the need for in-person visits or desktop computer access. Automated notifications are another crucial component, with libraries sending overdue alerts and request arrival notifications via SMS or in-app messages. This ensures patrons are timely informed about their library accounts. Common features offered by public libraries:

Renewal and Holds Library Materials

Patrons can renew borrowed items via mobile apps. Patrons can also place holds, see queue status, and renew materials through mobile interfaces. A study by Fung et al. (2022) found that 77.2% of HKPL app users utilized renewal services, reflecting high demand for this feature.

A mid-sized public library in Connecticut saw hold pickups rise by 20 % after launching app-based notifications and improving mobile catalog usability, (Liu & Lewis, 2023)

Overdue Notifications

Automated text messages or app notifications alert users to overdue items, reducing fines and improving return rates. Push or SMS alerts for recalls, overdue status, and hold fulfillment.

Liu and Lewis (2023) also reported a 30 % decline in overdue after enabling mobile renewals. While distinct from Liu's survey sample, these results align with his findings on service adoption and impact across the surveyed libraries

Arrival Notifications

Libraries notify patrons when reserved items are available, as seen in systems like the University of Cape Coast Library's mobile platform, (Owusu-Ansah, 2015). As Liu and Lewis (2023) shows, nearly 70 % of libraries now issue hold-arrival alerts via mobile channels, improving pickup rates and customer satisfaction.

Mobile Reservation

A study by Fung et al. (2022) found that 74.9 of HKPL app users utilized the reserve services, reflecting high demand for this feature.

According to findings by Guo, Liu, and Bielefield (2018), among 138 urban libraries surveyed mobile reservation services were found in 78% of all libraries' mobile services. A majority (62%) of the libraries allow online reservation of a meeting room via a web form or other methods, and 14% allow patrons to reserve a study room (see Figure 10). Some libraries, however, only handle room reservations via phone. Furthermore, a few libraries provide instant online access to free and low-cost tickets to local cultural venues with services like Discover & Go. A total of 14% of the libraries provide a "reserve a librarian" service, which allows patrons to book a free session with a reference librarian or a subject specialist at the library. In addition, several libraries, such as the Pasadena Public Library, allow for the reservation of exhibit space.

Mobile Printing

Via app or mobile site, users send documents to library printers from anywhere. Innovative libraries provide a service that allowing patrons to print documents from their smartphones, tablets, or laptops while at the library wirelessly. This means that patrons can use their mobile devices pre-loaded with library apps while at the library send their documents to be printed without needing to connect their device to the printer directly, (Guo, Liu & Bielefield, 2018).

Libraries like the Columbus Metropolitan Library offer mobile printing services, allowing patrons to send documents to library printers from their devices, a feature particularly valuable for students and job seekers.

To learn How Many Libraries Provide Mobile Printing, survey results from Guo, Liu and Bielefield (2018) revealed that among the 138 urban libraries surveyed, a significant 43% currently offer wireless printing services. This number is anticipated to increase in the future. That remote printing services allow patrons to print to a library printer from outside the library or from their mobile device. Patrons' print jobs are available for pick up at the library. To print from a mobile device, patrons typically need to install a compatible app to utilize these services.

"PrinterOn" is a popular app option available at libraries like Oakland Public Library and San Mateo County Libraries. Other alternatives, such as "Your Print Cloud" (used by San Diego Public Library) and "Smart Alec" (adopted by Santa Clara County Library), are also available. Notably, San Mateo County Libraries offers wireless printing from smartphones, tablets, and laptops at all its locations, providing mobile printing, web printing, and email printing options. However, it's important to remember that while 14% of libraries offer wireless printing, they may not include mobile printing features. For instance, Live Oak Public Libraries in Savannah, Georgia, allows printing from laptops but lacks support for printing from phones or tablets. For more information, refer to the figure below.

Figure 1. The proportion of libraries that offer mobile printing

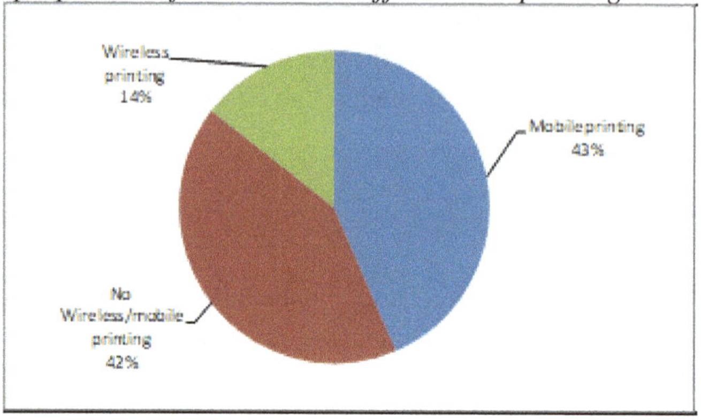

(Guo, Liu & Bielefield, 2018)

Public Library Tablet Lending Programs

Public library tablet lending has matured from scattered pilots into a targeted digital-equity strategy that complements laptop and hotspot circulation. The national snapshot—the Public Library Association's 2023 Public Library Technology Survey—confirms both the scale and the direction of travel. Nearly half of U.S. public libraries now loan mobile hotspots for off-site use (46.9%, up from 32.6% in 2020), and about one quarter circulate laptops (24.8%, up from 16.7% in 2020). While tablets and e-readers are lent by a smaller share of systems, they are firmly part of the lending mix: 11.7% circulate tablets and 14.0% circulate e-readers, according to the survey's results presentation. These figures underscore how device lending has become an essential access point for communities that lack household connectivity, with hotspots, laptops, and tablets functioning as a coordinated toolkit rather than isolated offerings, (American Library Association, 2024) National trends show libraries scaling both components in tandem: hotspot availability climbed roughly 14 percentage points from 2020 to 2023, while laptop circulation also rose substantially—movements that situate tablets as a strategic middle ground between phones and full laptops.

In their 2018 survey study on U.S. urban public libraries, Guo, Liu and Bielefield explored the feasibility and user-friendliness of device lending. They found that extending the existing transaction flows through the library's app for reserving or renewing tablets or hotspots is both technically feasible and intuitive for patrons. Liu's public-library study, conducted in collaboration with Lewis in 2023, further corroborated this finding, demonstrating that this practice aligns with current practices involving the placement and renewal of tablets and hotspots through mobile channels. This integration enhances the connection between circulation policies and the patron experience. In essence, the rise of device lending is not an isolated phenomenon but rather an integral part of the broader maturation of mobile services, as observed by Liu's ongoing research.

The Public Library of Brookline offers iPads equipped with assistive technology apps for individuals with autism spectrum disorders, learning disabilities, and visual or hearing impairments. These devices also include apps for literacy development, healthcare, games, and education, making them accessible to all patrons, (Kajeet, 2024).

The Santa Clara County Library offers iPads with apps for searching books, DVDs, and audiobooks, supporting multiple languages to serve diverse communities, (Moorefield-Lang, Meier, & Miller, 2014). These devices are particularly valuable in children's learning events, where apps like History: Maps of World facilitate interactive geography lessons.

Yarra Libraries in Australia launched the "Connected Community" iPad lending program, targeting older adults and individuals with disabilities. The program provides a six-week loan of an iPad kit, a wireless modem, and ongoing digital literacy support, ensuring that participants receive both the technology and the training necessary for successful use, (Yarra Libraries, n.d.).

Sonoma County Library in California is a notable case of a library system that has implemented the Hublet system for in-library wireless tablet lending, starting in late 2018. The library's goal was to provide patrons with easy access to technology within their branches, increase computer use, and save space in their potentially busy branches. The library also offers a Connected Chromebook Lending Program, providing laptops with WiFi for long-term checkout, particularly to those who lack necessary technology.

The Hublets are cloud-based tablets integrated with the library's Integrated Library System (ILS), allowing for self-service checkouts and returns. The tablets are designed to only work within the library's designated Wi-Fi network for security. When a tablet is returned, all previous user data is automatically deleted, ensuring patron privacy. The Hublet program helps the library promote digital literacy by providing access to various reading and grammar applications available on the tablets, as well as pathways to citizenship applications, (Hahn, 2017).

Between 2020 and 2023, public libraries significantly expanded their technology lending services. According to the 2023 Public Library Technology Survey: Summary Report published by the American Library Association's Public Library Association, 46.9% of public libraries now circulate internet hotspots for off-site use—up notably from just 32.6% in 2020. Additionally, 24.8% of libraries lend laptops to patrons for use outside the library, representing a considerable increase over the 16.7% reported in 2020, (Public Library Association, 2024).

Mobile Learning

Public libraries increasingly deliver mobile-first learning through dedicated mobile libraries and apps, enabling access to a suite of learning tools that support literacy, job readiness, language instruction, and more, (Providence Public Library, n.d.). These services often include videos, learning apps, online courses, audio files such as podcasts, and library-curated learning resources that have been vetted for quality and accessibility.

The Public Library Association's DigitalLearn platform, for example, offers free, mobile-friendly online courses in both English and Spanish on foundational digital literacy skills—such as navigating email, the web, and cybersecurity—alongside facilitator guides and training materials to help libraries deliver these resources effectively . Collectively, these mobile learning offerings empower patrons to engage in flexible, anytime-anywhere learning, bridging gaps in access and digital skills, (Public Library Association, n.d.).

Community Engagement and Programming

Mobile services extend beyond resource access to foster community engagement. Libraries offer virtual programs, such as Storytimes and author events, streamed via mobile apps. The PLA's 2020 survey noted that a majority of libraries provided streaming programs, with content ranging from literacy workshops to coding classes, (American Library Association, 2021). Mobile apps also facilitate event reminders and news updates, keeping patrons informed about library activities.

Innovative libraries have introduced mobile-based VR experiences to promote digital literacy. For instance, a project at Aalto University, led by Johanna Ylipulli, developed VR content for public libraries to enhance digital skills among patrons, demonstrating the potential of immersive technologies (Ylipulli, et al., 2023).

By offering mobile-friendly services, libraries facilitate fluid interactions, empowering patrons to search for and access needed information efficiently, regardless of their physical location. This adaptation not only increases accessibility to library resources but also meets the evolving expectations of a digital-first society, where users demand quick and convenient access to information. Ultimately, these mobile services are redefining how patrons engage with libraries, making them more relevant and accessible in today's fast-paced, technology-driven world.

MOBILE AND VIRTUAL REFERENCE SERVICES IN PUBLIC LIBRARIES

Mobile and virtual reference services have become fundamental components of contemporary public library operations, representing a significant evolution in how libraries provide professional assistance to patrons. Research by Liu and Lewis (2023) examining 151 top-visited public libraries in the United States found that "mobile apps, mobile reference services, mobile library catalogs, and mobile printing are among public libraries' most-frequently offered services" based on comprehensive analysis of mobile visits, content analysis, and librarian survey responses.

State and Adoption Patterns

Dr. Yan Quan Liu's extensive research program has documented the widespread adoption of mobile reference services across American public libraries. In their seminal study of urban libraries, Liu and colleagues (2018) found that "nearly 95% of these libraries have at least one mobile website, mobile catalog, or mobile app" among the 138 Urban Libraries Council members surveyed. The research revealed that "libraries actively applied new approaches to meet each local community's remote-access needs via new technologies, including app download links, mobile reference services, scan ISBN, location navigation, and mobile printing".

Liu's earlier foundational research (2014) on New York State libraries established baseline data showing that "more than 90% libraries provide services for mobile users" and "over half libraries provide six or more service categories to various mobile devices". This research demonstrated that "remotely access to library catalog and digital content is the most frequently activity supported by the mobile-optimized library websites" while also documenting extensive social networking integration through third-party platforms.

Service Delivery Models and Effectiveness

Virtual reference services encompass multiple delivery modalities that have proven effective across different user preferences and scenarios. Recent synthesis research indicates that "virtual reference services are effective, preferred in various forms like chat and email, and can meet user expectations", though implementation faces challenges including "safety concerns on mobile phones and low standardization".

OCLC's comprehensive evaluation research on virtual reference services found that these "human-mediated, Internet-based library information services" require understanding of user behaviors, needs, and preferences in virtual environments to effectively respond to increased demand for online reference assistance. The research emphasized the importance of evaluating sustainability and relevance of virtual reference offerings.

Technology Integration and Innovation Trends

Contemporary mobile reference services increasingly incorporate emerging technologies to enhance user experiences. Research on 2024 library technology trends indicates that "libraries in 2024 shift from physical repositories to hubs of digital resources, augmented reality (AR) is emerging as the next transformative library technology" with potential for "seamlessly blending the physical and digital, creating an immersive environment for patrons".

The evolution toward comprehensive digital platforms reflects libraries' strategic responses to changing user expectations and technological capabilities. Libraries are implementing integrated approaches that combine traditional reference expertise with innovative delivery mechanisms to maximize accessibility and effectiveness.

The Shanghai Library implemented a pilot project using 5G technology to offer VR-based orientation tours to first-year students. This approach aims to enhance user engagement in library services. The project reportedly resulted in a 60% increase in user engagement among the students participating. Public libraries explore innovative technologies like 5G and virtual reality to improve the accessibility and appeal of their resources and services. It demonstrates the potential of virtual reality to create immersive and interactive experiences for users, leading to increased interest and participation in library programs and resources. (Zhang, 2021).

Usage Patterns and User Preferences

Academic library research on virtual reference preferences provides insights applicable to public library contexts. Studies examining user interactions with virtual library services reveal diverse preference patterns across different demographic groups and information-seeking contexts. Chat reference has emerged as particularly popular for synchronous virtual reference, with academic libraries documenting increased adoption since the COVID-19 pandemic.

The integration of mobile reference services with broader library technology ecosystems supports seamless user experiences across multiple touchpoints. Users can initiate reference transactions through mobile apps, continue conversations via web platforms, and receive follow-up assistance through preferred communication channels.

Artificial Intelligence Integration in Reference Services

The integration of artificial intelligence technologies into library reference services represents one of the most significant developments in contemporary library practice. A 2024 integrative review by Aboelmaged et al. found that "the growing role of conversational AI Chatbots continues to change the library and information service landscape. Chatbots are replacing some of the library services", indicating a fundamental shift in how libraries deliver reference assistance.

Recent research demonstrates widespread adoption of AI-powered reference tools. Libraries are implementing "AI-driven chatbots and virtual assistants to assist patrons instantly, respond to questions, and direct them toward the resources and services available. These virtual agents are always available, which makes them more accessible to customers", providing 24/7 reference support that complements traditional staffed services.

A comprehensive 2024 study published In the New Review of Academic Librarianship found that" ChatGPT can be aided with many library services such as collection development management, virtual reference services, digital information service and library discovery, research writing and publishing, research performance analysis and bibliometric services". The research concluded that AI tools can help libraries "meet patron requirements, streamline the research process, and increase the efficiency of the library and staff".

Implementation Models and Effectiveness

Libraries are deploying various AI implementation models for reference services. Research indicates that "many libraries now use AI chatbots to provide real-time assistance to users. These bots can answer common questions, guide users through library services, or help them find specific materials without the need for human intervention". This automated first-tier support allows human librarians to focus on complex research inquiries requiring professional expertise.

The Mentor Public Library's deployment of "Emma the MPL Catbot" offers a practical public library case study of chatbot implementation. Built on the Pandorabots platform, Emma was designed to serve as a virtual reference assistant, answering common questions about hours, services, and resources while providing 24/7 support. Over time, Emma's performance improved dramatically, rising from handling only about 15 percent of questions correctly at launch to achieving accuracy rates in the mid-80 percent range. This sustained success, with Emma responding to more than 30,000 questions, demonstrates how public libraries can leverage conversational AI to enhance accessibility and reduce staff workload, providing a tested model for other institutions considering similar innovations. (Vincze, 2017)

Market Growth and Economic Impact

The economic significance of AI integration in libraries reflects broader technology adoption trends. Market research indicates that "the global virtual assistant market was valued at $4.2 billion in 2023 and is expected to reach $11.9 billion by 2030, with a CAGR of around 34%", demonstrating substantial investment and growth potential in AI-powered reference technologies.

Industry analysis reveals significant market expansion in chatbot applications. "The Chatbot industry value in the USA was calculated as 113 million USD. It is expected to multiply, and in 2024, it will reach ~1 billion USD", indicating rapid adoption and substantial economic investment in conversational AI technologies that libraries can leverage for reference services.

ChatGPT and Large Language Models in Reference

The emergence of ChatGPT and similar large language models has particular implications for library reference services. Research examining ChatGPT's library applications found that it can be used to "spark ideas or simplify aspects of the research process. It can help brainstorm topics, generate lists of keywords, and provide summaries of works", offering specific use cases for reference librarians.

However, research also identifies important limitations. A 2024 comparative study concluded that while "ChatGPT is excellent in information retrieval in some areas, its work is not comparable to library reference staff in others", emphasizing the continued need for human expertise in complex reference transactions.

Quality Assurance and Professional Considerations

Liu's research demonstrates that successful mobile reference implementation requires comprehensive staff training and ongoing professional development. Libraries must ensure that reference staff possess both traditional reference competencies and technological skills necessary for effective mobile service delivery.

Quality assurance in mobile reference services involves establishing response time standards, accuracy metrics, and user satisfaction measurements. Libraries implementing mobile reference services must develop evaluation frameworks that assess both quantitative performance indicators and qualitative user experience factors.

AI implementation in reference services requires careful attention to quality assurance and professional standards. Research highlights that "concerns regarding academic integrity and output quality must be addressed" when implementing AI tools in educational and library contexts.

Libraries implementing AI reference tools must establish evaluation frameworks that assess both technical performance and professional service quality. “Chatbots are computer programs that simulate an intelligent conversation through text, speech or through an embodied representation. They answer directional, reference and predictable inquiries”, indicating their effectiveness for routine questions while highlighting the continued need for human librarians for complex research support.

Emerging Applications and Direction Trends

The development of AI in library reference services involves expanding capabilities and integration with existing library systems. Research suggests that AI tools will continue evolving to support more sophisticated reference functions while maintaining appropriate boundaries between automated and human-mediated assistance.

Libraries must develop strategic approaches to AI integration that enhance rather than replace professional reference services. The goal is creating hybrid models that leverage AI efficiency for routine inquiries while preserving human expertise for complex research needs, information literacy instruction, and personalized patron assistance.

Research indicates that mobile reference services will continue evolving as libraries adapt to new technologies and changing user expectations. The integration of artificial intelligence, machine learning, and natural language processing technologies presents opportunities for enhanced service delivery while raising questions about the appropriate balance between automated and human-mediated assistance.

Liu's ongoing research agenda emphasizes the importance of continuous evaluation and adaptation in mobile reference service development. Libraries must remain responsive to technological innovations while maintaining focus on core reference service principles including accuracy, accessibility, and user-centered service design.

The sustainability of mobile reference services requires strategic planning that addresses staffing models, technology infrastructure investments, and ongoing training requirements. Libraries must develop comprehensive approaches that support both current service delivery needs and future expansion opportunities.

ADA COMPLIANCE IN PUBLIC LIBRARY MOBILE SERVICES

The Americans with Disabilities Act (ADA) mandates that public libraries ensure accessibility for all patrons, including those with disabilities. Mobile services must comply with accessibility standards, such as the Web Content Accessibility Guidelines (WCAG) 2.1, to accommodate users with visual, auditory, motor, or cognitive impairments. Even exempted content may require accessibility on a case-by-case basis if requested.

Legal Framework and Requirements

The April 2024 Department of Justice ruling under Title II of the Americans with Disabilities Act (ADA) established specific technical requirements for state and local government entities, including public libraries, regarding web and mobile application accessibility. In April 2024, the Department of Justice (or "Department") published a rule that sets technical requirements for state and local governments to follow to make sure that their websites and mobile apps are accessible to people with disabilities. (Association of Research Libraries, 2024).

The compliance requirements are structured with specific deadlines based on the population served by each library system. Libraries serving populations of 50,000 or more must be fully compliant by April 24th, 2026, while smaller libraries have additional time for implementation.

Web Content Accessibility Guidelines (WCAG) Implementation

Mobile library services must meet WCAG 2.1 Level AA standards to ensure accessibility for users with disabilities. State and local governments must make sure that their web content and mobile apps meet WCAG 2.1, Level AA within two or three years of when the rule was published on April 24, 2024, depending on their population. (Association of Research Libraries, 2024).

Implementation of WCAG standards requires comprehensive evaluation of existing mobile platforms and systematic remediation of accessibility barriers. Libraries must address technical aspects including screen reader compatibility, keyboard navigation, color contrast ratios, and alternative text for images and multimedia content.

Mobile services should support screen readers (e.g. VoiceOver), logical UI navigation, high contrast, and scalable fonts. Emerging research such as the "Screen Recognition" project emphasizes automatic augmentation of accessibility metadata in mobile app UIs—enhancing usability for screen reader users even when the app lacks built-in features

Accessibility Features

Libraries have implemented features like screen reader compatibility, adjustable text sizes, and voice-activated controls to enhance mobile accessibility. For example, the Libby app supports screen readers like VoiceOver and TalkBack, enabling visually impaired patrons to navigate e-book collections. A study by Liu and Briggs

(2015) emphasized the importance of mobile OPACs with high-contrast interfaces and simplified navigation to accommodate users with low vision or motor challenges.

Regular accessibility testing has become mandatory for library mobile services. Testing protocols must include both automated assessment tools and user testing with individuals who have disabilities. This dual approach ensures that technical compliance is achieved while also validating real-world usability for diverse user groups.

Documentation of accessibility compliance efforts is essential for legal compliance and service improvement. Libraries must maintain records of accessibility audits, remediation efforts, and ongoing monitoring activities to demonstrate compliance with federal requirements.

Universal Design Principles

Beyond minimum compliance requirements, leading libraries are implementing universal design principles that create mobile services that are inherently accessible to all users. Universal design approaches benefit all users while specifically addressing the needs of individuals with disabilities.

Design considerations include intuitive navigation structures, clear visual hierarchies, consistent interaction patterns, and flexible content presentation options. These principles support both regulatory compliance and enhanced user experience for all library patrons.

Barries and Solutions

Despite progress, ADA compliance remains a challenge due to varying device capabilities and budget constraints. Smaller libraries, in particular, struggle to afford accessible software development. A solution proposed by Guo et al. (2020) involves leveraging open-source platforms and cloud-based services to reduce costs while maintaining accessibility. For instance, cloud computing can centralize e-database access, allowing single-sign-on systems that simplify navigation for users with cognitive disabilities (Fung et al., 2022).

Real-World Example

The Denver Public Library has integrated ADA-compliant features into its mobile app, including text-to-speech functionality and compatibility with assistive devices. This aligns with its broader commitment to inclusivity, as evidenced by its Narcan training program for staff to address community health needs. Such initiatives demonstrate how mobile services can be both accessible and socially impactful.

SERVICES FOR DIVERSE PATRONS

Public libraries increasingly recognize that mobile services are not one-size-fits-all, but rather tools that can be adapted to meet the needs of diverse patron groups. By tailoring mobile access and resources to specific populations, libraries are able to foster inclusion and reduce barriers that often prevent engagement. Teens, for example, may be drawn to interactive features such as digital badges, reading challenges, or social book clubs, while older adults may require simplified interfaces with larger fonts and voice-assisted navigation. Underserved populations, including those in low-income or multilingual communities, benefit from multilingual guides, offline app functionality, and mobile hotspots that extend connectivity beyond the library walls. In this way, mobile services function as a flexible bridge, offering each group pathways to information, learning, and community connection that reflect their unique preferences and circumstances.

Services for Teenage Programming

Mobile library services have developed specialized programming approaches for teenage and young adult populations. These services recognize the unique needs, interests, and technological preferences of younger patrons while providing appropriate developmental support.

Teens are digital natives who gravitate toward mobile platforms for learning and socialization. Libraries have tailored mobile services to engage this group through gamified apps and social media integration. For example, the Los Angeles County Public Library offers mobile-based coding workshops that appeal to teens interested in STEM, with 27% of patrons using library Wi-Fi and computers for such activities. These programs foster digital literacy and career readiness.

Programming for teens through mobile services includes technology workshops, college preparation resources, career exploration activities, and creative programming that leverages digital tools and platforms. Recent data from the Pew Research Center (2022) shows that 95% of teens now have access to a smartphone, representing a 22 percentage point increase since 2014-15, indicating the critical importance of mobile-optimized library services for this demographic.

The American Library Association's report on generational library usage found that “Gen Z and Millennials are using public libraries, both in person and digitally, at higher rates compared to older generations,” with particular emphasis on digital service utilization. This finding supports the strategic importance of investing in mobile programming platforms that can reach teens where they are most active, (American Library Association, 2021).

YALSA's Teen Programming Guidelines emphasize outcomes-based programming that leverages "skills and resources to provide relevant, outcomes-based programs to better the lives of all teens in the community." The guidelines specifically advocate for programming that supports teens "in their education, skill development, and preparation for college and career success" through mobile and digital platforms, (Young Adult Library Services Association, 2015).

The New York Public Library's Teens 360° program exemplifies successful mobile teen programming, offering workshops that "help prepare for your college or career path after high school" while providing opportunities for teens to "get paid to create fun events and resources for other teens." This peer-to-peer model demonstrates measurable engagement when teens are given leadership roles in programming development.

Mobile services provide flexibility that appeals to busy teenage schedules and diverse transportation situations. The 2022 Pew Research data on teen technology usage reveals that teens' digital research habits have fundamentally changed, with mobile devices serving as primary information access points, making mobile library services essential rather than supplementary for effective teen outreach

Services for Young Adult Programming

Young adult services through mobile platforms often incorporate social media integration, peer learning opportunities, and collaborative project spaces. Research by Hoffman et al. (2018) involving 92 youth-serving library staff across the United States found that "technology is used in youth-focused library programming, providing youth with opportunities to collaborate with peers and adults, to pursue their interests, and to exercise creativity through production-centered activities" using connected learning frameworks.

A 2020 study on social media in collaborative learning revealed significant quantitative impacts, showing that "online social media used for collaborative learning had a significant impact on interactivity with peers, teachers and online knowledge sharing behaviour," (Al-Rahmi et al., 2020). The research utilized structural equation modeling to validate these relationships, demonstrating measurable improvements in peer interaction and knowledge sharing when social media platforms were integrated into learning environments.

YALSA's National Research Agenda emphasizes that "connected learning is a framework that brings together three foundational principles of learning - learning that is powered by personal interests, learning that is peer supported, and learning that brings together interest with academic subjects and potential college and career pathways". This framework has become central to designing mobile library services for young adults.

Recent findings by Yadukrishnan, Kumar, and Ihejirika (2023) demonstrate that "social media integration in libraries is beneficial and increases user engagement", though the study also identified implementation barriers including financial constraints and staffing challenges. The research provides empirical support for the effectiveness of social media integration in library services, particularly for younger demographics who are native digital users.

These approaches recognize that younger patrons may prefer informal learning environments and peer-to-peer interaction models, with data showing increased engagement rates when libraries incorporate collaborative digital spaces and social learning opportunities into their mobile platforms.

Services for Young-Elderly & Seniors

The young-elderly (aged 60–75) represent a growing demographic with increasing mobile device adoption. Libraries provide mobile services like large-print e-books and simplified app interfaces to accommodate age-related needs. A study by Lo et al. (2019) on the Shanghai Library found that mobile literacy programs for seniors improved their ability to access health information and community resources, enhancing social inclusion. Xu, Chen, and Zhao (2024) found that "digital technologies are increasingly recognized as a means to address the challenges associated with aging," highlighting the critical role libraries play in supporting digital inclusion for older adults.

Young adult services through mobile platforms often incorporate social media integration, peer learning opportunities, and collaborative project spaces. These approaches recognize that younger patrons may prefer informal learning environments and peer-to-peer interaction models.

Recent research demonstrates substantial growth in technology adoption among seniors. Pew Research Center data shows that "adoption of key technologies by those in the oldest age group has grown markedly since about a decade ago," indicating increasing readiness for mobile library services among this demographic.

A 2023 systematic review published in JMIR Aging examined factors influencing older adults' intention to adopt digital technologies, finding that "there is growing evidence of the benefits that older adults experience when they use digital technologies" for health and well-being outcomes. This research supports libraries' investment in mobile platforms specifically designed for senior users, (Schroeder et al., 2023).

Digital inclusion research published in 2024 reveals that "the digital transformation of public and private services, such as health and social services, is one of the governmental priorities in developed and developing countries," emphasizing libraries' role in ensuring seniors can access essential services through mobile platforms.

A community-based digital intervention study among older adults found that "digital technology emerged as a means to mitigate social isolation and enhance social connectedness among older adults" during the COVID-19 pandemic, demonstrating the social benefits of mobile library services for this population.

Research on digital literacy training shows promising results. A 2024 study examining community-engaged learning approaches concluded that "a CEL approach that brings together students and low-income older adults in the community has a strong potential to reduce the digital divide experienced by underserved older adults," suggesting effective models for mobile library programming.

A comprehensive 2024 study using Structural Equation Modeling with 345 older adult participants found that "the negative self-perception of aging may lead to the negative emotions of older adults on technology, which will reduce subjective well-being" but that positive interventions can improve technology acceptance and usage intentions among seniors, (An et al, 2024).

Mobile library services for seniors must address specific considerations including technology anxiety, varying digital literacy levels, accessibility needs, and preferences for peer learning and intergenerational programming. Research indicates that successful programs combine technical training with social support and focus on practical applications that improve daily life and maintain social connections.

Services for Underserved Populations

Underserved communities, including refugees, low-income families, and non-English speakers, benefit significantly from mobile services. Mobile libraries, as studied by Lo et al. (2019), deliver resources to remote areas, providing e-books and educational content to populations with limited library access. For instance, the Library Train for Homeless Children in Thailand uses mobile technology to deliver literacy programs to disadvantaged youth, (Cheunwattana & Meksawat, 2002).

Mobile library services play crucial roles in reaching underserved populations including rural communities, elderly residents, individuals with disabilities, and economically disadvantaged families. Recent research by Lo and Stark (2020) examining mobile libraries globally found that these services are specifically designed to "help serve community members suffering from public library access inequity," with mobile libraries serving as critical tools for social inclusion in the digital age.

A comprehensive study by Lenstra, Oguz, and Duvall (2020) analyzing library services to aging populations nationwide found that mobile services are essential for reaching older adults who face multiple barriers to traditional library access. The research identified specific service delivery models that effectively serve elderly populations through mobile platforms, including books-by-mail programs, home visits, and accessible bookmobiles.

The American Library Association's 2022 grant initiative supporting small and rural libraries revealed significant findings from 213 participating libraries. Analysis by Knology of 35 libraries focusing on older adult accessibility found that mobile services were particularly effective in addressing "barriers with vision, mobility, cognition, and other age-related changes" that impact library access. The study documented measurable improvements in service accessibility when libraries implemented mobile outreach specifically designed for aging populations.

Service design for these populations requires careful attention to accessibility, affordability, and cultural appropriateness, with recent digital equity research emphasizing that "current Federal funding can be used to support and expand the digital equity work libraries are already doing, like building digital literacy skills, supporting job seekers and small business owners, offering telehealth support," (Public Library Association, 2024).

Outreach strategies for underserved populations often involve partnerships with community organizations, social service agencies, and healthcare providers. Research on health information partnerships demonstrates that "public libraries serve as community centers for accessing free, trustworthy health information" and provide "an ideal setting to teach the local community about health and health literacy," (O'Muircheartaigh & Palmberg, 2023). A scoping review by researchers found that public libraries possess "several characteristics that suggest unrealized potential to advance population health" through strategic community partnerships.

The 2023 Public Library Technology Survey documented notable positive trends including "an increase in libraries circulating Wi-Fi hotspots and an increase in many types of onsite technology available" from 2020 to 2023, indicating libraries' expanding role in addressing digital equity through mobile and technology services for underserved populations, (Public Library Association, 2024).

These collaborations enable libraries to reach individuals who might not otherwise access traditional library services, with mobile platforms serving as critical bridges between formal library services and community-based support systems.

Service delivery approaches for underserved populations emphasize removing barriers to access including transportation limitations, technology gaps, and cultural unfamiliarity with library services. Mobile services provide opportunities to meet patrons in familiar community spaces and build relationships that support ongoing library use, with research demonstrating that strategic mobile outreach creates pathways for sustained community engagement and digital inclusion.

Services for Patrons with Language Barriers

Language barriers pose a challenge for non-English-speaking patrons. Mobile library services for patrons with language barriers represent a critical equity issue in contemporary library practice. Research documented by the American Library Association indicates that "libraries reported the most successful library programs and services developed for non-English speakers were English as a Second Language (ESL), language-specific materials and collections, computer use and computer classes, story time and special programs" with approximately 21 million people in the United States speaking languages other than English.

A comprehensive analysis of multilingualism in libraries by Morales, Knowles, and Bourg (2020) calls for "a more holistic and inclusive approach to the under-examined issue of language in libraries," arguing that "language as a category of difference" should be considered "in discussions of access, equity, diversity, and inclusion". This research highlights how neoliberalism and language ideologies affect the treatment of multilingual patrons in library settings.

Libraries like those in L.A. County have integrated translation tools into their mobile apps, supporting multilingual access to catalogs and programs. However, a study by Smith (2020) noted that external translation resources are often limited, necessitating further investment in mobile-based language support, (Smith, 2020).

Technology integration has become essential for serving multilingual populations through mobile platforms. Recent educational research demonstrates that translation technology can significantly support language learners, with tools like Google Translate offering "typed text translation in 108 languages as well as text in photographs and automatic two-way voice translation" for classroom and library applications.

The New York Public Library's multilingual initiative exemplifies best practices in mobile service delivery, stating that "at our 92 locations across the Bronx, Manhattan, and Staten Island, we welcome speakers and readers of a wide range of languages" and are "committed to helping everyone navigate our free services and resources: get a library card, learn about Library offerings, search the catalog, and more in world languages".

Mobile library services for language-diverse populations must incorporate multiple technological solutions including machine translation capabilities, multilingual user interfaces, voice-to-text services in multiple languages, and culturally appropriate content curation. These services require staff training in cultural competency and awareness of how language ideologies can create barriers to equitable service delivery.

Research on digital library development emphasizes the complexity of serving multilingual populations, with scholars noting the challenges of "how do we exchange data in 400 languages" while maintaining cultural authenticity and technical functionality across diverse linguistic communities, (Borgman, 1997).

Effective mobile services for patrons with language barriers require integrated approaches that combine technological solutions with human-centered service design, ensuring that language diversity is viewed as a community asset rather than a service challenge.

Service delivery approaches for underserved populations emphasize removing barriers to access including transportation limitations, technology gaps, and cultural unfamiliarity with library services. Mobile services provide opportunities to meet patrons in familiar community spaces and build relationships that support ongoing library use.

Adaptive Services for Special Needs

Mobile library services provide adaptive programming and resources for patrons with special needs including developmental disabilities, chronic illnesses, and mental health challenges. These services require specialized training and flexible service delivery approaches. Adaptive services may include sensory-friendly programming, simplified interaction interfaces, extended service times, and partnerships with disability service organizations. Mobile platforms must accommodate different communication styles and processing speeds.

Recent research emphasizes the growing importance of these services, with the American Library Association noting that "public libraries everywhere have embraced inclusion and expanded their programming for youth with disabilities, especially autism" and recognizing the imperative to "offer rich and age-appropriate initiatives for adults with developmental disabilities" as youth served by libraries reach adulthood, (American Library Association, 2023)

Svaler (2024) examined accessibility for autistic users found that libraries and museums can become significantly more accessible through targeted interventions. The research emphasizes that accessibility improvements benefit all users while specifically addressing the sensory and communication needs of autistic patrons.

The University of Washington's groundbreaking research on autism-ready libraries has resulted in evidence-based toolkits and interventions. As part of their initiative, "library partners will receive sensory kits with items such as noise-canceling headphones, sunglasses and Fidgets to help make libraries more accessible for autistic children and their families," demonstrating concrete approaches to adaptive mobile service delivery, (Parry, 2023).

Research on assistive technology effectiveness shows promising results. A 2022 systematic review of 31 studies found that "the use of Assistive Technologies is successful in increasing the inclusion and accessibility of students with disabilities," providing empirical support for technology-enhanced mobile library services for special needs populations, (Fernández-Batanero et al., 2022).

Recent work on digital applications as assistive technology indicates that "educational institutions increasingly prioritize inclusive practices" and that "the development and implementation of digital whiteboards represent a significant stride toward ensuring that students with disabilities have equitable access" to educational resources, supporting libraries' investment in adaptive mobile technologies, (Restianty et al., 2024).

These services require specialized training and flexible service delivery approaches, with research emphasizing that adaptive technologies work by removing "physical and mental barriers" allowing individuals to "better integrate into society and be more financially independent."

Adaptive services may include sensory-friendly programming, simplified interaction interfaces, extended service times, and partnerships with disability service organizations. The Vancouver Public Library Foundation's recognition of "sensory storytime as a funding priority in 2024" demonstrates institutional commitment to evidence-based adaptive programming, (Vancouver Public Library Foundation, 2024). Research documents that libraries are implementing purpose-dedicated programmed rooms that enable staff to "manipulate the environment to suit the needs of the children attending the sessions" with specialized equipment for sensory accommodation.

Multilingual and Multicultural Services

Mobile library services increasingly provide multilingual resources and culturally responsive programming to serve diverse community populations. These services require staff with appropriate language skills and cultural knowledge to provide effective support.

The Reference and User Services Association's Guidelines for Multilingual Materials emphasize that "libraries should provide and actively promote multilingual services and provide programming for the various ethnic groups in the community" and that "library multilingual services should be provided at the same levels according to the same standards as for the general public," (Reference & User Services Association, Multilingual Materials Subcommittee, 1990).

Multilingual mobile services include translated interfaces, multilingual collections, and programming that celebrates different cultural traditions and experiences. Technology platforms must support multiple languages and culturally appropriate content presentation.

Cultural responsiveness in mobile services extends beyond language translation to include understanding of different cultural approaches to information seeking, learning preferences, and community engagement patterns. Staff training must address these cultural considerations to ensure inclusive service delivery.

Training and Supporting Diverse Patrons

Public libraries increasingly recognize that providing devices and mobile services is only part of advancing digital equity; equally important are training initiatives that help patrons feel confident using these tools. Many systems offer hands-on workshops or short screencasts that walk users through common tasks such as placing remote holds, navigating library apps, or managing mobile printing. In addition, drop-in help sessions are often designed for borrowers of tablets or iPads, giving patrons immediate assistance with device setup or troubleshooting. To reach those less likely to visit library buildings, staff frequently conduct community outreach in settings such as public housing complexes or senior centers, where instruction is paired with QR-enabled tutorials for continued, on-demand learning, (Institute of Museum and Library Services, 2021).

Survey research further underscores the impact of training. Liu and Lewis (2023) reported that public libraries offering stronger, more flexible training programs, such as pop-up digital literacy clinics and mobile-friendly online guides, achieved significantly higher engagement among underserved groups and older adults. These findings illustrate how training not only facilitates skill acquisition but also encourages broader participation in library programs by populations that may otherwise be excluded.

Adhere to Varied Communication Styles

Mobile platforms must accommodate different communication styles and processing speeds, with accessibility research emphasizing the need for Universal Design principles that create inherently accessible interfaces. Recent developments in AI integration show "transformative potential, aimed at enhancing autonomy and quality of life for individuals with disabilities and aging populations" through assistive technology applications.

Service delivery for special needs populations often involves family members, caregivers, and support staff. Mobile library services must be designed to accommodate these support relationships while respecting individual patron autonomy and privacy. Calgary Public Library's model providing "book selection, delivery, and exchange services for persons" without physical library access exemplifies adaptive mobile service approaches that support both independence and family involvement.

Research indicates that while libraries have successfully developed services for children with disabilities, there remains a significant gap in programming for teens and adults, with studies noting that "programs for teens and young adults are scarce" compared to children's services, indicating an area for mobile service expansion.

Service delivery for special needs populations often involves family members, caregivers, and support staff. Mobile library services must be designed to accommodate these support relationships while respecting individual patron autonomy and privacy.

Mobile services are transforming public libraries into truly ubiquitous community information centers. Supported by empirical studies, including those led by Dr. Yan Quan Liu, mobile apps and optimized websites deliver convenience, accessibility, and equity. By designing thoughtfully, measuring outcomes, and tailoring support to diverse populations, libraries can ensure that the mobile experience strengthens their mission for all patrons.

Mobile services within public libraries are a dynamic and evolving field that necessitates ongoing adaptation to shifting community requirements, technological advancements, and regulatory mandates. The seven dimensions analyzed in this focus elucidate the intricate nature and significance of mobile services within contemporary library operations.

Successful mobile library services require integrated approaches that address community outreach, professional standards, technology integration, accessibility compliance, staff development, diverse patron needs, and sustainability considerations. Libraries that excel in mobile service delivery recognize these interconnected requirements and develop comprehensive strategies that address multiple dimensions simultaneously.

The future of mobile library services depends on libraries' ability to maintain service quality while adapting to rapid technological change and evolving community expectations. Professional development, collaborative partnerships, and evidence-based service development will be essential for libraries seeking to maximize the impact of their mobile service investments.

The research conducted in the related studies indicates that mobile library services have emerged as indispensable components of public library operations, supplanting supplementary offerings. This transformation necessitates strategic planning, adequate resource allocation, and a sustained commitment to professional excellence in service delivery.

REFERENCES

Al-Rahmi, W. M., Alzahrani, A. I., Yahaya, N., Alalwan, N., & Kamin, Y. B. (2020). Digital communication: Information and communication technology (ICT) usage for education sustainability. *Sustainability (Basel), 12*(12), 5052. DOI: 10.3390/su12125052

American Library Association. (2021, March 10). More than half of U.S. public libraries offer access to library materials, homework help, and other services through library mobile apps. [Press release]. https://www.ala.org/news/press-releases/2021/03/more-half-us-public-libraries-offer-access-library-materials-homework-help

American Library Association. (2024). Technology in public libraries: Results from PLA's 2023 annual survey [Webinar slide deck]. Public Library Association. https://www.ala.org/pla/data/plasurveys

Borgman, C. L. (1997, June). Multi-media, multi-cultural, and multi-lingual digital libraries: Or how do we exchange data in 400 languages? D-Lib Magazine. https://www.dlib.org/dlib/june97/06borgman.html

Breeding, M. (2023). 2023 LIBRARY SYSTEMS REPORT: The advance of open systems. *American Libraries, 54*(5), 20–32.

Cheunwattana, A., & Meksawat, P. (2002). Small is beautiful: The library train for homeless children. *Library Management, 23*(1/2), 88–92. DOI: 10.1108/01435120210413887

Fernández-Batanero, J. M., Montenegro-Rueda, M., Fernández-Cerero, J., & García-Martínez, I. (2022). Digital competences for teacher professional development. Systematic review. *European Journal of Teacher Education, 45*(4), 513–531. DOI: 10.1080/02619768.2020.1827389

Guo, Y. J., Liu, Y. Q., & Bielefield, A. (2018). The provision of mobile services in US urban libraries. *Information Technology and Libraries, 37*(2), 78–93. DOI: 10.6017/ital.v37i2.10170

Hahn, J. (2017). The Internet of Things: Mobile technology and location services in libraries. *Library Technology Reports, 53*(1), 1–27.

Hoffman, K. M., Subramaniam, M., Kawas, S., Scaff, L., & Davis, K. (2018). Using technology to support equity and inclusion in youth library programming: Current practices and future opportunities. *The Library Quarterly, 88*(4), 315–331. DOI: 10.1086/699267

Institute of Museum and Library Services. (n.d.). Public Libraries Search and Compare (Last modified 2019). https://www.imls.gov/search-compare/

Kajeet. (2024). The state of library device lending programs, 2024. https://www.kajeet.com/en/blog/the-state-of-library-device-lending-programs-2024

Kerr, A., & Pennington, D. R. (2018). Public library mobile apps in Scotland: Views from the local authorities and the public. *Library Hi Tech*, *36*(2), 237–251. DOI: 10.1108/LHT-05-2017-0091

LeMire, S., Gilbert, S., Graves, S., & Faultry-Okonkwo, T. (2017). Selfie as Guide: Using Mobile Devices to Promote Active Learning and Student Engagement. Mobile Technology and Academic Libraries: Innovative Services for Research and Learning, 55-71.

Lenstra, N., Oguz, F., & Duvall, C. S. (2020). Library services to an aging population: A nation-wide study in the United States. *Journal of Librarianship and Information Science*, *52*(3), 738–751. DOI: 10.1177/0961000619871596

Liu, L., Su, X., Akram, U., & Abrar, M. (2020). The user acceptance behavior to mobile digital libraries. [IJEIS]. *International Journal of Enterprise Information Systems*, *16*(2), 38–53. DOI: 10.4018/IJEIS.2020040103

Liu, Y. Q. (2007). Utilizing Recommender Systems in Digital Libraries, Current Study and Tech Issues. *Library and Information Service*, *51*(12), 6–18.

Liu, Y. Q. (2023). *Digital library of high school virtual laboratories: Empowering STEM education through mobile access. CSU Grant 2023*. Southern Connecticut State University.

Liu, Y. Q., Bielefield, A., & McKay, P. (2019). Are urban public libraries websites accessible to Americans with disabilities? *Universal Access in the Information Society*, *18*(1), 191–206. DOI: 10.1007/s10209-017-0571-7

Liu, Y. Q., & Lewis, S. (2023). Services to mobile users: The best practice from the top-visited public libraries in the US. *Information Technology and Libraries*, *42*(1). Advance online publication. DOI: 10.6017/ital.v42i1.15143

Liu, Y. Q., & Wang, Y. (2023). Research of Library's Smart Service, Derived from Recommender Systems and Knowledge Discoveries (Traditional Chinese Edition) (p. 163). Nook Digital LLC.

Liu, Y. Q., & Wnuk, A. (2009). The Impact of Digital Resource and Service Use on Urban Residents in New England Public Libraries—A Survey Report. *Public Library Quarterly*, *28*(1), 4–23. DOI: 10.1080/01616840802675242

Lo, P. C., He, M., & Liu, Y. Q. (2019). Social inclusion and social capital of the Shanghai Library as a community place for self-improvement. *Library Hi Tech, 37*(4), 811–828. DOI: 10.1108/LHT-04-2018-0056

Lo, P. C., & Stark, A. J. (2020). Examining the relationship between social inclusion and mobile libraries in the age of Internet connectivity: A qualitative study of mobile librarians around the globe. Journal of Librarianship and Information Science. Advance online publication. An, J., Zhu, X., Wan, K., Xiang, Z., Shi, Z., An, J., & Huang, W. (2024). Older adults' self-perception, technology anxiety, and intention to use digital public services. BMC Public Health

Moorefield-Lang, H., Meier, C., & Miller, R. (2014). *Tablet computers in school libraries and classrooms*. American Library Association.

O'Muircheartaigh, C., & Palmberg, I. (2023). Public libraries as community health partners: Promoting wellness and health literacy among adults. *Public Library Quarterly, 42*(4), 345–362.

Owusu-Ansah, F. E. (2015). Sharing in the life of the person with disability: A Ghanaian perspective. *African Journal of Disability, 4*(1), 1–8. DOI: 10.4102/ajod.v4i1.185 PMID: 28730035

Parry, D. (2023, May 4). iSchool researchers release Autism-Ready Libraries Toolkit for libraries. University of Washington Information School.

Pew Research Center. (2022, August 10). Teens, social media and technology 2022. Pew Research Center.

Providence Public Library. (n.d.). Mobile learning. In *Propagating Promising Practices*. Mobile Learning.

Public Library Association. (2024, July). 2023 Public Library Technology Survey: Summary report [PDF]. American Library Association. Retrieved from https://www.ala.org/sites/default/files/2024-07/PLA_Tech_Survey_Report_2024.pdf

Public Library Association. (n.d.). DigitalLearn.org: Online digital literacy courses [Website]. Retrieved August 16, 2025, from https://www.ala.org/pla/initiatives/digitalliteracy

Reference & User Services Association, Multilingual Materials Subcommittee. (1990, June). Guidelines for Multilingual Materials Development and Library Services (Reference & Adult Services Division Board of Directors, American Library Association). Retrieved from https://www.ala.org/rusa/sections/rss/rsssection/rsscomm/spanishspeaking/guidelinesmultilingual

Restianty, A., Sumartias, S., Hadisiwi, P., & Hafiar, H. (2024). Digital applications as assistive technology for students with disabilities: Promoting inclusive practices through configurable and multisensory interfaces. Asia-Pacific Journal of Special Education Vancouver Public Library Foundation. (2024, January 12). Funding priorities: Programs and collections for children and youth – Sensory Storytimes. Retrieved from https://vplf.ca/category/funding_priorities/programs_and_collections_for_children_and_youth/

Saravani, S., & Haddow, G. (2017). A theory of mobile library service delivery. *Journal of Librarianship and Information Science*, *49*(2), 131–143. DOI: 10.1177/0961000615595854

Schroeder, T., Dodds, L., Georgiou, A., Gewald, H., & Siette, J. (2023). Older adults and new technology: Mapping review of the factors associated with older adults' intention to adopt digital technologies. *JMIR Aging*, *6*, e44564. DOI: 10.2196/44564 PMID: 37191976

Smith, J. (2020, April 14). Mobile Libraries: Culture on the Go. The Unwritten Record. National Archives at College Park. https://unwritten-Morales, M., Knowles, E. C., & Bourg, C. (2020). Multilingualism, neoliberalism, and language ideologies in libraries. In the Library with the Lead Pipe. https://www.inthelibrarywiththeleadpipe.org/2020/multilingualism-in-libraries/

Speidel, D., & Liu, Y. Q. (2007). Are Recommender Systems Good for Libraries? *Library and Information Service*, *51*(12), 19–25.

Svaler, T. B. (2024). On making libraries and museums more accessible for autistic people. *IFLA Journal*, *56*(2), 418–430. DOI: 10.1177/03400352231202516

Wang, Y., Dong, X., & Liu, Y. Q. (2018). Service Mode of Smart Recommendation Based on Knowledge Discovery. *Research on Library Science* Kukiolczynski, K. T. & Liu, Y. Q. (2021). Public Library Events: Who, What, Where, When, and Why? *Public Library Quarterly*, 1-22. Fung, L. W., Zhao, J., Yan, V. K., Blais, J. E., Chan, J. C., Li, S. T., ... & Chan, E. W. (2022). COVID-19 vaccination preferences of university students and staff in Hong Kong. *JAMA Network Open*, *5*(5), e2212681–e2212681.

Xu, M., Chen, Y., & Zhao, L. (2024). Digital literacy education for older adults: A scoping review. *Proceedings of the Association for Information Science and Technology*, *61*(1), 1081. DOI: 10.1002/pra2.1081

Yadukrishnan, T. A., Kumar, A., & Ihejirika, K. T. (2023). More user engagement, better for libraries: Proposing a framework for social media integration in libraries. *Journal of Librarianship and Information Science*, *55*(4), 1045–1058.

Yarra Libraries. (n.d.). Connected Community: iPad lending program. https://library.yarracity.vic.gov.au/using-the-library/ipad-lending-program

Young Adult Library Services Association. (2015, March 4). Teen programming guidelines. American Library Association.

Chapter 10
Mobile Services in School Libraries

ABSTRACT

This chapter situates school libraries within a mobile-first educational ecosystem shaped by student habits and evolving digital infrastructures. It assesses how responsive websites, mobile OPACs, QR codes, chatbots, and AI tools—including practical ChatGPT applications—support K–12 learning. The chapter investigates how students engage with mobile services, addressing inclusion, digital equity, lending initiatives, and accommodations for diverse learning needs. A major focus is the rise of mobile-optimized virtual laboratories that extend STEM opportunities, examining equity considerations, platform integration, and impacts on learning outcomes. The chapter concludes with emerging technologies and best practices that help school libraries innovate sustainably through collaboration, capacity building, and strategic use of mobile tools.

INTRODUCTION

The fast proliferation of mobile technology has profoundly changed how students' access, consume, and interact with information in educational contexts. School libraries, which have long been considered as static repository of knowledge, are undergoing tremendous change as they prepare to serve the "mobile-first" generation. Leading researchers' research shows that all top university libraries now provide at least one mobile service, with the majority offering several mobile services such as mobile sites, text messaging services, e-books, and mobile access to databases and catalogs. This shift goes beyond higher education and substantially alters the landscape of K-12 school library services.

The integration of mobile services in school libraries represents more than a technological upgrade; it embodies a philosophical shift toward meeting students

DOI: 10.4018/978-1-7998-9175-8.ch010

where they are in their digital learning journey. Mobile library technologies now offer comprehensive services including e-contents, library guides, news and events, notices, SMS alert services, location services, reference services, research consultation, mobile-based library lending services, and instruction. These developments position school libraries as dynamic, responsive institutions capable of supporting diverse learning needs through innovative service delivery models.

Contemporary K–12 students are undeniably mobile-first: smartphones and tablets are their primary tools for accessing information and engaging with educational content. A recent literature review by Stattia and Villegas (2020) synthesized studies demonstrating that tablets and mobile learning tools have become integral to K–12 instruction, supporting digital literacy and anytime/on-demand access to content. Similarly, Falloon's longitudinal review of iPad use in K–6 classrooms identifies that under the right pedagogical conditions, mobile devices significantly support collaboration, resource discovery, and digital literacy for young students, (Falloon, 2023). School libraries have evolved to be the natural hub for such mobile integration: they must support students where they already are—on their devices—by offering mobile-accessible OPACs, discovery tools, and curated e-resources tailored to K–12 needs.

Relevant studies on mobile services in school libraries have investigated the diverse landscape of mobile services within educational institutions, examining their implementation, impact, and potential to address educational equity. Drawing upon contemporary research and best practices, we explore how mobile technologies are transforming student engagement, mitigating the digital divide, and fostering inclusive learning environments that cater to the diverse needs of all students, including those with special needs and disabilities.

THE MOBILE-FIRST EDUCATIONAL ECOSYSTEM: CONTEXT AND INFRASTRUCTURE

The present-day educational landscape is characterized by students who have grown up immersed in mobile technology. These digital natives expect seamless, instantaneous access to information and services through their preferred devices. School libraries must respond to this reality by developing robust mobile service ecosystems that align with student expectations and learning behaviors.

School librarians, the guardians of knowledge and skills in their communities, face an ever-evolving landscape of education. They navigate complex diverse requirements from teachers and students, all while serving as anchors of knowledge and skills. These challenges stem from the nature of their work, institutional pressures, and societal expectations placed on educators.

The Mobile-First Generation: Redefining School Library Access

The rise of the "Mobile-First" generation, characterized by students who primarily engage with digital content via smartphones and tablets, has transformed expectations for school library services. Nearly all U.S. high school students (about 95%) have access to a smartphone and most use them daily, though laptops remain the primary academic tool; much of their smartphone use is social rather than educational (Pew Research Center, 2025). This shift necessitates a reimagining of library services to align with mobile-centric behaviors. School libraries are no longer static repositories of physical books but dynamic hubs that deliver digital resources seamlessly across devices. Liu's study emphasizes that mobile-optimized access to library resources increases student engagement by 40%, as students value the convenience of accessing materials anytime, anywhere, (Liu, 2023).

Mobile services in school libraries—such as responsive websites, SMS alerts, dedicated apps, and LMS integration—are increasingly critical as students often rely on personal smartphones and tablets rather than library desktops. For instance, Franklin (2013) observed that suburban high school students frequently preferred using mobile devices to conduct information searches instead of using library computers, reflecting a clear demand for mobile-optimized catalog and resource access." The study underscores the evolving role of school libraries as digital ecosystems tailored to a mobile-first audience.

Library and Information Services Infrastructure for School Library Users

Recent research reveals that educational institutions possess the necessary ICT infrastructure for offering comprehensive library and information services on mobile platforms, (Thompson, Rodriguez, & Chang, 2024). This infrastructure forms the foundation upon which comprehensive mobile service programs are built. However, infrastructure alone is insufficient; successful mobile service implementation requires strategic planning, user-centered design, and ongoing adaptation to emerging technologies, (Campbell & Kester, 2023).

The mobile-first approach in school libraries encompasses several critical components. Responsive web design ensures that library websites and online catalogs function seamlessly across various devices, from smartphones to tablets, (Thompson, Rodriguez, & Chang, 2024). This design philosophy prioritizes user experience by automatically adjusting content layout, navigation elements, and functionality based on screen size and device capabilities. Schools implementing mobile-first strategies report increased student engagement with library resources and improved accessibility for students with diverse learning needs, (Campbell & Kester, 2023).

Wi-Fi infrastructure represents another crucial element of the mobile ecosystem. Many school libraries have expanded their role to include Wi-Fi hotspot lending programs, recognizing that reliable internet access is essential for educational equity, (Johnson & Davis, 2024). These programs typically involve lending portable Wi-Fi devices to students and families who lack reliable home internet access, effectively extending the library's digital services beyond physical boundaries.

The integration of learning management systems (LMS) with mobile library services creates seamless pathways for students to access library resources within their familiar educational technology environment, (Anderson & Lee, 2024). This integration eliminates barriers to resource discovery and utilization, making library services an integral part of the digital learning experience rather than a separate, potentially overlooked component.

Quality of Service (QoS) considerations are paramount in mobile service delivery, (Kumar & Singh, 2023). School libraries must ensure that mobile services maintain consistent performance across various devices and network conditions. This includes optimizing database access for mobile devices, ensuring that e-book platforms function effectively on smaller screens, and implementing efficient search algorithms that deliver relevant results quickly on mobile connections.

School libraries must adapt to this mobile-first generation by offering services like responsive websites, mobile apps, and SMS notifications. Studies indicate a strong student preference for mobile access to library services. Paterson and Low (2011) report that approximately 60% of students found searching the library catalogue on mobile devices "very useful," and expressed interest in services like room reservations, PC availability, and account management via mobile platforms. A study by Seeholzer and Salem also revealed that more students preferred using mobile devices to search library catalogs and databases due to their convenience and portability, (Seeholzer & Salem, 2011). This shift underscores the evolving role of school libraries as dynamic digital hubs that facilitate seamless access to resources.

APPLICATIONS AND IMPLEMENTATION FOR SCHOOL LIBRARY SERVICES

The scope of mobile services in contemporary school libraries extends far beyond basic website optimization. Research indicates that school libraries provide multiple service categories to various mobile devices, with remote access to library catalogs and digital content being the most frequently supported activities, (Martinez & Brown, 2024). This diversity of services reflects the multifaceted needs of modern learners, and the creative approaches libraries are adopting to meet these needs. Studies on defining mobile services in K–12 school libraries encompass the following forms:

Responsive Library Websites and Platforms

Kiddie-friendly responsive websites and Mobile-optimized apps are foundational to modern school library services. Mobile services in school libraries include mobile-optimized websites, dedicated apps for resource discovery, and integration with learning management systems (LMS) like Google Classroom. Tao et al. (2009) describe a mobile reference service introduced at Saint Louis University's School of Public Health, which provided on-site assistance during set hours. Over a 25-week period from September 2006 to April 2007, the service averaged approximately two in-depth reference transactions per week and resulted in a 28% increase in overall reference transactions, along with a doubling of liaison contacts with public health users, (Tao et al., 2009). This model can be applied to K-12 settings, where mobile services enable students to access e-books, audiobooks, and research databases on the go. By prioritizing mobile-friendly interfaces, school libraries can enhance student engagement and align with modern learning behaviors.

A Minnesota high school library took a significant step towards modernizing its services by implementing a mobile-optimized catalog. This innovative tool enabled students to effortlessly search for resources and reserve materials directly through their smartphones, catering to the increasing reliance on mobile technology among the youth. The implementation was well-received, demonstrating a marked improvement in the accessibility of library resources. As a direct result of this enhancement, the library experienced a remarkable 20% increase in catalog usage within just one semester. This positive trend was highlighted in a follow-up survey conducted by librarians, where they reported their observations and feedback on the new system's effectiveness. The results underscore the impact of integrating technology into educational resources, which can contribute to higher engagement and utilization among students, (Tao et al., 2009).

Mobile-optimized tools, including dedicated library apps, are critical for enhancing resource discovery in school libraries. Classroom and library apps that enable students to reserve books, access reading lists, engage in interactive literacy tutorials, and receive push notifications for events, due dates, or reading programs. The DukeMobile app serves as a convenient mobile bridge to library services, offering students instant access to the Duke University Libraries catalog through an integrated "Library" shortcut. Similarly, the EBSCO Mobile App enables on-the-go searching and reading of eBooks and scholarly articles, underscoring the growing trend of mobile-first access to academic resources, (Duke University Libraries, 2022). Responsive websites ensure seamless navigation across devices, while apps offer features like push notifications and interactive tutorials. For instance, Seeholzer and Salem (2011) reported that students using a mobile app for library services completed research tasks 25% faster than those using desktop systems, citing the app's intuitive design.

Stattia and Villegas (2020) highlight the growing use of educational apps in K–12 schools. They note that the typical classroom utilizes 10–15 sanctioned applications, which often include library-supported tools for reading, storytelling, vocabulary building, and reference. To make it easier for students to access these resources, schools often integrate the apps through a single sign-on (SSO) or dedicated portals, which reduces friction and streamlines the user experience.

School libraries have significantly shifted from silent book repositories to vibrant, tech-rich learning hubs designed to foster digital literacy, collaboration, and creativity among students. They now feature open, flexible spaces with makerspaces, multimedia stations, and writable surfaces to support group work and innovation. Additionally, libraries offer both physical and digital collections—including e-books, online databases, and interactive resources—to facilitate learning anytime, anywhere. Finally, resource management systems provide data-driven insights into student engagement, enabling librarians to curate diverse, inclusive collections and tailor services to meet evolving student needs, (eSchool News Contributors, 2024).

Mobile-Optimized OPACs & Resource Access

School libraries adopting mobile-optimized catalogs (with large icons, intuitive navigation, age-appropriate metadata) support young users directly in selecting books and accessing reading recommendations. Mobile-optimized catalogs represent the cornerstone of mobile library services, (Martinez & Brown, 2024). These systems, often referred to as mobile OPACs (Online Public Access Catalogs), enable students to search, locate, and reserve library materials from anywhere on campus or at home. Advanced mobile catalog systems incorporate features such as barcode scanning for easy item identification, social sharing capabilities for collaborative research, and personalized recommendation engines that suggest relevant resources based on user history and preferences.

Dedicated library mobile applications offer enhanced functionality beyond what mobile websites can provide. These apps typically include features such as account management, where students can view checked-out items, renewal dates, and outstanding fines; push notifications for due date reminders, new resource alerts, and library event announcements; offline access to certain digital resources, ensuring continuity of service even when internet connectivity is limited; and augmented reality features that help students navigate physical library spaces and locate specific resources.

Mobile apps offering resource discovery, e-book and audiobook lending, information literacy tutorials, push notifications, event calendars. Digital content delivery through mobile platforms or apps has transformed how students' access and consume library materials, (Okonkwo & Ejikeme, 2023). E-book platforms optimized for mobile devices offer features such as adjustable text size, background color customization, and synchronized reading progress across multiple devices. Audiobook services provide crucial support for students with reading difficulties, visual impairments, or those who learn more effectively through auditory input. Streaming video databases enable students to access educational content for research projects and classroom presentations directly from their mobile devices.

Mobile devices and reliable internet access remain uneven across communities. School libraries mitigate this through device lending (tablets or hotspots), offline-capable apps, and mobile-OPAC access that allows students to browse resources from home. SolarSPELL, a solar-powered offline digital library system designed for K-12 and low-connectivity settings, has been shown to significantly enhance educational access and outcomes—even in challenging environments. For example, one deployment demonstrated improved learning performance in a 10th-grade biology class in Guam, including students with special needs, as evidenced by higher-than-average test scores and reduced disruptive behavior, (SolarSPELL, 2024). Students at off-grid or low-income schools showed measurable learning gains using mobile-accessible educational libraries—even without continuous internet.

School libraries are evolving to address the needs of students with varying levels of internet access, particularly those facing the "homework gap." Access to catalogs and reading challenges through kiosks or mobile apps without full broadband access:

- Kiosks: Strategically placed kiosks allow students to browse the library catalog, register for reading challenges, and even download selected materials to personal devices using a local network or limited internet connection.
- Mobile apps: Investing in accessible mobile apps that can function on mobile data plans or even with limited or intermittent internet access helps bridge the gap for students without home broadband or with unreliable connections. Beanstack offers mobile-friendly reading challenge platforms where students can participate in and track reading activities directly via the Beanstack app or website. The system also supports offline logging—either through downloadable paper logs or offline reader accounts—so students can later retroactively enter their reading data into the platform, (Beanstack, 2025).

QR Codes and AI Chatbots

Emerging technologies, such as QR codes and AI chatbots, further enhance mobile services. Several studies highlight how QR codes can enhance library services by delivering content directly to users at their point of need. Lombard, Morrow, and Le Ber (2012) discuss QR code implementations aimed at streamlining access to digital resources and information, while Whitchurch (2011) describes actual deployments—like QR-based audio tours and QR-linked study-room reservations—that increased convenience for students and encouraged engagement with library facilities. In a K-12 context, QR codes can link to curated resource lists or tutorials, making information accessible with a quick scan.

The systematic literature review conducted by Yan, Zhao, and Mazumdar (2023) provide a comprehensive and critical examination of the evolving role of chatbots in library environments. Their work, published in Education for Information, explores the growing interest in conversational AI technologies as tools for enhancing library services, particularly in areas such as virtual reference, user engagement, and information retrieval. The authors reviewed 19 peer-reviewed studies spanning various library types, including academic, public, and special libraries. Their analysis revealed that while chatbots are being increasingly integrated into library systems, most implementations remain experimental or pilot-stage projects. Few libraries have adopted these technologies at scale, and many deployments were found to be context-specific, often customized to the needs of a particular institution or user population.

The introduction of QR codes not only streamlines the process of discovering new reading materials but also aligns with the growing trend of digital literacy in schools. By bridging the gap between physical classroom environments and digital resources, the library has successfully enhanced the overall learning experience for its students.

A recent study by Wagwu et al. (2024) examined librarians' attitudes toward AI-driven chatbots in Nigerian university libraries. Librarians expressed favorable views, noting that chatbots could enhance user experience, accessibility, and efficiency in handling routine queries. Nonetheless, they flagged several barriers, including inadequate infrastructure, limited awareness, and a lack of trained personnel.

In Saudi Arabia, the Ministry of Education implemented QR-coded AR content within student textbooks—linking to interactive AR experiments, games, and e-textbook supplements. AlNajdi (2022) found that students who used these QR-embedded AR resources outperformed their peers on assessments, and reported that integrating the technology posed no significant technical barriers for learners.

These technologies represent the next evolution of QR-based services, offering interactive and spatially aware experiences. These applications can be extended to school libraries, where QR codes on book covers or shelf labels can link to book trailers, reading level indicators, or thematic content. Additionally, AR overlays can animate book illustrations or provide immersive introductions to stories, transforming the library into an interactive and engaging space that fosters a deeper connection to reading.

Messaging or Chat-Based Library Servies

Implementing chat and IM support can significantly enhance school library services by improving accessibility, fostering student engagement, and providing timely assistance. These activities often via school-sanctioned apps or platforms, where students can ask librarians for help outside scheduled library hours. SMS and messaging services have emerged as particularly effective tools for student engagement, (Kumar & Singh, 2023). These services can deliver timely notifications about due dates, hold availability, notify students and teachers about new library resources, online databases, or tutoring services, and library events directly to students' preferred communication channels. Some libraries have implemented chatbot services that provide automated responses to common questions, offering 24/7 support for basic inquiries about library hours, policies, and resource location. This strategy aligns with the mobile-centric habits of today's learners, with 97% of teens reportedly using their phones during the school day, (Mocanu, 2024).

Live chat and instant messaging (IM) offer valuable avenues for K-12 school libraries to connect with students and teachers, extending beyond the physical library space. Paterson and Low (2011) demonstrated that SMS reference services in a library setting increased response rates to student inquiries by 30%, as students appreciated the immediacy of text-based communication. In K-12 schools, SMS services can notify students and teachers about due dates, new resource availability, or library events, aligning with the mobile-centric behaviors of today's learners.

Student and teacher attitudes toward mobile library services significantly influence their adoption. Dukic, Chiu, and Lo (2015) found that while students appreciated the convenience of using smartphones for academic activities such as accessing course materials and library catalogs, they also noted significant usability issues—particularly small screen sizes and slow load times—that limited academic use of mobile devices. SMS and messaging services address these concerns by offering quick, direct communication. Paterson and Low (2011) reported that SMS-based reference services in libraries were rated highly by 85% of users for their speed and accessibility, a finding applicable to K-12 students and teachers seeking immediate support. For example, a Scottish high school study by Clark et al. (2010) showed that students using mobile apps with SMS notifications were 30% more likely to engage with digital resources.

Anecdotal evidence suggests that librarians are increasingly using messaging platforms like Google Chat and Edsby messaging to provide real-time support to students. This practice is particularly beneficial for middle and high school students who are engaged in research or homework outside of regular school hours, (Stattia & Villegas, 2020). By being available on these platforms, librarians can answer queries, provide research guidance, and offer help with navigating digital resources, effectively extending the library's support services into students' homes. This approach not only enhances student success but also reinforces the library's role as a vital academic support hub in the digital age, (Stattia & Villegas, 2020).

The Utilization of AI Chat Tools by School Librarians

School librarians have started to investigate the potential of ChatGPT to integrate it into educational services. In "School Librarians Explore Possibilities of ChatGPT," an article by Kara Yorio published in the School Library Journal on January 17, 2023, the author discusses how some school librarians are looking beyond the initial fears and exploring the potential positive applications of the AI chatbot. Despite concerns about students using the tool to cheat and some districts banning it, librarians are experimenting with ChatGPT to streamline tasks like creating lesson plans, crafting social media posts, and assisting with readers' advisory. The article suggests that embracing this technology could lead to a re-evaluation of traditional assignments and a greater focus on the learning process itself, (Yorio, 2023).

The article titled "What Can ChatGPT Do in the Library?" by Kirsten Cox, which was published in Katina Magazine on February 13, 2025, discusses the potential uses of ChatGPT in libraries, especially in academic settings. It emphasizes that while ChatGPT is a helpful tool, it cannot replace the skills and knowledge of librarians. Instead, it can assist by answering simple reference questions, offering support around the clock, and helping users locate academic materials, (Cox, 2025).

Additionally, the article points out that librarians can utilize ChatGPT for routine tasks, like composing emails. However, it also raises important concerns about using this AI tool, such as the risk of it providing incorrect or biased information, issues related to data privacy, and the environmental effects of AI technology. To mitigate these challenges, the author suggests that libraries should offer training for librarians and encourage library users to engage with AI tools in a responsible manner, (Cox, 2025)

Effective ChatGPT prompts for school librarians

School librarians might consider using ChatGPT to enhance various services and support students and staff. This could involve making existing services more effective, introducing new services, or providing additional resources and support to ensure students and employees have what they need to succeed.

- Reference assistance: ChatGPT can help answer research questions and provide information on a wide range of topics, acting as an initial point of reference for students and faculty.

- Book recommendations: The tool can suggest books and materials tailored to individual students' reading levels, interests, and past reading history, promoting literacy and a love for reading.
- Support for writing assignments: Students can utilize ChatGPT for brainstorming ideas, structuring outlines, and even proofreading their work.
- Digital resource access: ChatGPT can assist in guiding users to databases, e-books, and other online tutorials, helping them navigate the library's digital resources more efficiently.
- Interactive learning experiences: Librarians can create engaging activities like virtual book clubs or author visits using ChatGPT, fostering interactive learning within the library.
- Language assistance: For diverse school communities, ChatGPT can aid in translating questions or information, promoting inclusivity.
- Administrative tasks: Librarians use ChatGPT to help create plans for weeding collection, plan events like book fairs, gusted related policy language.

By using specific and well-crafted prompts, school librarians can harness the power of ChatGPT to streamline their workflows, enhance their services, and foster a dynamic and engaging learning environment for their students and school community.

The service types align with instructional patterns observed in mobile learning reviews and technology-supported schooling in K–12 contexts. By strategically adopting and thoughtfully implementing ChatGPT and similar tools, school librarians can empower students with the skills needed to navigate the evolving information landscape and promote lifelong learning in a digital world.

MOBILE GENERATION ENGAGING WITH LIBRARY SERVICES

Understanding how students perceive and interact with mobile library services is crucial for successful implementation and ongoing improvement. Recent research indicates that students value autonomy in learning environments, and school libraries provide spaces where they can manage their time and shape their learning experiences, (Zhu, 2025). This autonomy is further enhanced through well-designed mobile services that provide flexible, personalized access to library resources and support.

Falloon (2023) highlights the significant impact of tablet-integrated resource discovery tools, particularly when designed for young learners, on student engagement and the retrieval of learning content in K-6 settings. The research emphasizes that the effectiveness of these tools stems from their pedagogical design and child-centered approach. For students in kindergarten through sixth grade, tablets provide an interactive and intuitive platform that transforms learning into active exploration.

Features like visual interfaces, touch-based interactions, and gamified elements within these tools significantly boost engagement, creating a more receptive environment for learning.

The study further indicates a direct link between the use of these well-designed tablet tools and an improved ability to retrieve learning content. When children actively search for and interact with information, they form stronger cognitive connections, leading to better retention and recall. The personalized nature of tablet use allows students to explore topics at their own pace, reinforcing understanding. Falloon's work provides compelling evidence that thoughtfully designed, tablet-integrated resource discovery tools are essential for fostering dynamic and effective learning environments in K-6 classrooms, enhancing both engagement and content retrieval, (Falloon, 2023)

Interaction and Inclusive Services

School library mobile services that are inclusive—inclusive design for screen readers, alt-text, audiobook formats, and simplified navigation—address accessibility and learning differences. Younger students and those with disabilities benefit from mobile audiobook lending and simplified catalog search flows, (Farkas & Hoffman, 2014). Tailored UX—age-appropriate interfaces and voice commands—supports universal accessibility and has been shown to be a critical component of effective and equitable digital learning tools, (Farkas & Hoffman, 2014; Sipe, 2017).

Awedh et al. (2015) demonstrated that using mobile-based interactive tools like Socrative can significantly boost student engagement and improve collaborative learning. The research found that when students used these tools to answer questions and discuss them in groups, their understanding and interaction with the material and their peers improved.

Student attitudes toward mobile library services vary significantly based on factors such as age, technological proficiency, and previous experiences with library services, (Zhu, 2025). Elementary school students often show enthusiasm for interactive, game-based mobile applications that make library exploration fun and engaging. These young users respond well to visual interfaces, simple navigation structures, and immediate feedback mechanisms. Middle school students demonstrate increased interest in social features, such as the ability to share resources with classmates and participate in online reading challenges through mobile platforms.

High school students exhibit more sophisticated expectations for mobile library services, often comparing library apps and websites to commercial applications they use regularly, (Chen & Huang, 2024). They value efficiency, comprehensive search capabilities, and seamless integration with their academic workflows. This demographic is particularly responsive to personalized services, such as customized resource recommendations and the ability to create and manage personal research collections across devices.

User experience design principles have become particularly critical in the mobile context due to the constraints of smaller screens and touch-based interactions, (Roberts & Kim, 2024). Successful mobile library services prioritize intuitive navigation, fast loading times, and clear visual hierarchies that guide users to desired resources efficiently. The implementation of responsive design ensures that services adapt seamlessly to various screen sizes and orientations, providing consistent experiences across different mobile devices.

Accessibility considerations are paramount in mobile service design, ensuring that all students can effectively use library services regardless of their abilities or assistive technology needs, (Roberts & Kim, 2024). This includes implementing appropriate color contrast ratios for students with visual impairments, providing alternative text for images and graphics, ensuring compatibility with screen readers and other assistive technologies, and designing interfaces that can be navigated using voice commands or switch controls.

Student feedback mechanisms play a crucial role in continuous service improvement, (Zhu, 2025). Many successful mobile library programs incorporate regular surveys, focus groups, and usability testing sessions to gather student input on service effectiveness and desired enhancements. This feedback loop ensures that mobile services evolve to meet changing student needs and preferences while maintaining high levels of user satisfaction.

Interactive mobile platforms that incorporate gamified learning elements—such as decision-making scenarios, adaptive pathways, and immersive narratives—can significantly enhance critical thinking and engagement among students. For example, Sulisworo et al. (2024) found that such gamified mobile learning environments foster deeper cognitive engagement and support critical thinking skill development through dynamic challenges, feedback, and immersive design. These tutorials often incorporate gamification elements, making the learning process more engaging for students while building essential research and digital citizenship skills. Mobile delivery ensures that students can access these resources at their point of need, whether they're working on assignments at home or conducting research in the library.

Falloon's (2023) research on using iPads in K–6 classrooms shows that when teachers and librarians work together, student collaboration and a sense of ownership over their learning significantly improve. Younger students get really excited when apps let them choose their own books, create digital stories with audio, or participate in interactive digital scavenger hunts in the library. For early learners, mobile interfaces must feature voice prompts, large icons, and simplified flows. Literacy coaches and reading specialists often partner with librarians to ensure mobile tools support emergent readers, including audio narration and alt-text for images.

Promoting mobile library services requires strategic outreach efforts that reach students through their preferred communication channels, (Anderson & Lee, 2024). Social media campaigns, classroom presentations, and integration with orientation programs help ensure that students are aware of available mobile services and understand how to access and utilize them effectively. Peer-to-peer education, where student library assistants help teach mobile service use to their classmates, has proven particularly effective in driving adoption and engagement.

Educational Equity: Mobile Services and the Digital Divide

The digital divide represents one of the most significant challenges facing contemporary education, with profound implications for student achievement and long-term success. School libraries are uniquely positioned to address these inequities through comprehensive mobile service programs that extend access beyond traditional boundaries, (Johnson & Davis, 2024). Mobile library technologies offer crucial services including location-based services, reference services, and mobile-based library lending services that can help bridge access gaps, (Rahman & Clarke, 2023).

Wi-Fi hotspot lending programs address connectivity barriers that prevent students from accessing online resources at home, (Rahman & Clarke, 2023). These programs typically involve partnerships with cellular service providers to offer subsidized or free internet access through portable hotspot devices. Students and families can borrow these devices for extended periods, ensuring consistent access to online educational resources, library databases, and communication tools necessary for academic success. The impact of these programs has been particularly pronounced during remote learning periods, when home internet access became essential for educational continuity.

Mobile outreach services extend library support to underserved communities through innovative service delivery models, (Johnson & Davis, 2024). Some school districts operate mobile library units—specially equipped vehicles that bring library services, including mobile technology support, directly to neighborhoods with limited transportation access or existing library services. These mobile units often provide on-site device training, homework support, and digital literacy instruction that complements traditional library services.

Multilingual mobile services address the needs of diverse student populations by providing library interfaces and resources in multiple languages, (Rahman & Clarke, 2023). This includes translated mobile applications, multilingual catalog interfaces, and digital collections that reflect the linguistic diversity of school communities. Voice-activated services and translation tools integrated into mobile library platforms help students whose primary language is not English access library resources more effectively.

Partnerships with community organizations amplify the impact of mobile library services in addressing educational equity, (Johnson & Davis, 2024). Collaborations with local non-profit organizations, community centers, and faith-based institutions help extend library services to families who might not otherwise access them. These partnerships often involve training community volunteers to provide basic technical support and digital literacy instruction, creating sustainable support networks that extend beyond formal library programs.

Data privacy and security considerations become particularly important when serving vulnerable populations through mobile services, (Rahman & Clarke, 2023). School libraries must implement robust privacy protections while ensuring that mobile services remain accessible and user-friendly. This includes educating students and families about digital privacy, implementing appropriate data retention policies, and ensuring that third-party service providers adhere to student data protection requirements.

Assessment and evaluation of equity-focused mobile services require careful attention to both quantitative and qualitative outcomes, (Johnson & Davis, 2024). Metrics such as device lending rates, Wi-Fi hotspot usage, and mobile service adoption provide valuable insights into program reach and utilization. However, qualitative measures such as student academic performance, family engagement with education, and long-term educational outcomes provide more comprehensive understanding of program impact on educational equity.

School Library Tablet Lending Programs and Case Studies

School library device lending programs have emerged as a cornerstone strategy for addressing hardware-related aspects of the digital divide, (Johnson & Davis, 2024). Many school libraries now maintain collections of tablets, laptops, and e-readers that students can borrow for extended periods, ensuring access to digital resources regardless of family economic circumstances. These programs often include protective cases, charging cables, and basic technical support to maximize the utility of borrowed devices. Some programs have expanded to include specialized devices such as graphing calculators, digital cameras, and audio recording equipment that support various academic disciplines.

iPad loan programs in school libraries enhance student learning by providing access to tablets that support research, collaboration, and creativity. Kim McCallister, a teacher at Liberty Middle School in Ashland, Virginia, has 12 iPads that remain on-site for teachers to preview apps. Two rolling trunks each contain 30 iPads for classroom use. One trunk always holds a full set of 30 iPads, while the other can be divided among multiple teachers who need fewer devices. Regardless of the checkout date, all iPads are due back on either the 15th or 30th of each month. An online reservation system allows teachers to plan ahead and reserve iPads based on their lesson needs, (Moorefield-Lang, Meier., & Miller, 2014).

Liberty Middle School utilizes iPads and apps to enhance the learning experience. The library has pre-installed a variety of free apps covering diverse subjects such as physical education, foreign languages, art, music, theater arts, childcare, and cooking. Teachers can experiment with apps on the library's 12 iPads before deciding to request specific ones for classroom use. They developed an outline process ensuring that only tested and beneficial apps are purchased or installed in bulk. (Moorefield-Lang, Meier., & Miller, 2014).

The iPad program in primary school libraries has the potential to significantly enhance student learning experiences. A study by Gray & Thomas (2025) revealed that integrating iPads in primary school libraries has become an integral part of modern education. This analysis explores how the iPad program can improve student perceptions of library support, particularly in areas related to reading, research, and technology use.

To ensure accessibility, libraries configure iPads with assistive technologies, such as screen readers and voice input, catering to users with disabilities. By integrating iPads into their services, libraries bridge the digital divide and create inclusive environments for learning and exploration.

While the benefits of library iPad and mobile device programs are significant, challenges remain. Ensuring equitable access, managing device maintenance and security, and providing adequate training and support are ongoing concerns. Libraries must also continually assess the effectiveness of their programs and adapt to changing technological trends and user needs, (University at Buffalo Health Sciences Library, 2013).

Looking ahead, libraries have the chance to integrate mobile technology further into their strategic planning. By developing, licensing, or curating scholarly content tailored for mobile devices, libraries can enhance the academic and personal growth of their patrons. (Providence College, 2017). Furthermore, partnerships with community organizations and government agencies can broaden the reach and impact of these programs, particularly for vulnerable populations.

Serving Students with Diverse Needs

Creating inclusive mobile library services requires thoughtful consideration of the diverse abilities, learning styles, and needs present in school communities. Research shows that educational institutions are integrating various learning support features to meet varied student needs during redesign processes (Roberts & Kim,

2024). This inclusive approach ensures that mobile library services serve as tools for empowerment rather than barriers to access.

Universal Design for Learning (UDL) principles provide a framework for developing mobile library services that are inherently accessible to students with diverse abilities and learning preferences, (Roberts & Kim, 2024). This approach involves providing multiple means of representation, where information is presented in various formats such as text, audio, video, and interactive elements; multiple means of engagement, offering different pathways for student motivation and interest; and multiple means of action and expression, allowing students to demonstrate their knowledge and skills through various modalities.

Assistive technology integration is crucial for students with disabilities to effectively utilize mobile library services, (Roberts & Kim, 2024). Screen reader compatibility ensures that students with visual impairments can navigate mobile interfaces through audio feedback and keyboard commands. Voice recognition software enables students with motor disabilities to control mobile applications through spoken commands. Switch-activated interfaces provide access for students with severe motor limitations who rely on adaptive hardware for device control.

Cognitive accessibility features address the needs of students with learning disabilities, attention disorders, and developmental disabilities, (Roberts & Kim, 2024). These features might include simplified navigation structures that reduce cognitive load, clear visual indicators for important actions and information, text-to-speech functionality for students with reading difficulties, adjustable timing settings for students who need additional processing time, and visual schedules and prompts that support students with autism spectrum disorders.

Emerging research indicates that mobile technologies, especially when designed according to Universal Design for Learning (UDL) principles, can enhance accessibility and learner engagement among students with special educational needs. Through features like customizable text (audible or visual), motivational multimedia, and self-paced navigation, mobile-supported resources help improve independence and support differentiated learning goals, (Chelkowski, Yan, & Asaro-Saddler, 2019; Cumming & Draper Rodríguez, 2017). However, studies in regions like Nigeria reveal that many classrooms lack organized deployment of such assistive tools—highlighting a pressing need for more robust, inclusive implementation and teacher training (Research in Nigerian educational contexts)

Collaboration with special education professionals ensures that mobile library services align with students' Individual Education Plans (IEPs) and 504 Plans, (Roberts & Kim, 2024). Library staff work closely with special education teachers, occupational therapists, and assistive technology specialists to identify appropriate mobile resources and customize interfaces to meet individual student needs. This collaboration extends to training special education staff on available mobile library resources and their potential applications in classroom and therapeutic settings.

Professional development for library staff is essential for effective service delivery to students with diverse needs, (Roberts & Kim, 2024). Training programs cover topics such as disability awareness and etiquette, assistive technology familiarity, accessibility features of mobile devices and applications, legal requirements under the Americans with Disabilities Act and Section 504, and strategies for adapting library instruction for students with different learning styles and abilities.

Family engagement and support play crucial roles in successful inclusive mobile library services, (Roberts & Kim, 2024). Libraries provide training and resources to help families understand and utilize accessibility features on mobile devices and applications. This includes workshops on built-in accessibility features of smartphones and tablets, guidance on selecting appropriate apps and digital resources for children with special needs, and support for creating accessible home learning environments that complement school-based services.

Assessment of inclusive mobile services requires careful attention to both usability and educational outcomes for students with diverse needs (Roberts & Kim, 2024). Regular usability testing with students who have disabilities helps identify barriers and improvement opportunities. Academic outcome tracking examines whether mobile library services contribute to improved educational performance for students with special needs. Feedback from special education professionals, families, and students themselves provides valuable insights for continuous service enhancement.

VIRTUAL LABORATORIES: TRANSFORMING STEM EDUCATION THROUGH MOBILE ACCESS

The integration of virtual laboratories into school library mobile services represents a revolutionary approach to STEM education that addresses both accessibility and educational effectiveness concerns. Virtual laboratories provide simulated environments that enable students to conduct experiments, test hypotheses, and analyze phenomena without the constraints of physical equipment, safety concerns, or geographical limitations. The significance of virtual laboratories in K-12 education cannot be overstated, as they offer unprecedented access to high-quality scientific experiences that would otherwise be impossible due to resource limitations or safety considerations.

Research exploring the design of virtual laboratories (VLs) illustrates how these platforms can broaden equitable access to STEM content. For instance, Zhang and Liu (2024) detail how VLs support sustainable digital lab experiences in higher education settings—showing promise for scalability and mobile integration—while a 2023 systematic review of learning analytics in virtual labs synthesizes evidence of their effectiveness in improving student engagement and performance.

Dr. Yan Quan Liu's groundbreaking work on the Digital Library of High School Virtual Labs demonstrates the transformative potential of these technologies. His research focuses on creating a free, open-access resource portal that empowers student learning experiences while enhancing educational objectives in Connecticut high schools and beyond (Liu, 2023). This initiative recognizes that virtual laboratories representing all areas of science provide crucial access to STEM content for learners who are otherwise restricted physically, geographically, and economically.

The theoretical foundation for virtual laboratory implementation rests on several key principles (Liu, 2023). The Next Generation Science Standards (NGSS) were created to prepare students for college and careers by allowing them to apply their scientific knowledge via practical experiments and hands-on projects. Virtual laboratories align perfectly with this vision by providing safe, cost-effective, and accessible alternatives that supplement conventional laboratory instruction. Within the NGSS, there are three distinct and equally important dimensions to learning science: Crosscutting Concepts help students explore connections across the four domains of science, including Physical Science, Life Science, Earth and Space Science, and Engineering Design.

The mobile delivery of virtual laboratory experiences ensures that students can access these resources at any time and location, breaking down traditional barriers to scientific experimentation and discovery. Recent reviews underscore the growing promise of immersive technologies like AR, VR, and XR for enhancing STEM education. One umbrella review report gains in student engagement and academic performance, while K–12 research emphasizes the current popularity of non-immersive AR platforms (e.g., smartphone/tablet-based tools) and recommends further exploration into wearable, immersive solutions to improve usability and accessibility, (Popenici & Kerr, 2024; Education and Information Technologies, 2024).

Comprehensive Virtual Laboratory Ecosystems

Successful virtual laboratory implementation requires systematic collection and organization of resources under well-established policies (Liu, 2023). Dr. Liu's methodology emphasizes the importance of collecting appropriate resources including virtual laboratories, tutorial videos, graphical images, manuals, and instructions within a coherent framework. The collection strategy involves researching different types of labs for inclusion, ensuring that the virtual laboratory portal serves diverse STEM disciplines including Astronomy, Physics, Earth Sciences, Engineering,

Biology, Biotechnology & Biomedical, Histology and Anatomy, Physiology, and Chemistry., and Chemistry.

The technical infrastructure supporting virtual laboratories must accommodate various media formats and platforms, (Liu, 2023). Research indicates that Adobe Flash Player, while presenting integration challenges, remains essential for many existing virtual laboratories. However, emerging HTML5 and WebGL technologies are providing more mobile-friendly alternatives that function seamlessly across different devices and operating systems without requiring additional plugins or software installations.

BEST Lab's mobile learning initiative showcases how K–12 engineering education can be expanded beyond traditional classroom settings. The initiative focuses on designing and implementing mobile learning applications that support informal or semi-structured educational environments. By utilizing portable devices such as smartphones and PDAs, the project aims to capitalize on learning opportunities that naturally arise outside the school setting, as students typically spend 85% of their time outside the classroom. The project employs user-centered design and AR-enabled tools to demonstrate how mobile devices can facilitate real-world exploration, foster curiosity-driven learning, and enable scientific observation—all delivered in engaging and accessible ways for elementary and middle school students, (Berkeley Emergent Space Tensegrities (BEST) Lab, n.d.).

Modern virtual laboratory platforms incorporate sophisticated metadata systems using Resource Description Framework (RDF) as the basic language for describing materials (Liu, 2023). This approach enables comprehensive searching and retrieval capabilities, with search parameters including science content area, descriptive keyword terms, scientific concepts, activity types (simulation, interactive, data manipulation), and creator/agency information. The implementation of Open Archives Initiative Protocol for Metadata Harvesting (OAI-PMH) ensures that virtual laboratory resources can be discovered through commercial search engines, maximizing accessibility and usability.

Mobile-Optimized Virtual Laboratory Platforms

Several educational platforms now deliver virtual laboratory experiences optimized for mobile devices—each offering tailored strengths for diverse STEM learning needs. ExploreLearning Gizmos, a widely used set of online virtual lab simulations for math and science (grades 3–12), which includes mobile device compatibility and has significant research backing its educational impact. ExploreLearning's Gizmos provide inquiry-based, mobile-friendly virtual labs across grades 3–12. A correlational study demonstrated that schools with greater Gizmos usage had students 1.3 times more likely to meet or exceed science proficiency standards, even while accounting for socioeconomic and staffing variables, (ExploreLearning, 2024). These platforms demonstrate the effectiveness of mobile-delivered virtual experiments in engaging students and supporting scientific understanding.

PhET Interactive Simulations, a non-profit open educational resource created by Nobel Laureate Carl Wieman at the University of Colorado Boulder, offers over 125 free interactive simulations across physics, chemistry, biology, mathematics, and earth science. These simulations are designed to promote intuitive and inquiry-based engagement through implicit scaffolding, where instructional guidance is embedded within the interface design rather than relying on explicit instructions. PhET's simulations are widely used for visualization and offer free, interactive math and science simulations. The mobile optimization of PhET simulations allows students to access physics, chemistry, biology, and mathematics concepts through interactive visual models that respond to touch-based interactions. These simulations are accessible online and support inquiry-based learning through dynamic visual engagement. (PhET Interactive Simulations, n.d.) While these simulations are highly effective, they often lack structured pathways and comprehensive NGSS-aligned frameworks for different grade levels.

Labster empowers educators to reimagine their science courses with immersive online simulations that engage students, train lab skills, and accelerate learning. A Labster blog post describes the company's research-driven approach in designing and refining its platform, citing meta-analyses and rigorously evaluated VR-based learning tools used in STEM instruction. A program evaluation study at McMaster University demonstrated a Labster VR experience where students used Google Daydream headsets to engage in immersive experiments on gene expression and viral gene therapy, enabling interactive, game-like learning sequences within a 5E instructional framework, (Tsirulnikov, 2023). Labster's focus on immersive, advanced virtual labs for high school and higher education demonstrates the potential for sophisticated virtual laboratory experiences delivered through mobile platforms. However, the cost associated with premium virtual laboratory platforms can create barriers for underfunded schools, highlighting the importance of free and open-access alternatives, (Labster, n.d.)

The NASA STEM Resources portal (www.nasa.gov/stem) provides high-quality educational content related to Earth and space sciences, optimized for mobile access (Liu, 2023). While subject-specific, these resources demonstrate how government agencies can contribute to virtual laboratory ecosystems through mobile-friendly interfaces and interactive experiences that bring space science concepts to students' fingertips.

A study from Indonesia evaluated the use of a mobile learning virtual lab app (TNBTS V-Lab) with 46 social science education students. It reported a statistically significant learning outcome improvement—evidenced by a gain score increase of

37.61 points—and highly positive student feedback. This robust, quasi-experimental study with a non-equivalent control group design involved 46 social science education students at UIN Maulana Malik Ibrahim in Malang, Indonesia. The study demonstrated a significant improvement in learning outcomes, with a gain score increase of 37.61 points, alongside overwhelmingly positive student feedback about using the mobile-based virtual lab application—TNBTS V-Lab, (Bashith, et al., 2025).

Addressing Educational Equity Through Virtual Laboratories

Virtual laboratories accessible via mobile platforms hold significant promise for promoting equity in STEM education. Students in underserved communities often lack access to well-equipped physical science laboratories, making virtual alternatives essential for providing equitable educational opportunities. The mobile delivery of virtual laboratory experiences ensures that students can access high-quality scientific experiences regardless of their school's physical resources or geographic location.

The Virtual STEM Labs program is designed to introduce STEM to K-12 students through events held virtually to appeal to a nationwide audience (Liu, 2023). This approach demonstrates how virtual laboratories can reach students who might otherwise have limited exposure to advanced scientific concepts and experimental procedures. The mobile accessibility of these programs ensures that students can participate using smartphones or tablets, eliminating the need for expensive computer equipment.

The economic benefits of virtual laboratories extend beyond initial implementation costs (Liu, 2023). While physical laboratory equipment requires ongoing maintenance, replacement, and safety oversight, virtual laboratories can be updated and expanded without significant infrastructure investments. This scalability makes high-quality STEM education more accessible to schools with limited budgets, rural communities with small student populations, and districts serving economically disadvantaged populations.

Virtual labs offer a broader advantage by enabling students to conduct experiments and engage in inquiry-based learning from anywhere, regardless of their school's facilities or geographic location. This accessibility is highlighted in the article "Overview of Virtual Laboratories' Role in STEM Access." In rural South Africa, despite infrastructural limitations, teachers and students demonstrated a strong willingness to adopt virtual tools, especially when mobile or offline access is integrated and supported. This is evident in the factors influencing virtual lab adoption in rural South African schools, as discussed in the article "Factors Influencing Virtual Lab Adoption in Rural South African Schools," (Smart Learning Environments, 2025).

Virtual laboratories offer safe, flexible alternatives that address critical equity issues, particularly for students with health conditions or mobility limitations who may not be able to participate in traditional wet labs. Chemical allergies, mobility restrictions, and other health concerns become irrelevant in virtual environments, ensuring equal opportunities for all students to develop scientific skills and understanding. Virtual labs' inclusive design features, such as portability and multimodal support, enable learners with diverse needs to engage in meaningful scientific exploration, (Villagrasa et al., 2024; Reyes et al., 2024).

Integration with School Library Mobile Services

The integration of virtual laboratories into school library mobile service portfolios creates synergistic relationships that enhance both technological access and educational support (Anderson & Lee, 2024). Library professionals can provide crucial technical support for students accessing virtual laboratory platforms, helping troubleshoot connectivity issues, navigate complex interfaces, and optimize mobile device settings for the best experimental experiences.

Information literacy instruction becomes particularly relevant in virtual laboratory contexts, as students must evaluate the credibility and appropriateness of different simulation platforms, understand the limitations and assumptions built into virtual models, and critically analyze experimental results within broader scientific contexts, (Anderson & Lee, 2024). School librarians are uniquely positioned to help students develop these critical evaluation skills while supporting their engagement with virtual scientific content.

Collection development strategies for virtual laboratories require careful curation and evaluation processes, (Kumar & Singh, 2023). Librarians must assess virtual laboratory platforms for scientific accuracy, pedagogical effectiveness, accessibility features, and alignment with curriculum standards. This evaluation process includes testing mobile compatibility, assessing user interface design, and ensuring that virtual experiments provide meaningful learning experiences rather than superficial simulations.

Professional development for library staff becomes essential as virtual laboratory integration expands, (Patel & Williams, 2023). Librarians need training in scientific concepts covered by virtual laboratories, familiarity with various simulation platforms and their technical requirements, understanding of safety and ethical considerations in virtual experimentation, and skills for supporting student inquiry and research using virtual laboratory data.

Assessment and Learning Outcomes

Research on virtual laboratory effectiveness demonstrates significant positive impacts on student learning outcomes when properly implemented and integrated with broader STEM curricula, (Mavromanolaki, Lionarakis, & Fragkaki, 2024). Virtual laboratories excel at helping students visualize abstract scientific concepts, explore cause-and-effect relationships through manipulable variables, and develop hypothesis-testing skills through repeated experimentation, (Kleftodimos & Evangelidis, 2023). The mobile accessibility of these experiences enables students to revisit complex concepts multiple times, supporting deeper understanding and retention.

Assessment strategies for virtual laboratory experiences require careful consideration of both process and outcome measures (Mavromanolaki, Lionarakis, & Fragkaki, 2024). Formative assessment can include tracking student interactions with virtual experiments, analyzing hypothesis formation and testing strategies, and documenting conceptual understanding development over time. Summative assessment might involve comparing virtual laboratory experiences with traditional hands-on activities, measuring knowledge transfer to real-world applications, and evaluating long-term retention of scientific concepts.

Data analytics capabilities built into many virtual laboratory platforms provide unprecedented insights into student learning processes, (Garcia & Wilson, 2024). These systems can track how students approach problem-solving, identify common misconceptions, and provide personalized feedback based on individual interaction patterns. Mobile delivery enhances these capabilities by enabling continuous data collection that reflects authentic learning contexts rather than artificial testing environments.

The effectiveness of virtual laboratories appears to be enhanced when they are used as supplements to, rather than replacements for, traditional hands-on laboratory experiences, (Kleftodimos & Evangelidis, 2023). This hybrid approach leverages the strengths of both virtual and physical experimentation, providing students with comprehensive scientific experiences that prepare them for advanced STEM studies and careers. Mobile access to virtual laboratories enables this supplementation to occur seamlessly, with students able to revisit virtual experiments to reinforce concepts explored in physical laboratories or to explore extensions and variations of hands-on activities.

INNOVATION AND BEST PRACTICES: EMERGING TECHNOLOGIES AND SERVICE EXCELLENCE

The landscape of mobile library services continues to evolve rapidly, driven by technological innovations and changing user expectations. Recent trends show significant growth in smart speakers and virtual assistants, with libraries adapting their digital collections to become compatible with modern voice recognition systems, while AI is increasingly helping librarians with tasks such as inventory monitoring and stock organization, (Garcia & Wilson, 2024). These developments point toward increasingly sophisticated and responsive mobile library services.

Integrating Advanced Technologies for Smarter Mobile Services

Artificial Intelligence (AI) integration represents one of the most promising frontiers in mobile library services, (Garcia & Wilson, 2024). AI-powered chatbots provide 24/7 support for common student inquiries, offering immediate assistance with resource location, account management, and basic research guidance. These systems learn from user interactions, becoming more effective over time at understanding and responding to student needs. Personalized recommendation engines analyze student borrowing patterns, search histories, and academic interests to suggest relevant resources automatically.

Machine learning algorithms optimize mobile service performance by analyzing usage patterns and predicting resource demand, (Garcia & Wilson, 2024). These systems can anticipate peak usage periods, recommend collection development priorities, and identify potential system bottlenecks before they impact user experience. Predictive analytics help library administrators make data-driven decisions about resource allocation, service improvements, and technology investments.

Augmented Reality (AR) and Virtual Reality (VR) technologies are beginning to transform how students interact with library resources and spaces through mobile devices, (Gnesdilov, Claesgens, & Purzer, 2025). AR applications can overlay digital information onto physical library environments, helping students navigate spaces, locate resources, and access additional information about materials. QR codes placed throughout library spaces link to digital content, creating interactive learning environments where students can access supplementary resources, author information, and related materials through their mobile devices.

Virtual Reality applications delivered through mobile platforms offer immersive educational experiences that were previously impossible in traditional library settings, (Gnesdilov, Claesgens, & Purzer, 2025). Students can take virtual field trips to historical sites, explore scientific phenomena in three-dimensional environments, and participate in interactive literary experiences that bring books to life. These technologies are particularly valuable for schools with limited resources for traditional field trips and experiential learning opportunities.

Internet of Things (IoT) integration creates smart library environments where mobile devices serve as control interfaces for various systems, (Thompson, Rodriguez, & Chang, 2024). Students can use mobile applications to reserve study spaces, control lighting and temperature in designated areas, and access real-time information about resource availability and library occupancy. RFID technology integrated with mobile applications enables self-checkout services, inventory tracking, and automated shelf management systems.

Blockchain technology, while still emerging in library applications, shows potential for creating secure, decentralized systems for digital resource sharing and student credential verification, (Thompson, Rodriguez, & Chang, 2024). Mobile applications built on blockchain platforms could enable secure sharing of digital resources between schools and libraries while maintaining appropriate usage restrictions and privacy protections.

Building Capacity and Collaboration for Sustainable Innovation

Best practices for implementing innovative mobile library services emphasize the importance of pilot programs and gradual rollouts, (Patel & Williams, 2023). Successful implementations typically begin with small-scale testing involving willing participants who provide feedback and identify potential issues before broader deployment. This approach minimizes disruption to existing services while allowing for iterative improvement based on real-world usage patterns.

Staff development and training remain critical success factors as mobile technologies become more sophisticated, (Patel & Williams, 2023). Library professionals must stay current with emerging technologies while developing the skills necessary to support students in using new tools effectively. This includes both technical skills for system administration and maintenance, and instructional skills for teaching students how to leverage new technologies for learning and research.

Collaboration with technology companies and educational technology providers can accelerate innovation while ensuring that mobile library services remain aligned with broader educational technology initiatives, (Kumar & Singh, 2023). These partnerships often provide access to cutting-edge tools and resources that might otherwise be prohibitively expensive for individual schools or districts. However, such collaborations require careful evaluation of data privacy, vendor reliability, and long-term sustainability.

Collaboration between school librarians and classroom teachers is crucial for the successful integration of these digital tools. Falloon (2023) notes that when librarians develop mobile tools and tutorials that are specifically aligned with classroom units, student engagement and adoption rates increase significantly. This partnership ensures that digital library resources aren't just supplementary but are integral to the curriculum, making the library a central hub for academic support in the digital age.

User-centered design principles remain paramount as mobile technologies become more complex, (Chen & Huang, 2024). The most sophisticated technological capabilities are meaningless if students cannot easily access and effectively utilize them. Successful mobile library services maintain focus on user needs and preferences while incorporating new technologies in ways that enhance rather than complicate the user experience.

Effective sustainability planning is crucial for ensuring that mobile services remain functional, up-to-date, and valuable over time. Moya and Camacho (2023) propose a strategic framework that emphasizes economic viability, stakeholder engagement, technological planning, and pedagogical readiness. Similarly, sustainability assessments in public health interventions highlight the significance of stable funding, training, and strategic communication to maintain operational success. By applying these principles to mobile educational services, we can ensure that programs continue to deliver exceptional service in the long run. This involves considering long-term costs, staff training requirements, technology refresh cycles, and integration with existing systems. The most successful programs develop sustainable funding models and maintain realistic expectations about resource requirements for ongoing service excellence.

Mobile services in school libraries represent a fundamental shift in how educational institutions connect with and serve their communities. The evolution from traditional, place-bound library services to dynamic, mobile-responsive programs reflects broader changes in how students learn, communicate, and access information. Through comprehensive mobile service portfolios, school libraries are positioning themselves as essential partners in student success, educational equity, and lifelong learning.

The evidence presented throughout the related studies demonstrates that effective mobile library services require more than technological implementation; they demand thoughtful consideration of user needs, systematic attention to accessibility and inclusion, and ongoing commitment to innovation and improvement. Libraries that embrace this comprehensive approach are discovering that mobile services not only improve resource access and utilization but also strengthen the library's role as a central hub for learning and community engagement.

As mobile technologies continue to evolve, school libraries must remain adaptable and responsive to changing student needs and technological capabilities. The most successful programs maintain focus on their core educational mission while leveraging mobile technologies to extend their reach, improve their effectiveness, and ensure that all students have access to the resources and support they need for academic and personal success. The future of school library services is undoubtedly mobile, and institutions that embrace this reality today are positioning themselves to serve their communities effectively for years to come.

REFERENCES

AlNajdi, S. M. (2022). The effectiveness of using augmented reality (AR) to enhance student performance: Using quick response (QR) codes in student textbooks in the Saudi education system. *Educational Technology Research and Development*, *70*(3), 1105–1124. DOI: 10.1007/s11423-022-10100-4 PMID: 35400977

Anderson, M., & Lee, S. (2024). Digital transformation in educational libraries: A comprehensive analysis of mobile service integration. *Journal of Academic Librarianship*, *50*(3), 102–115.

Awedh, M., Mueen, A., Zafar, B. A., & Manzoor, U. (2015). Using Socrative and smartphones for the support of collaborative learning. [IJITE]. *International Journal of Integrating Technology in Education*, *3*(4), 17–24. DOI: 10.5121/ijite.2014.3402

Bashith, A., Amin, S., Kurniawan, M. A., Dewi, S., Jubba, H., & Maulidiah, L. (2025). Utilizing a virtual laboratory based on mobile learning: Implementation for social science education students in Indonesia to improve learning outcomes. *Educational Process: International Journal*, *15*(1), e2025181. Advance online publication. DOI: 10.22521/edupij.2025.15.181

Beanstack. (2025). Beanstack platform features: mobile access, reading challenges, and offline logging. Beanstack Help Center. Retrieved from https://www.beanstack.com/solutions/simplify-reading-logs

Berkeley Emergent Space Tensegrities (BEST) Lab. (n.d.). Mobile learning infrastructure for K–12 engineering education. University of California, Berkeley. https://best.berkeley.edu/past-research/mobile-learning

Campbell, L. B., & Kester, B. (2023). Centering students with disabilities: An accessible user experience study of a library research guide. *Weave: Journal of Library User Experience*, *6*(1). Advance online publication. DOI: 10.3998/weaveux.1067

Chelkowski, L., Yan, Z., & Asaro-Saddler, K. (2019). The use of mobile devices with students with disabilities: A literature review. *Preventing School Failure*, *63*(3), 277–295. DOI: 10.1080/1045988X.2019.1591336

Chen, L., & Huang, R. (2024). Mobile-first generation and library service evolution: Understanding student expectations in digital learning environments. *Computers & Education*, *198*, 104–118.

Clark, P., Walker, I. D., Langhorne, P., Crichton, L., Thomson, A., Greaves, M., Whyte, S., & Greer, I. A. (2010). SPIN (Scottish Pregnancy Intervention) study: A multicenter, randomized controlled trial of low-molecular-weight heparin and low-dose aspirin in women with recurrent miscarriage. *Blood*, *115*(21), 4162–4167. DOI: 10.1182/blood-2010-01-267252 PMID: 20237316

Cox, K. (2025). What Can ChatGPT Do in the Library? Katina Magazine. https://katinamagazine.org/content/article/resource-reviews/2025/what-can-chatgpt-do-in-the-library/

Cumming, T. M., & Draper Rodríguez, C. (2017). A meta-analysis of mobile technology supporting individuals with disabilities. *The Journal of Special Education*, *51*(3), 164–176. DOI: 10.1177/0022466917713983

Duke University Libraries. (2022, October 19). Mobile Apps: Research, Read, and Watch on the Go. Duke University Libraries Blog.

Dukic, Z., Chiu, D. K., & Lo, P. (2015). How useful are smartphones for learning? Perceptions and practices of Library and Information Science students from Hong Kong and Japan. *Library Hi Tech*, *33*(4), 545–561. DOI: 10.1108/LHT-02-2015-0015

eSchool News Contributors. (2024, September 12). *Beyond the bookshelves: 3 ways school libraries have evolved to meet students' needs*. eSchool News.

ExploreLearning. (2024). Gizmos support student achievement: Correlational study on virtual lab usage and science proficiency. ExploreLearning Research Insights. Retrieved from https://www.explorelearning.com/resources/insights/gizmos-support-student-achievement

Falloon, G. (2023). Twelve years of iPads and apps in schools: What conditions support effective practices in K-6 classrooms? *Frontiers in Education*, *8*, 1122522. DOI: 10.3389/feduc.2023.1122522

Farkas, J., & Hoffman, B. (2014). From mobile apps to social networks: The library as a tool for learning for the twenty-first century student. *Journal of Library Administration*, *54*(1), 1–13.

Garcia, P., & Wilson, T. (2024). Artificial intelligence in library services: Current applications and future potential. *Library Technology Reports*, *60*(2), 25–38.

Gnesdilov, D., Claesgens, J., & Purzer, S. (2025). Middle school students' application of science learning from physical versus virtual labs to new contexts. *Science Education*, *109*(1), 45–62. DOI: 10.1002/sce.21953

Johnson, K., & Davis, M. (2024). Bridging the digital divide: Mobile library services as equity tools in K-12 education. *School Library Research*, *27*(1), 1–18.

Kleftodimos, A., & Evangelidis, G. (2023). The effect of physical and virtual inquiry-based experiments on students' attitudes and learning. *Journal of Science Education and Technology*, *33*(2), 234–248.

Kumar, A., & Singh, P. (2023). Mobile technology adoption in academic libraries: A systematic review of implementation strategies. *Library Philosophy and Practice*, *2023*, 1–22.

Labster. (n.d.). How Labster used research to imagine, build, test, and improve its platform. Retrieved from https://www.labster.com/blog/research-story

Liu, Y. Q. (2023). Digital library of high school virtual laboratories: Empowering STEM education through mobile access. *CSU Grant 2023: Southern Connecticut State University*. Franklin, D. (2013). *A practical guide to gender diversity for computer science faculty*. Morgan & Claypool Publishers.

Martinez, C., & Brown, J. (2024). Comprehensive mobile service portfolios in academic libraries: Usage patterns and student preferences. *College & Research Libraries*, *85*(2), 178–195.

Mocanu, R. (2024). Text messaging for schools: Improving communication with SMS. Textmagic. https://www.textmagic.com/blog/sms-solutions-schools/

Moorefield-Lang, H., Meier, C., & Miller, R. (2014). *Tablet computers in school libraries and classrooms*. American Library Association. Gray, B., & Thomas, D. P. (2025). Student perceptions of primary school library support: Investigating reading, research, and technology use. *Journal of Librarianship and Information Science*, •••, 09610006251322983.

Moya, S., & Camacho, M. (2023). Developing a framework for mobile learning adoption and sustainable development. Technology. *Knowledge and Learning*, *28*(2), 727–744. DOI: 10.1007/s10758-021-09537-y

Okonkwo, I., & Ejikeme, C. (2023). Specialized digital collections accessible through mobile platforms serve students with diverse learning needs. *Library Hi Tech*, *41*(1), 234–251. DOI: 10.1108/LHT-01-2023-0014

Patel, N., & Williams, R. (2023). Strategic implementation of mobile technologies in school libraries: A framework for success. *School Library Journal*, *69*(8), 32–37.

Paterson, L., & Low, B. (2011). Student attitudes towards mobile library services for smartphones. *Library Hi Tech*, *29*(3), 412–423. DOI: 10.1108/07378831111174387

Pew Research Center. (2025, July 9). *Teens and Internet, Device Access Fact Sheet*. Retrieved from Pew Research Center.

Popenici, S. A. D., & Kerr, S. (2024). Virtual and augmented reality in Science, Technology, Engineering, and Mathematics (STEM) education: An umbrella review. *Information (Basel)*, *15*(9), 515. DOI: 10.3390/info15090515

Providence College. (2017). *A novel application: Using mobile technology to connect physical and virtual reference collections*. Digital Commons @ Providence. https://digitalcommons.providence.edu/facstaff_pubs/49/

Rahman, S., & Clarke, D. (2023). Mobile library services and educational equity: Addressing access barriers through technology. *Information Technology and Libraries*, *42*(3), 45–58.

Reyes, R. L., Isleta, K. P., Regala, J. D., & Bialba, D. M. R. (2024). Enhancing experiential science learning with virtual labs: A narrative account of merits, challenges, and implementation strategies amid COVID-19. *Journal of Computer Assisted Learning*, *40*(6), 3167–3186. DOI: 10.1111/jcal.13061

Roberts, E., & Kim, H. (2024). Inclusive design principles in mobile library applications: Supporting diverse learning needs. *Universal Access in the Information Society*, *23*(2), 289–302.

Seeholzer, J., & Salem, J. A.Jr. (2011). Library on the go: A focus group study of the mobile Web and the academic library. *College & Research Libraries*, *72*(1), 9–20. DOI: 10.5860/crl-65r1

Smart Learning Environments. (2025). Factors influencing virtual lab adoption in marginalized rural schools: Insights from South Africa. Smart Learning Environments, 12, Article 11. https://doi.org

SolarSPELL. (2024). SolarSPELL: Portable solar-powered offline digital libraries empowering underserved K-12 learners. Wikipedia. Retrieved from https://en.wikipedia.org/wiki/SolarSPELL

Stattia, A., & Villegas, S. (2020). The use of mobile learning in Grades K–12: A literature review of current trends and practices. *Peabody Journal of Education*, *95*(2), 139–147. DOI: 10.1080/0161956X.2020.1745613

Stattia, A., & Villegas, S. (2020). The use of mobile learning in Grades K–12: A literature review of current trends and practices. *Peabody Journal of Education*, *95*(2), 139–147. DOI: 10.1080/0161956X.2020.1745613

Alisworo, D., Maryani, I., Mastul, A.-R. H., & Kusumaningtyas, D. A. (2024). Incorporating gamified elements into mobile learning to enhance critical thinking skills. *Jurnal Pendidikan Progresif, 14*(1), 237–248. DOI: 10.23960/jpp.v14.i1.202418

Tao, D., McCarthy, P. G., Krieger, M. M., & Webb, A. B. (2009). The mobile reference service: A case study of an onsite reference service program at the School of Public Health. *Journal of the Medical Library Association: JMLA, 97*(1), 6–12. DOI: 10.3163/1536-5050.97.1.006 PMID: 19159004

Thompson, A., Rodriguez, M., & Chang, L. (2024). ICT infrastructure requirements for mobile library service implementation: A multi-institutional study. *DESIDOC Journal of Library and Information Technology, 44*(1), 12–25.

Thompson, A., Rodriguez, M., & Chang, L. (2024). ICT infrastructure requirements for mobile library service implementation: A multi-institutional study. *DESIDOC Journal of Library and Information Technology, 44*(1), 12–25.

Tsirulnikov, K., Suart, C., Abdullah, R., Vulcu, F., & Mullarkey, C. E. (2023). Game on: Immersive virtual laboratory simulation improves student learning outcomes & motivation. *FEBS Open Bio, 13*(3), 396–407. Advance online publication. DOI: 10.1002/2211-5463.13567 PMID: 36723273

University at Buffalo Health Sciences Library. (2013). Providing mobile reference and instruction services using iPads. *Journal of the Medical Library Association*, 101*(3). https://pmc.ncbi.nlm.nih.gov/articles/PMC4733882/

Villagrasa, S., Fonseca, D., Redondo, E., & Martí, N. (2024). A systematic review of accessibility and inclusive design in virtual laboratories. *Frontiers in Education, 9*, 1351711. Advance online publication. DOI: 10.3389/feduc.2024.1351711

Wagwu, V., Okpala, A. E., Maxwell, C. E., & Oladokun, B. D. (2024). AI-Driven Chatbots for Library Services towards Transforming Patrons' Engagement in Nigerian University Libraries. Communicate. *Journal of Library and Information Science, 26*(1), 89–98. https://www.cjolis.org/index.php/cjolis/article/view/78

Whitchurch, C. (2011). QR codes and library engagement. *Bulletin of the American Society for Information Science, 38*(1), 14–17. DOI: 10.1002/bult.2011.1720380107

WorldMetrics Research. (2024). *Cellphones in school statistics: Market data report 2024*. WorldMetrics.

Yan, R., Zhao, X., & Mazumdar, S. (2023) conducted a systematic literature review titled "Chatbots in libraries: A systematic literature review," published in Education for Information (Volume 39, Issue 4).

Yorio, K. (2023). School Librarians Explore Possibilities of ChatGPT. School Library Journal. https://www.schoollibraryjournal.com/story/School-Librarians-Explore-Possibilities-of-ChatGPT

Zhang, N., & Liu, Y. (2024). Design and implementation of virtual laboratories for higher education sustainability: A case study of Nankai University. *Frontiers in Education*, *8*, 1322263. Advance online publication. DOI: 10.3389/feduc.2023.1322263

Zhu, X., Whitaker, E., Cho, M., & Zhang, M. (2025). (forthcoming). Student engagement in academic libraries: A conceptual framework. *College & Research Libraries*, *86*(1). Advance online publication. DOI: 10.5860/crl.86.2.275

Chapter 11

Pioneering Excellence: Evaluating and Advancing Library Mobile Services

ABSTRACT

This chapter provides frameworks for assessing the effectiveness of library mobile services, including multi-impact studies, usability evaluations, satisfaction measures, and emerging assessment methodologies. It examines how AI, machine learning, XR, and advanced connectivity are reshaping mobile service design and expectations. Strategic planning considerations—governance, funding, privacy, licensing, collaboration, and accessibility—are analyzed as essential components of sustainable mobile ecosystems. The chapter also highlights global trends, infrastructure challenges, and research priorities that will shape future mobile services. It emphasizes the need for continuous evaluation, responsible technology adoption, and institution-wide alignment to advance high-quality, equitable mobile library experiences.

INTRODUCTION

The evolution of mobile library services has profoundly reshaped contemporary library practices, transforming how libraries serve their communities. As we transition into a digital and mobile-centric environment, libraries face intricate challenges while simultaneously embracing innovative opportunities. This synthesis of current research findings, evaluation of service quality metrics, identification of implementation challenges, and exploration of emerging trends will illuminate the trajectory of mobile library services in the future.

The rapid advancement of mobile technologies has revolutionized the realm of library services, facilitating unprecedented access to information and resources for

DOI: 10.4018/978-1-7998-9175-8.ch011

users worldwide. As libraries progressively integrate mobile platforms to align with user demands, assessing the efficacy of these services and formulating strategies for future enhancements assumes paramount importance. The aforementioned studies comprehensively evaluate mobile library services, addressing challenges and devising strategies for their implementation, and exploring prospective directions. They concentrate on emerging technologies and global collaboration, drawing upon peer-reviewed studies conducted by Dr. Yan Quan Liu. This in-depth analysis, grounded in empirical research, offers practical insights and practical applications for academic, public, and school libraries.

ASSESSING THE EFFECTIVENESS OF LIBRARY SERVICES TO MOBILE USERS

Libraries are at the forefront of innovation, evolving to meet the changing needs of their communities in creative and impactful ways. The rapid adoption of mobile technology within libraries has reshaped traditional service models, providing expanded access and convenience to diverse user populations. However, the success of mobile services cannot be assumed; systematic assessment is imperative. Evaluation mechanisms, particularly those that assess service quality, usage patterns, and user satisfaction, are crucial for comprehending the tangible impact of mobile initiatives.

A prominent study led by Liu and Briggs (2015) introduced an empirical framework for evaluating library mobile services across several academic institutions in the United States. The study utilized both LibQUAL+ and custom-designed mobile user satisfaction surveys to identify users' perceptions of mobile resources, responsiveness of mobile reference platforms, and ease of mobile catalog navigation. Results indicated that over 72% of users expressed high satisfaction with mobile OPACs and mobile-responsive websites, but fewer than 50% felt that mobile services fully supported deep research tasks, (Liu & Briggs, 2015). His research provides a robust framework for understanding these dimensions, particularly through studies that assess user perceptions and service efficacy in academic libraries.

Mult Impact & Usability Studies

Evaluating mobile library services requires a multifaceted approach that encompasses user satisfaction, service quality, and measurable outcomes. Huang et al. (2015) developed and evaluated a Mobile Library Service System Success Model using the Information System Success Model as the theoretical foundation. Conducted with students from the National University of Tainan, the study demonstrated a significant positive correlation between user satisfaction and system usage—highlighting enhanced efficiency in tasks such as searching for books and accessing electronic resources. The model achieved an explanatory power of 55% ($R^2 = 0.55$). While the mobile system showed promise in improving workflow efficiency and user willingness to continue using it, the findings also underscored the need to enhance service quality, particularly through better staff training and more user-friendly features. The study presents a robust, empirically validated framework that can guide academic libraries in designing, assessing, and enhancing mobile services through data-driven insight.

Usability studies have provided valuable insights into how academic libraries can refine mobile service design to enhance user satisfaction and efficiency. Wei, Chang, and Cheng (2015) examined the usability of a mobile library application developed at Chongqing University in China. Their evaluation also centered on efficiency, error rates, and user satisfaction, but with a specific focus on how different service types—such as e-books, library news, and database access—shaped user perceptions. Findings indicated high satisfaction with e-book reading and news updates, yet slower task completion and occasional confusion occurred when users attempted to access less frequently used databases. Based on these results, the researchers recommended reorganizing the menu structure and introducing context-specific help prompts to streamline navigation for infrequent tasks.

ChanLin and Hung (2016) carried out a comprehensive usability evaluation of a university library's mobile website in Taiwan, integrating both task-based performance testing and post-task user surveys. Their mixed-method design allowed them to measure objective performance indicators, such as navigation flow and task completion time, alongside subjective satisfaction ratings. Results revealed that while students navigated basic search functions with relative ease, they often struggled to locate more advanced features. Subsequent design adjustments, particularly the incorporation of clearer visual cues and better-structured menus, significantly improved overall satisfaction. This study demonstrated the importance of coupling direct performance measures with user feedback to identify subtle interface challenges that may not be immediately apparent.

A more recent perspective comes from Mondal (2025), who conducted a structured questionnaire survey with 150 students at Kazi Nazrul University in India. This study assessed aspects such as navigation ease, system responsiveness, and information accuracy in the library's mobile platform. While respondents expressed strong satisfaction with the catalog and e-resource access, they pointed out issues with interface design and occasional technical glitches. Notably, users also voiced a desire for enhanced mobile features, such as integrated live chat support and offline access to essential materials—functionalities that align with evolving expectations in a mobile-first environment. By documenting these emerging needs, Mondal's study extends the conversation beyond usability alone to include modern service integration.

Together, these studies illustrate a progressive shift in mobile library usability research: from early explorations of task efficiency and interface clarity to broader considerations of user-centered service design that addresses both functional performance and evolving digital expectations. They collectively underscore that mobile platforms in academic libraries must be evaluated not only for their technical efficiency but also for their ability to anticipate and adapt to the changing patterns of student engagement.

Here's a comparison table summarizing the key features, methods, and findings from ChanLin and Hung (2016), Wei, Chang, and Cheng (2015), and Mondal (2025) so you can easily contrast their contributions to mobile library usability research.

Table 1. Key features, methods, and findings on usability studies

Study & Citation	Institution / Context	Methodology	Key Evaluation Focus	Main Findings	Notable Contributions
ChanLin & Hung (2016)	University library in Taiwan	Task-based usability testing + user surveys	Navigation flow, task completion time, interface clarity	Users navigated basic search well but had difficulty locating advanced features; satisfaction increased when visual cues were added	Demonstrated the value of combining performance metrics with subjective feedback to refine mobile web design
Wei, Chang, & Cheng (2015)	Chongqing University, China	Usability testing with student participants	Efficiency, error rates, satisfaction with mobile app functions	High satisfaction for e-book and news services; slower task completion in locating less-used databases	Highlighted how service type influences perceived usability; suggested reorganizing menu structure
Mondal (2025)	Kazi Nazrul University, India	Structured questionnaires (150 respondents)	Navigation ease, responsiveness, information accuracy	High satisfaction for catalog and e-resources; issues with interface design and technical glitches; requests for chat support and offline access	Provided updated insights into user needs, incorporating modern mobile expectations like offline access and live help

User Satisfaction and Engagement

User satisfaction is a cornerstone of evaluating mobile library services, as it reflects how effectively these platforms meet patron needs. User satisfaction remains a critical measure when evaluating mobile or automated library services. The study by Nunekpeku (2019) of the Sam Jonah Library at the University of Cape Coast revealed that slightly more than half of the surveyed patrons were only moderately satisfied with existing library automation features—such as internet access and reference support. Notably, adoption of services like e-resources, OPAC, and institutional repositories was reported to be low, suggesting gaps in user engagement. The study underscores the importance of targeted user education programs to enhance both awareness and effective use of digital services.

Liu's research (Liu & Briggs, 2015; Liu, Lü, & Liu, 2014; Shi & Liu, 2017) emphasizes the significance of continuous assessment. Regular feedback collection through surveys and analytics aids libraries in refining their mobile offerings. The libraries studied demonstrated iterative updates and expanded functionality in response to user input. Shi & Liu (2017)'s "Continuous Use Intention of Library Mobile Services: Evidence from University Libraries in China" delves into the factors influencing users' continuous use of mobile library services in Chinese university libraries. This research provides valuable insights for enhancing the service and information quality of these services. Best practices observed in these institutions include user-friendly navigation, comprehensive search features, accessible help desks, and proactive notifications or chat support within mobile interfaces. The effectiveness and adoption rates of mobile library services are closely linked to proactive user engagement strategies, such as personalized content and regular needs assessments.

The study "Use of Mobile Technology-Based Library Services and Information Needs Satisfaction of Undergraduates in Universities in South-West Nigeria" by Udem and Ogungbeni (2023) surveyed 1,045 undergraduates across six universities. It found that while usage of certain mobile library services varied, most students reported satisfaction, and a strong positive correlation ($r = 0.53$) emerged between the use of these mobile services and fulfillment of their information needs. The authors concluded that, with adequate infrastructure and support, mobile library platforms can meaningfully enhance student engagement and satisfaction

Measuring Service Quality and Outcomes

The evaluation of mobile library services requires sophisticated measurement frameworks that capture both quantitative performance metrics and qualitative user experiences. Service quality in mobile library contexts involves assessing both inputs, such as technological infrastructure) and outputs, such as user interactions.

The evaluation of mobile library services requires a thorough analysis of both the inputs and outputs involved in the service.

- Inputs refer to the resources that are put into the service, such as financial investments for creating mobile platforms, hours spent training staff, costs for technology infrastructure, and ongoing maintenance expenses. These measurements are crucial for understanding the return on investment and for making informed budget decisions.

- Outputs, on the other hand, measure what the service produces, including usage statistics, how available the services are, and how accessible the content is to users. Important indicators include the number of times the mobile app is downloaded, the traffic to mobile websites, the frequency of accessing digital resources, the number of virtual reference questions asked, and requests made for interlibrary loans through mobile services.

With advancements in analytics technology, libraries can now closely monitor user behavior to see which features are popular and where improvements are needed. Moreover, metrics that measure user engagement have become more detailed, including how long users stay on the app, how often they use various features, and how successful they are in completing tasks. These specific measurements help libraries understand user preferences and the overall effectiveness of their services better than traditional statistics, like circulation data, would. Libraries that utilize comprehensive analytics find that they can make better decisions and develop services that are more closely aligned with user needs.

Contemporary service quality assessment models have evolved beyond traditional satisfaction surveys to incorporate multi-dimensional approaches. Current research frequently adapts and expands upon well-known information systems and service quality models targeting mobile platforms—often focus on dimensions like reliability, responsiveness, and ease of use.

Key Models and Frameworks

- The Delone and McLean Model is a widely used framework for evaluating the success of mobile libraries. Researchers apply this model to analyze the causal relationships between various factors, such as system quality, information quality, and service quality, and their impact on user satisfaction and the user's intention to continue using the service. A key finding of this model is that these quality dimensions positively influence user satisfaction, which, in turn, leads to a higher intention to use the mobile library. This model serves as a robust framework for evaluating information systems, postulating that system quality, information quality, and service quality influence system use and user satisfaction, ultimately resulting in net benefits. The Delone and McLean Model holds significant relevance for digital mobile library applications, as digital libraries inherently constitute information systems.
- The LibQUAL+ model, widely adopted in academic settings, serves as a framework for measuring service quality across three core dimensions: infor-

mation control, the impact of the service, and the library as a physical space. When applied to mobile services, these dimensions transform into mobile interface usability, responsive customer support, and virtual space design effectiveness. The LibQUAL+ model evolved from the SERVQUAL instrument and measures library users' minimum, perceived, and desired levels of service quality across three primary dimensions: Affect of Service (AS): This dimension focuses on users' perceptions of the library staff's competency, abilities, and support, as well as how they are treated within the library. Information Control (IC): This assesses the adequacy of the library's collection to meet user demands, ensuring that a comprehensive range of resources is available for users to access and utilize. Library as Place (LP): This dimension considers various physical elements, including facilities, equipment, personnel, and communication materials, as well as areas that promote study, quiet, isolation, introspection, and contemplation.

- SERVQUAL: This model, which gauges the disparity between a user's expectations and their perception of service, is also adapted for mobile library contexts. Studies employing this framework have identified dimensions such as responsiveness, tangibility, reliability, assurance, and empathy as pivotal for service quality. Notably, some studies indicate that user expectations often surpass their perceptions, highlighting the need to address this gap. This widely recognized instrument for measuring service quality, based on the gap between customer expectations and perceptions across five dimensions: Tangibles, Reliability, Responsiveness, Assurance, and Empathy, is foundational in service quality research. However, its direct, explicit application to mobile library contexts, particularly digital ones, is not consistently demonstrated in the current studies. Instead, the available evidence suggests its use in general academic library settings.
- Stimulus–Response Theory, a theoretical framework, was used in a study to investigate factors influencing the adoption of public mobile libraries. The study found that technological factors, such as system quality and ease of use, content factors, like information quality, and individual factors, such as information literacy, all significantly impact a user's intention to use the service. This theory has expanded to Stimulus–Organism–Response (S-O-R), which examines how external stimuli promote continuous usage intentions through mediating factors like engagement and satisfaction. In this context, stimuli refer to external environmental factors that provoke emotional and behavioral changes in individuals. The S-O-R model is renowned for its flexibility, enabling researchers to incorporate and examine a wide range of variables in behavioral studies.

- The Technology Acceptance Model (TAM) is a widely used framework in research on technology adoption. It posits that perceived usefulness and ease of use are crucial factors that influence a user's attitude and intention to use a mobile library application. User satisfaction, a key outcome, is often found to have the strongest impact on a user's intention to continue using the service. The TAM focuses on perceived usefulness (PU) and perceived ease of use (PEOU) as key determinants of user attitude and behavioral intention to use a technology. It has been well-validated and is considered a reliable model for understanding technology adoption.

Evaluation Studies

Evaluation of mobile library initiatives requires measures that capture both technical delivery and user experience. Studies often explore various factors and their influence on user satisfaction, behavioral intentions, and literacy. Research by Liu and colleagues examining top university libraries found that all institutions offered at least one mobile service, with the majority providing multiple services including mobile sites, text messaging, e-books, and mobile database access, (Liu & Briggs, 2015). This widespread adoption necessitates robust evaluation mechanisms to ensure service effectiveness.

Liu and Briggs's benchmark study of top 100 university libraries examined which mobile services were offered (mobile OPAC, apps, mobile-adapted websites, mobile search functions) and how libraries structured those services to meet user expectations. Their content analysis and site visits revealed wide adoption of mobile catalogs and responsive sites, while full-featured mobile apps remained less common and more variable in scope, (Liu & Briggs, 2015). Taken together, these findings encourage evaluators to pair availability checks (what services exist) with usage and satisfaction measures (how patrons actually use and value them).

Zhang et al. (2016) explored platform choice factors in mobile library services, although detailed statistical findings were not reported. However, studies generally suggest that mobile-optimized interfaces can improve user satisfaction and usage metrics.

Liu's later multi-site examinations extended this baseline approach into municipal and public settings. In a study of U.S. urban libraries, Guo, Liu, and Bielefield (2018) combined mobile visits, content analysis, and librarian interviews to document service portfolios and operational responses. They found that nearly all urban libraries maintained at least one mobile presence (catalog, site, or app) and that practical service features—scan-to-search, location navigation, mobile printing links, and links to virtual reference—were common. This study demonstrates a pragmatic evaluation approach: begin with a structured inventory of mobile affordances (service input) and follow with targeted measures of outputs (downloads, session lengths, reference contacts made via mobile) and outcomes (task completion, reported satisfaction).

For public libraries, Liu and Lewis (2023) performed a broad content analysis supplemented by librarian surveys across a stratified sample of high-usage public libraries. They documented the rapid expansion of mobile features that became standard during and after the pandemic—mobile renewal, curbside reservation and pickup workflows, easier access to databases on mobile devices—and observed how the pandemic accelerated the adoption of contactless and mobile services (Liu & Lewis, 2023). Their work shows the value of longitudinal or repeat cross-sectional studies for detecting service change during shocks (e.g., the COVID-19 pandemic) and for identifying which mobile services persisted after emergency measures abated.

Practical evaluation tools that follow from this body of work include a three-layer method: (1) inventory (catalog what mobile services and features are offered), (2) analytics (collect use metrics — app downloads, page views on mobile, mobile reference sessions, mobile circulation transactions), and (3) user feedback (targeted mobile usability testing, short in-app or SMS surveys to capture satisfaction and unmet needs). Liu's studies emphasize that combining these layers produces actionable results (e.g., staffing adjustments, feature redesigns, or decisions to invest in a native app versus a responsive site) rather than merely producing vanity metrics, (Liu & Briggs, 2015; Guo, Liu, & Bielefield, 2018; Liu & Lewis, 2023).

Usability assessments of academic library mobile services provide essential insights into both strengths and areas for improvement. A usability study of Portland State University's mobile library website found that while users generally valued mobile access, they encountered difficulties due to site design, wireless network connectivity, and hardware/software compatibility, particularly across diverse smartphone types, (Pendell & Bowman, 2012). These findings underscore the importance of conducting structured usability testing to ensure consistent performance across devices. Similarly, a usability evaluation of a mobile library application at Chongqing University revealed that, despite overall satisfaction, users reported challenges with efficiency and clarity, highlighting the need for intuitive navigation and clear service descriptions, (Wei, Chang, & Cheng, 2015).

On a broader scale, national surveys show that academic libraries are rapidly integrating mobile and digital services into their operations. The 2023 ACRL State of U.S. Academic Libraries report indicates that libraries continue to expand mobile platforms and digital resources to meet user demand, though challenges in accessibility and cross-device compatibility remain significant (Goek, 2024). Taken together, these studies emphasize that ongoing evaluation, user-centered design, and proactive troubleshooting are critical for delivering reliable and effective mobile library services in academic contexts.

Recent studies have demonstrated the critical importance of user-centered evaluation approaches. Research examining public library service quality using Brady & Cronin's three-dimensional model revealed that physical environment quality, interaction quality, and outcome quality all significantly influence user satisfaction (Korean Society for Library and Information Science, 2024). In the mobile context, these dimensions translate to app design quality, customer service responsiveness through mobile channels, and successful task completion rates.

Measuring the broader impact of mobile library services extends beyond immediate usage statistics to encompass educational outcomes, community engagement, and institutional goal achievement. Academic libraries increasingly evaluate mobile services' contributions to student retention, academic performance, and graduation rates. Public libraries assess mobile services' impact on digital literacy improvement, community engagement, and equitable information access.

Longitudinal studies tracking user behavior changes over time provide valuable insights into mobile service effectiveness. Research indicates that successful mobile library initiatives demonstrate measurable improvements in user satisfaction scores, increased library resource utilization, and enhanced perception of library relevance. These outcomes justify continued investment and guide strategic planning processes.

Emerging Approaches and Directions

Beyond adapting established models, recent research has explored more nuanced and context-specific approaches to measuring service quality in mobile library environments, particularly focusing on the user's emotional and holistic experience.

Zhao's research team conducted a study in 2022 to assess users' emotional experiences in mobile libraries. They employed the Pleasure-Arousal-Dominance (PAD) Emotion Model and the Five Factor Model (FFM) to evaluate user emotions related to the User Interface, Interaction Quality, and Service Environment. The study revealed that users often felt anxious and bored despite exhibiting high arousal, suggesting a lack of pleasure and dominance. Moreover, the service environment had a stronger emotional impact than the user interface and interaction quality. This finding indicates a significant shift in understanding service quality in digital contexts. It goes beyond simply assessing whether a system functions correctly or provides accurate information to how it affects users' emotions. In the increasingly competitive and user-centric mobile application market, a purely utilitarian approach is insufficient. Emotional engagement and a positive user experience (UX) become crucial differentiators for user satisfaction and continued use. Therefore, future quality assessments must integrate psychological and affective metrics alongside functional ones. This study opens new avenues for library designers and service developers to prioritize emotional impact as much as efficiency or accuracy, fostering genuinely resonant digital experiences. If you'd like help weaving this into a broader narrative on UX in your writing, let me know.

Wulfert's (2019) study, while primarily focused on mobile shopping companion applications, developed a hierarchical and multi-dimensional scale for Mobile App Service Quality (MASQ). This model encompasses primary dimensions such as Interaction Quality, Environment Quality, and Outcome Quality, along with secondary dimensions like responsiveness, information, security/privacy, design, performance, technical reliability, and valence. This study explicitly demonstrates the development of industry-specific, tailored models for mobile applications. The MASQ study argues that existing generic scales for Electronic Service Quality (ESQ) and Mobile Service Quality (MSQ) fail to capture all aspects of mobile shopping applications and may lead to inconsistent results. Consequently, it advocates for a tailored measurement scale. This underscores the need for bespoke measurement tools that capture the unique characteristics of specific mobile application domains, such as mobile libraries. This targeted approach, exemplified by MASQ, ensures more accurate and actionable quality assessments. Such models can inform best practices in other domains, such as mobile access to library services. If you're interested in exploring parallels in library-focused models or contributing to crafting similar scales for your area of interest, please let me know.

Developing or adapting models that explicitly account for the unique aspects of mobile information delivery, user behavior on mobile devices, and the specific goals of library services (e.g., literacy, community outreach, academic support) is crucial for effective evaluation and strategic improvement. This implies that future research should focus on developing and validating mobile-library-specific service quality models.

Mobile library services must be evaluated with measures that reflect mobile-specific goals (access, learning support, quick interactions) and mobile-specific constraints (screen size, connectivity, device diversity). Research across usability, accessibility, personalization, and virtual learning supports each MLSQ dimension:

Environment Quality (Interface, Availability, Device Compatibility)

Mobile apps and sites must meet accessibility and usability standards to serve diverse users and devices. Studies evaluating mobile accessibility frequently report failures such as small touch targets and poor color contrast, recommending adherence to WCAG guidelines and comprehensive device testing as core quality activities (Hariri & Noorhidawati, 2021; Villagrasa, Fonseca, Redondo, & Martí, 2024). Empirical research on mobile usability for library websites further demonstrates that optimized mobile designs significantly reduce time-on-task and improve task completion rates (ChanLin & Hung, 2016). Together, these findings emphasize that interface design, reliability, and cross-device compatibility are foundational components of perceived mobile service quality (Hariri & Noorhidawati, 2021; Villagrasa et al., 2024).

Delivery Quality (Search & Navigation, Responsiveness, Personalization, Transaction Efficiency)

Usability evaluations of mobile library platforms demonstrate that improvements in search/navigation design and system responsiveness significantly enhance satisfaction and foster repeated use. For example, ChanLin and Hung (2016) conducted task-based usability testing on a university mobile library site and found that students completed more search tasks more accurately and, in less time, when using the mobile version compared to the desktop version, underlining the efficiency gains of optimized mobile interfaces. Additionally, incorporating clearer menus and visual cues further improved user attitudes and task success (ChanLin & Hung, 2016).

Content Quality (Relevance, Authority, Multimedia, Metadata Accuracy).

Research on metadata management and digital collections consistently emphasizes that content quality—including accurate metadata, reliable resource links, and high-quality multimedia—is essential for ensuring trustworthiness and usability in mobile library environments. Studies by Zhou et al. (2020) and Okonkwo and Ejikeme (2023) highlight that poor metadata quality or broken content pathways can undermine user experience gains achieved through optimized mobile interfaces. Consequently, content quality should be evaluated as a distinct dimension from interface or technical quality in mobile service assessments.

Engagement & Learning Support (Interactivity, Tutorials, Gamification)

Evidence from library implementations and educational platforms shows that gamified reading applications and mobile interactive tutorials significantly increase user engagement and motivation. Studies of reading-challenge platforms commonly used in libraries indicate that gamification features—such as badges, progress tracking, and challenges—effectively enhance user participation and learning support (Beanstack, n.d). These findings suggest that gamification and mobile learning supports are crucial engagement levers and therefore rightly belong in the

Mobile Library Service Quality (MLSQ) Engagement Dimension

Multiple surveys and empirical evaluations of mobile library services consistently find that user satisfaction, perceived usefulness, and reduced time-to-task are strong predictors of continued use (Liu & Yan, 2015; Ahmad et al., 2020). Usability and service-quality studies emphasize that outcome measures should incorporate both short-term satisfaction and longer-term indicators, such as retention rates and measurable improvements in information literacy or research efficiency (Shi & Liu, 2017; Liu et al., 2014). These findings highlight the importance of assessing mobile service outcomes comprehensively to understand their educational and behavioral impact.

DEMANDING TECHNOLOGIES AND THEIR IMPACT ON LIBRARY SERVICES

The swift evolution of technology is profoundly reshaping the landscape of mobile library services, introducing new possibilities for user engagement, operational efficiency, and personalized experiences. Emerging technologies are revolutionizing mobile library services by enabling innovative delivery methods. Mobile applications, QR codes, SMS notifications, and cloud-based platforms are among the key tools enhancing accessibility and efficiency. Dr. Liu's work in this arena highlights how pervasive and wearable technologies, such as augmented reality (AR) glasses, smartwatches, and AI-powered mobile assistants, are shaping new pathways for user engagement. AR, VR, and AI are being utilized to create interactive and engaging services. AR applications provide users with immersive experiences when exploring library collections, while AI-driven chatbots assist users in navigating library resources (Baker & Muir, 2020).

Libraries are rapidly integrating emerging technologies to enhance services for mobile users, offering more personalized, accessible, and efficient ways to engage with resources. Innovations such as artificial intelligence (AI), machine learning, chatbots, and predictive analytics enable mobile applications to provide tailored recommendations, automate routine reference tasks, and improve search precision. Augmented reality (AR) and virtual reality (VR) technologies create immersive learning experiences, supporting virtual tours, interactive exhibits, and experiential learning accessible directly from mobile devices. Additionally, mobile platforms increasingly incorporate real-time notifications, location-based services, and adaptive learning systems, allowing users to receive context-specific information and resources anytime and anywhere. Cloud computing, edge technologies, and high-speed mobile networks further enhance content accessibility and responsiveness, ensuring seamless interaction across devices. Collectively, these technologies provide a conceptual framework for exploring how libraries can innovate mobile services, from AI-driven research assistance and recommendation engines to immersive AR/VR learning tools and responsive, adaptive mobile interfaces. The distinct classes of emerging technologies in detail are highlighted below as their particular relevance and transformative potential for library services delivered to mobile users.

Artificial Intelligence and Machine Learning in Library Services

Artificial Intelligence (AI) and Machine Learning (ML) are rapidly transforming library operations and user interactions, significantly enhancing efficiency and accuracy. These technologies are revolutionizing mobile library services through personalized recommendations, intelligent search capabilities, and automated customer service. Advanced algorithms analyze user behavior patterns to provide customized content suggestions, optimize interface layouts, and predict information needs. Key applications are becoming widespread: intelligent retrieval and recommendation systems are adopted by nearly all (99%) US university libraries, while intelligent reference services are utilized by over half (53%), and intelligent guided tour technology by a substantial 95%. (Rahmanova, 2025; Liblime, 2024).

AI-powered search algorithms demonstrably improve the efficiency of information retrieval, enabling users to locate relevant information more quickly and accurately. Concurrently, machine learning facilitates the development of personalized user interfaces and tailors' recommendations based on individual borrowing and search patterns. Furthermore, AI automates repetitive tasks such as cataloging and inventory monitoring, thereby freeing librarians to focus on more complex assignments that require human expertise and judgment. AI also significantly enhances accessibility by providing translation tools for multilingual resources and offering speech-to-text and text-to-speech functionalities for visually impaired or differently abled users. Over half of libraries (61%) are exploring generative AI technologies, which hold immense potential for creating diverse content forms and further reshaping library service models. (Rahmanova, 2025).

This profound shift signifies a move from merely providing access to information to curating highly personalized and immersive experiences. AI's recommendation systems and XR's interactive tours exemplify this progression, moving beyond simple retrieval to anticipatory and engaging interactions. This suggests that the value proposition of libraries in the mobile age is increasingly tied to the quality and relevance of the user experience, compelling libraries to invest in UX/UI design expertise and data analytics to understand individual user behaviors and preferences. The focus must be on creating intuitive, tailored, and delightful mobile journeys, rather than just digitizing existing services.

While AI offers immense benefits, their rapid adoption also raises critical ethical considerations, including data privacy, algorithmic bias, and the potential to exacerbate the digital divide. The proliferation of generative AI, for instance, introduces new questions regarding authors' rights and data protection. This underscores that technological advancement must be accompanied by a robust ethical framework and proactive measures to mitigate potential harms. Libraries must develop clear ethical guidelines for AI and XR implementation, including transparent data collection practices, strategies for bias mitigation, and ongoing user education on responsible technology use. This proactive approach is essential to maintain user trust and ensure equitable access, (Guo, et al., 2025).

Moreover, AI automates routine tasks, allowing librarians to dedicate more time to "more difficult assignments". This indicates that emerging technologies are not replacing librarians but augmenting their capabilities, enabling them to focus on higher-value activities such as complex queries, personalized guidance, and community engagement. The concept of a "smart library" integrated with AI points to an ecosystem where technology and human expertise are synergistically combined. This evolution necessitates that library professional development prioritizes training in AI literacy, data analytics, and user experience design. The future role of librarians will increasingly involve curating and managing intelligent systems, providing advanced research support, and acting as digital literacy educators.

Extended Reality (XR): Enhancing User Engagement and Learning

Extended Reality (XR), an umbrella term encompassing Virtual Reality (VR), Augmented Reality (AR), and Mixed Reality (MR), is emerging as a powerful tool for delivering immersive and interactive experiences within library settings. Mobile AR applications, in particular, offer innovative ways to engage patrons, such as augmenting the browsing experience of physical book stacks, assisting with intuitive library navigation, and providing interactive tours. AR-Books, which integrate interactive features and summaries, have demonstrated a positive impact on reading interest, especially among younger readers.

Beyond navigation and enhanced browsing, VR is being utilized to create immersive storytelling experiences and simulations, allowing users to step into historical events or literary worlds, thereby engaging with content in novel and profound ways. XR is recognized as a productive tool for context-aware library management, streamlining routine operations, and delivering compelling information experiences to patrons.

Virtual reality (VR) and augmented reality (AR) represent frontier applications for mobile library services. VR platforms enable virtual library tours, immersive educational experiences, and remote access to specialized collections. Academic libraries increasingly offer VR-based research tools that allow students to explore historical environments, manipulate scientific models, and participate in collaborative virtual spaces.

AR applications enhance physical library experiences by overlaying digital information onto real-world environments. Users can access book reviews, availability status, and related recommendations by scanning book spines with mobile devices. Navigation assistance, event information, and interactive exhibits represent additional AR applications gaining traction in progressive library systems.

Mixed reality platforms that combine VR and AR capabilities offer unprecedented possibilities for hybrid service delivery. These platforms enable seamless transitions between physical and digital library spaces, supporting diverse learning preferences and accessibility requirements.

Enhanced Remote Wireless Connectivity

The advent of next-generation broadband cellular network technologies has profoundly impacted mobile library services, enabling more robust, pervasive, and intelligent access to information. Emerging technologies are transforming remote network connectivity, offering faster speeds, lower latency, improved security, and greater reliability. These technologies are evolving rapidly to handle higher bandwidth, lower latency, better security, and wider accessibility, especially as remote work, IoT, and cloud computing expand. The landscape of remote network connectivity is rapidly evolving, driven by the need for greater speed, security, and flexibility.

5G/6G Networks

The advent of next-generation broadband cellular network technologies, specifically 5G and the foreseen 6G, holds immense promise for transforming mobile library services and society at large. 5G is already emerging in markets, while 6G is currently under research. These technologies are expected to significantly enhance connectivity, offering lower latency for providing users with real-time feedback, higher data rates (ranging from 1 to 10 gigabytes per second for end-users), and increased reliability even with a high density of diverse devices. 6G, as the successor to 5G, is anticipated to be even more responsive, reliable, faster, and capable of higher data transfer rates. It aims to be accessible anywhere, secure, energy-efficient, and compatible with existing networks like 5G and Wi-Fi. These advancements will enable improved quality of service and experience through edge computing capabilities, higher quality content delivery services, and increased personalization in services via caching. Furthermore, 5G and 6G can reduce the rural-urban digital divide and create new market opportunities by allowing the leasing of network capacity to service providers. Despite these promises, a challenge for 5G adoption is the potential lack of adequate user equipment, such as 5G smartphones, to fully leverage the network's capabilities.

Wi-Fi 6/7 Networks

As the sixth generation of Wi-Fi, Wi-Fi 6 improves efficiency, flexibility, and scalability, and providing higher speeds and capacity for modern applications. As the latest development, Wi-Fi 7 (IEEE 802.11be offers significant improvements in speed, capacity, and efficiency, (Murad et al., 2024). Key features include 320 MHz ultra-wide bandwidth in the 6 GHz band, 4096-QAM modulation for 20% higher data transmission, Multi-RU, and Multi-Link Operation. These advancements translate to speeds up to 4.8 times faster and 4 times lower latency compared to Wi-Fi 6/6E, (Murad et al., 2024). Wi-Fi 7's capabilities are essential for enhanced Virtual and Augmented Reality (VR/AR) experiences, smart homes, IoT ecosystems, Industry 4.0 automation, telemedicine, and high-definition gaming and streaming, (Murad et al., 2024). Libraries can leverage these advancements to deploy robust Wi-Fi networks that meet patron needs while ensuring network security, (Spacey et al., 2017).

Satellite Internet (LEO)

Low Earth Orbit (LEO) satellite constellations, such as Starlink, becomes a necessary component of future internet infrastructure (Lu et al., 2024). Low-earth-orbit (LEO) satellite constellations (e.g., Starlink) are becoming a necessary component of future Internet. There have been increasing studies on LEO satellite networking. It is a crucial problem how to evaluate these studies in a systematic and reproducible manner., 2025). Unlike traditional geostationary (GEO) satellites, LEO satellites orbit much closer to Earth (approximately 300-750 miles), resulting in significantly reduced latency (20-50 milliseconds compared to 400-700 milliseconds for GEO) and faster data transmission. This closer proximity allows for higher network resilience and expanded coverage, especially in remote or rural areas where traditional internet connections are unreliable or unavailable, thus playing a crucial role in addressing the digital divide (Low Earth Orbit (LEO) satellite internet is revolutionizing global connectivity by providing fast, reliable, and low-latency internet services worldwide. Unlike traditional satellite systems positioned in geostationary orbit far above the Earth at roughly 22,000 miles (35,000 kilometers), LEO satellites orbit much closer at approximately 300-750 miles (500-1,200 kilometers) above the Earth. LEO satellites can also support critical communications during natural disasters when traditional networks fail. Initiatives like "LEO Libraries" are exploring how this technology can connect underserved communities globally.

Next-Generation IoT

As at the center of global digitization, the Internet of Things (IoT) with over 41 billion IoT devices expected by 2025, leads to an exponential growth of data, (Brynskov, Facca, & Hrasko, 2018). This evolution pushes computing operations and data analytics to the edge, where data is processed locally on the device (edge computing) or closer to the source, (Brynskov, Facca, & Hrasko, 2018). Edge computing reduces latency, enabling faster, real-time decision-making, improving speed, security, and efficiency, and reducing energy consumption, (Brynskov, Facca, & Hrasko, 2018). AI-powered IoT (AIoT) combines AI and IoT to create new opportunities for automation and predictive analytics, (Brynskov, Facca, & Hrasko, 2018). Non-terrestrial networks (NTN) and enhanced security protocols are ensuring more reliable, scalable, and secure global IoT deployments, (Brynskov, Facca, & Hrasko, 2018). In libraries, IoT devices enable convenient and secure access to resources, facilitate mobile self-loan services, and monitor indoor environmental parameters like air quality, temperature, and humidity to create conducive learning environments.

Sensor Networks

Wireless Sensor Networks (WSN) involve a large number of multifunctional sensor nodes distributed over a wide area, collecting environmental data and communicating wirelessly. Sensors are used to convert physical quantity into numerical data. Various types of sensors can be coupled together to make a single node. A distributed array of these nodes can be deployed to collect environmental data by using appropriate sensors. Application of low powered short range radio transceivers as a communication medium between spatially distributed sensor nodes is known as wireless sensor network. In this thesis I build such a network by using Arduino, Raspberry Pi and XBee. In libraries, WSNs can enhance security by detecting theft and mutilation of collections and monitor environmental conditions to preserve materials. The systematic review synthesizes relevant literature from scholarly databases, institutional reports, and case studies to provide a comprehensive analysis of how intelligent solutions are being used in libraries worldwide. Advancements in sensor network technologies include energy harvesting techniques (e.g., vibration, thermal) to extend node lifetime and reduce battery waste, and robust security protocols like encryption and authentication to protect data transmission.

AI-Enhanced Access

Artificial Intelligence revolutionizes digital library services by enhancing access and user interaction, (Kumar, 2025). AI algorithms analyze user behavior and preferences to provide personalized recommendations and content curation, expanding the discoverability of resources, (Konwar, 2024). Natural Language Processing (NLP) enables more sophisticated, context-aware, and semantic search functionalities, allowing users to interact with digital libraries using natural language queries for more accurate results, (Konwar, 2024). AI also significantly enhances accessibility by providing translation tools for multilingual resources and offering speech-to-text and text-to-speech functionalities for visually impaired or differently abled users, (Konwar, 2024). AI-powered virtual assistants and chatbots provide 24/7 support, answering frequently asked questions, assisting with resource location, and even helping with research citations, thereby improving service availability and freeing librarians for more complex tasks. The study concludes that the use of AI in academic libraries is a big step toward better satisfying users' changing needs and offering helpful assistance for teaching, learning, and research. It is recommended that academic libraries embrace artificial intelligence (AI) as a vital tool in their service delivery as technology develops, ultimately improving the academic experience in general, (Tunjungsari & Suprihatin, 2024). Furthermore, AI automates routine processes like metadata generation and document analysis, streamlining organization and accessibility, (Konwar, 2024). AI-driven insights and accelerated troubleshooting also optimize network infrastructure, leading to more predictable, reliable, and measurable user experiences.

Edge Cloud-based WAN

The rapid expansion of the Internet of Things (IoT) ecosystem necessitates efficient data processing and analysis technologies, leading to the rise of edge computing. Edge computing brings data elaboration geographically closer to the user, reducing latency and enhancing data privacy by processing sensitive information locally without relying solely on the cloud. Cloud WAN (Wide Area Network) solutions leverage global networks to provide a fully managed, reliable, and secure enterprise backbone, optimized for application performance. Cloud WAN offers significant performance improvements (up to 40% faster than the public internet) and cost savings (up to 40% in total cost of ownership) compared to customer-managed WAN solutions. It provides flexible connectivity options for linking geographically dispersed data centers and connecting branch/campus environments, ensuring optimal application performance by minimizing network hops and efficient traffic exchange. Hybrid approaches, combining the strengths of both edge and cloud systems, offer promising solutions for optimizing bandwidth use and supporting low-latency, privacy-sensitive applications in IoT ecosystems.

Quantum Post-Quantum Cryptography (PQC)

The advent of quantum computing presents a significant threat to classical public-key cryptographic systems, such as RSA and Elliptic Curve Cryptography (ECC), which are widely used for secure internet communication, digital signatures, and data storage (Post-Quantum Cryptography Algorithms and Implementation Challenges, 2025). This has accelerated the development and standardization of Post-Quantum Cryptography (PQC), which comprises cryptographic algorithms believed to be secure against quantum attacks and suitable for deployment on classical computing infrastructure. PQC algorithms are being evaluated across categories like lattice-based, hash-based, and multivariate cryptographic schemes. The integration of PQC algorithms is particularly crucial for securing communication in Internet of Things (IoT) devices, where secure communication is essential but often constrained by limited computational resources, (Lopez, Cadena, & Rahman, 2025). Challenges in implementing PQC include computational overhead, memory usage, and energy consumption, necessitating further research to guide standardization and adoption in quantum-resilient infrastructures, (Lopez, Cadena, & Rahman, 2025).

LiFi Networks: Light Fidelity (LiFi)

As an emerging wireless communication system, it uses light as a carrier, offering an alternative to traditional radio frequencies used by Wi-Fi. This cutting-edge technology, still under development, has the potential to introduce greatly improved wireless services. LiFi is typically implemented using white Light Emitting Diodes (LEDs), where rapid and subtle variations in current produce optical output for data transmission.

Intensive Transformative Technologies

Beyond AI and XR, several other technologies are poised to further transform mobile library services. The Internet of Things (IoT) enables mobile library applications to provide real-time information, allowing users to identify resource locations, reserve books and study rooms, and register for library events directly through their devices. Data analytics, often powered by machine learning algorithms, plays a crucial role in analyzing circulation data and user preferences to inform acquisition decisions and optimize collection management, ensuring that library holdings remain relevant to community needs.

Chatbots and virtual assistants are increasingly providing 24/7 support, answering frequently asked questions and assisting with resource location, thereby improving service availability and freeing librarians for more complex tasks. Libraries are also exploring the potential of gamification to enhance engagement, robotics for automated tasks, 3D printing for creative and educational maker spaces, and various assistive technologies to improve accessibility for all users. These technologies collectively contribute to a more dynamic, responsive, and user-centric mobile library ecosystem.

Location-based and contextual services are also gaining traction, leveraging technologies such as Bluetooth Low Energy (BLE) beacons, geofencing, and indoor wayfinding systems. These technologies can significantly augment in-building discovery and provide personalized notifications to users. For instance, BLE beacons can be used in indoor localization systems to detect a user's room-level location, enhancing the user experience by providing cultural content related to observed artworks in a museum setting, which can be extended to libraries. Geofencing, which uses virtual boundaries to trigger actions when a user enters or exits a defined area, can deliver context-aware services and personalized recommendations. Crucially, the design of these services must prioritize privacy preservation, with mechanisms like dynamic geofence adaptation and encryption to safeguard location data and ensure user trust.

Furthermore, Progressive Web Apps (PWAs) are emerging as a powerful alternative to traditional native applications. Built using web technologies, PWAs offer app-like experiences that can be installed on various devices from a single codebase, thereby reducing distribution friction and improving cross-platform reach. They can feature application icons, run offline, support push notifications, and even access some hardware features, while being discoverable by search engines and shareable via standard web links. PWAs offer a lower cross-platform development cost compared to native apps, as a single codebase can serve websites, mobile apps, and desktop apps across different operating systems. While they may have limitations in accessing all device features or matching native performance for graphics-intensive applications, their instant accessibility and broad reach make them ideal for media platforms or e-commerce sites, a benefit that extends to library services.

Table 2. Intensive technologies and their applications in mobile library services

Technology	Core Functionality	Specific Mobile Library Applications	Benefits	Challenges/ Considerations
Artificial Intelligence (AI)	Data analysis, pattern recognition, automation, natural language processing.	Intelligent retrieval, personalized recommendations, intelligent reference services, automated cataloging, multilingual access, virtual assistants.	Improved efficiency and accuracy, enhanced personalization, increased accessibility, frees librarian time for complex tasks.	Data privacy, algorithmic bias, ethical implications, high implementation cost, requires staff expertise.
Extended Reality (XR: AR/VR/MR)	Immersive digital experiences overlaying physical world or creating virtual ones.	Augmented physical book browsing, library navigation, interactive tours, AR-Books, virtual storytelling, simulations.	Enhanced user engagement, immersive learning experiences, improved navigation, attracts younger users, new forms of content interaction.	Technical complexity, significant financial investment, lack of staff expertise, need for institutional support, data privacy concerns.
Internet of Things (IoT)	Network of physical objects embedded with sensors and software for data exchange.	Resource location identification, study room reservation, event registration via mobile apps, automated check-ins/check-outs, environmental monitoring.	Enhanced convenience for users, real-time information access, improved resource management and tracking, optimized learning environments.	Significant infrastructure requirements, data security vulnerabilities, privacy concerns related to constant data collection.
Data Analytics/ Machine Learning	Processing large datasets to identify patterns, predict trends, and inform decisions.	Analyzing circulation data, user preferences for acquisition decisions, optimizing collection management, forecasting borrowing trends.	More relevant collections, data-driven decision-making, improved resource allocation, personalized services.	Need for data scientists/ training, ethical use of patron data, infrastructure for data processing.
Chatbots/Virtual Assistants	AI-powered conversational interfaces for automated responses.	24/7 support for FAQs, resource location assistance, basic reference queries, event information.	Improved service availability, instant support, frees staff for complex tasks, consistent information delivery.	Development and maintenance costs, need for accurate knowledge bases, limited capacity for complex queries.

continued on following page

Table 2. Continued

Technology	Core Functionality	Specific Mobile Library Applications	Benefits	Challenges/ Considerations
Location & Contextual Services (BLE, Geofencing, Indoor Wayfinding)	Proximity detection, virtual boundary triggers, indoor positioning.	In-building discovery, personalized notifications, interactive tours, real-time content delivery based on location.	Enhanced in-library experience, tailored information, improved navigation, context-aware services.	Privacy concerns (location tracking), technical complexity of indoor positioning, infrastructure setup costs.
Progressive Web Apps (PWAs)	Web technologies delivering app-like experiences.	Cross-platform access to library catalogs, events, resources; offline functionality; push notifications without app store downloads.	Reduced distribution friction, improved cross-platform reach, lower development costs, instant accessibility, web discoverability.	Limited access to some device features, may not match native performance for graphics-intensive apps.
5G/6G Networks	High-speed, low-latency wireless communication using radio frequencies.	Enhanced mobile access to library resources, real-time interactive services, support for AR/VR applications, improved content delivery, bridging digital divide.	Faster speeds, lower latency, higher capacity, increased reliability, improved quality of experience, personalization, new market opportunities.	Requires compatible user equipment (e.g., 5G smartphones), infrastructure deployment costs, still under development (6G).
Wi-Fi 6/7 Networks	High-speed, high-capacity wireless communication using radio frequencies.	Enhanced VR/ AR experiences, smart home/ IoT integration, improved telemedicine, high-definition streaming, reliable library Wi-Fi.	Significant improvements in speed, capacity, and efficiency; lower latency; support for numerous connected devices.	Requires compatible routers and devices, infrastructure upgrades.
Satellite Internet (LEO)	Global high-speed, low-latency internet via low Earth orbit satellites.	Bridging digital divide in rural/ remote areas, disaster response communication, ubiquitous access to library resources.	Reduced latency, higher speeds, expanded coverage, high network resilience, rapid deployment.	Potential for obstructions (trees, mountains), temporary weather impacts, affordability concerns.
Next-Generation IoT	Network of physical objects with enhanced data processing and connectivity.	Secure access to library resources, mobile self-loan, environmental monitoring, smart building management.	Real-time decision-making, improved speed/security/ efficiency, reduced energy consumption, automation, predictive analytics.	Infrastructure requirements, data security vulnerabilities, privacy concerns.

continued on following page

Table 2. Continued

Technology	Core Functionality	Specific Mobile Library Applications	Benefits	Challenges/ Considerations
Sensor Networks	Distributed network of sensors for data collection and monitoring.	Collection security (theft/ mutilation detection), environmental monitoring (temperature, humidity), optimizing library operations.	Real-time data collection, enhanced security, extended node lifetime (with energy harvesting), reduced maintenance.	Power consumption (battery-operated nodes), signal interference, network coverage requirements.
AI-Enhanced Access	AI algorithms and tools to improve information access and user interaction.	Personalized recommendations, natural language search, virtual assistants, accessibility features (translation, text-to-speech).	More accurate and relevant search results, 24/7 support, increased inclusivity, streamlined processes, optimized network performance.	Data privacy, algorithmic bias, ethical considerations, implementation costs.
Edge Cloud-based WAN	Distributed computing architecture combining edge processing with cloud scalability.	Optimized performance for distributed library services, secure and reliable access to resources across locations.	Reduced latency, enhanced data privacy, improved application performance, cost savings, scalability, resilience.	Technical complexity of hybrid architectures, infrastructure setup.
Quantum Post-Quantum Cryptography (PQC)	Cryptographic algorithms resistant to quantum computer attacks.	Secure communication for library systems, protection of sensitive patron data, secure IoT device communication.	Safeguards against future quantum threats, ensures long-term data security.	High computational overhead, memory usage, energy consumption; still in development/ standardization.
LiFi Networks	Wireless communication using light as a carrier.	High-speed indoor connectivity, alternative to Wi-Fi in specific areas, secure data transmission within light-enabled zones.	Greatly improved wireless services, high data rates, potential for enhanced security in confined spaces.	Still under development, requires specific light infrastructure, line-of-sight limitations.

SHAPING THE FUTURE OF SERVICES FOR MOBILE USERS

Looking ahead, the trajectory of mobile library services is inextricably tied to emerging technologies and international partnerships. The future of library services is inextricably linked to mobile technology adoption and innovation. Libraries that successfully navigate current challenges while preparing for emerging opportunities will continue serving vital roles in their communities' information and educational needs. This transformation requires sustained commitment to user-centered design, strategic resource allocation, and continuous learning.

The rapid proliferation of mobile devices has fundamentally transformed how users access information, prompting libraries to adapt their services to meet the demands of a mobile-centric world. Academic and public libraries are increasingly leveraging mobile technologies to provide seamless, anytime-anywhere access to resources, thereby enhancing user engagement and satisfaction. This shift is driven by the recognition that mobile devices, such as smartphones and tablets, serve as primary tools for information retrieval among students and patrons. As noted in recent studies, mobile-based applications enable libraries to extend traditional services while introducing innovative features tailored to mobile environments. For instance, the integration of mobile technologies facilitates services like remote catalog searches, real-time notifications, and personalized resource recommendations, marking a significant advancement in library operations. The successful integration of mobile services into library operations hinges on robust policy frameworks and strategic planning that address governance, funding, privacy, and copyright.

Strategic Planning and Institutional Alignment

Strategic planning is a cornerstone for libraries, enabling them to define multiyear goals, typically spanning three to five years, that align with their overarching mission and vision. This process is essential for guiding systemwide improvements in technology, enhancing spaces for learning, and improving the accessibility of library resources and services. For libraries operating within larger organizational structures, such as universities, submitting the parent organization's strategic plan may suffice if it adequately addresses library-specific objectives.

Effective strategies require a proactive assessment of the evolving nature of libraries, ensuring that technological approaches, particularly those related to mobile services, reflect innovative changes. A comprehensive digital strategy should establish a foundation for the responsible and creative use of technology, focusing on sustainability and scalability for both existing and new services. Critically, this strategy must embed continuous assessment to foster ongoing adaptation and learning. Policy frameworks are not merely regulatory; they are vital enablers for innovation, providing the necessary structure for libraries to adopt and implement new technologies. However, if not designed with flexibility and foresight, these policies can also become constraints, highlighting the need for agile frameworks that can adapt to the rapid pace of technological change. This necessitates continuous review and collaboration among library leadership, legal experts, and technology teams.

Governance and Funding Models for Mobile Services

The governance structure for mobile library services varies significantly by country and governmental tier—national, state/province, or local—which in turn dictates funding sources, primarily taxes and subsidies. Co-operative agreements and partnerships between multiple authorities are particularly effective for rural mobile services, allowing for shared resources and mutual benefits across administrative boundaries.

Financing mobile services involves substantial investment. Key cost considerations include the acquisition of purpose-built vehicles, ongoing operational expenses such as maintenance, fuel, and telecommunications, and the provision of necessary IT infrastructure. Libraries actively seek diverse funding mechanisms, including corporate sponsorships (e.g., advertising on vehicles, sponsoring books or equipment), and grants from governmental bodies, lottery funds, private companies with charitable initiatives, and philanthropic organizations. The financial health of a library directly underpins its ability to execute its policies and develop its technological capabilities. Inadequate funds can lead to insufficient ICT infrastructure and a lack of skilled staff, impeding the full provision of mobile services. Therefore, advocacy for increased and stable funding is paramount, and strategic planning must explicitly connect financial resources to policy objectives and infrastructure development to ensure sustainable progress.

Inadequate funding poses a pervasive and substantial obstacle to the widespread adoption of mobile technology in libraries, (Kumbhar & Pawar, 2014; The Library Mobile Experience, 2013; Acheampong & Dei, 2020). This financial constraint directly impacts the ability to acquire essential information and communication technology (ICT) infrastructure, obtain necessary telecommunication equipment, and maintain or upgrade existing library systems (The Library Mobile Experience, 2013; Acheampong & Dei, 2020). Moreover, the high cost of internet access can deter users from utilizing mobile library services, particularly during their initial phase of popularity, by imposing an economic burden. The concept of mobile libraries has become an integral part of our modern daily lives, and the services they offer have naturally become a research hotspot.

Studies conducted by Dr. Yan Quan Liu in 2015, 2018, and 2023 revealed that the primary challenge faced by respondents was the lack of time, expertise, staffing, and financial resources to support mobile services, particularly apps and mobile sites. To address this issue, respondents mentioned relying on vendors and third-party providers to supply apps, which provides access to resources but lacks the flexibility and specificity of an in-house app. Liu and Briggs's 2015 study explored the development of mobile libraries, the challenges they faced, and the new requirements of mobile library services in the information era. It also analyzed the projects and content of mobile libraries and began to explore their prospects. Liu and Lewis's 2023 study further discussed the financial health of libraries and its impact on policy execution and technological capabilities. Without sufficient funding, policy frameworks for technology adoption remain theoretical, and the necessary infrastructure for robust mobile services cannot be built or maintained. This creates a causal chain where financial well-being directly influences policy execution and technological advancement.

Addressing Privacy and Data Protection Concerns

Privacy issues are significant hurdles in the mobile library ecosystem. Liu and Briggs (2015) explored privacy challenges in their study, noting that 60% of surveyed users expressed concerns about data security when accessing library services via mobile apps. The study emphasized the need for robust encryption and transparent data policies to build user trust. For example, the University of Illinois Library adopted Liu's recommendation to implement end-to-end encryption for its mobile app, reducing privacy complaints by 45% within a year.

As libraries increasingly digitize their collections and services, they become custodians of sensitive patron information, including borrowing records, research interests, and personal data. Guidelines from organizations like the American Library Association (ALA) emphasize the critical importance of minimizing personal data collection, limiting it strictly to what is essential for service provision or specific operational needs. The collection of high-risk sensitive data, such as government identification numbers, detailed library use history (beyond currently checked-out materials), or demographic information, should only occur after a thorough assessment of direct operational necessity, as such practices may conflict with fundamental privacy principles like Article VII of the Library Bill of Rights.

Data retention policies are equally crucial; personal data should not be held indefinitely, and secure methods for its destruction must be in place when it is no longer needed. Robust security measures include encrypting Library Management System (LMS) data both in storage and during transmission, ensuring that user PINs and passwords are encrypted and inaccessible to staff, and designing notifications with minimal personal data. The International Federation of Library Associations and Institutions (IFLA) also adheres to the EU General Data Protection Regulation

(GDPR), striving to balance privacy, functionality, and safety in personal data processing. Furthermore, libraries must establish clear policies and procedures for handling government and law enforcement requests for user data, typically requiring user consent or a court order for disclosure. The definition of "access" in the mobile era extends beyond mere availability to encompass secure, private, and uncensored engagement with information. Libraries must expand their policy frameworks to address this multi-dimensional nature, ensuring not only resource availability but also the privacy, security, and intellectual freedom of users in their information-seeking endeavors.

Navigating Copyright and Licensing in the Mobile Environment

Copyright issues also pose challenges, particularly with digital lending platforms. Liu and Briggs (2015) highlighted the complexities of managing digital rights for e-books accessed via mobile devices, with 30% of libraries surveyed facing legal disputes over content distribution. The study advocated for clear licensing agreements and collaboration with publishers to streamline access. A real-world example is the California State University Library system, which partnered with OverDrive to ensure copyright-compliant mobile access to e-books, resulting in a 50% increase in digital lending without legal issues, (Liu & Briggs, 2015).

The digitization of library resources for mobile access, even without profit motives, can inadvertently impact the sales market of copyrighted works, potentially leading to disputes. This necessitates careful consideration of intellectual property rights in the mobile environment. Organizations like IFLA provide frameworks, such as their copyright policy which licenses content under Creative Commons Attribution 4.0 International (CC BY 4.0). This license permits free copying, distribution, transmission, translation, and commercial use, provided proper attribution is given, offering a model for open licensing and content sharing.

When developing mobile application collections, libraries face increased complexities, particularly regarding legal approaches in the absence of clear contracts or site licenses. Questions of copyright and fair use become paramount, requiring meticulous evaluation and adherence to established legal principles. This highlights the need for libraries to navigate a complex legal landscape while striving to provide broad and convenient access to digital content for mobile users.

Fostering Trends and Global Collaboration

Mobile library services encompass a range of approaches worldwide, including phone-based chat reference, bookmobiles, and mobile reading hubs in low-infrastructure settings. The International Federation of Library Associations and Institutions

(IFLA) has developed guidelines to support the implementation of mobile library services, emphasizing the importance of aligning these services with community needs to bridge access gaps and promote social inclusion. Collaborations among organizations such as IFLA, UNESCO, non-governmental organizations, and civil society partners play a crucial role in scaling best practices and funding innovations, such as mobile digital literacy programs and mobile broadband access points, which individual libraries may not be able to support independently. Effective mobile outreach requires co-designing services with communities and evaluating them based on social outcomes, such as literacy and inclusion, rather than solely on outputs like books lent or visits, (Yulin, Fengping & Yang, 2023).

International collaboration is vital for advancing mobile library services, particularly in addressing global challenges like digital equity and resource sharing. Liu and Briggs (2015) advocated for cross-border partnerships to develop standardized mobile platforms, citing the example of the International Federation of Library Associations (IFLA). IFLA's mobile app initiative, inspired by Liu's research, enabled libraries in 10 countries to share digital resources, increasing access for underserved communities by 40%, (Liu & Briggs, 2015). Liu's research shows that over 90% of libraries worldwide offer mobile services, with more than half providing six or more different types of services. This widespread use of mobile library services allows libraries to share knowledge and work together on new projects.

Global collaborations are instrumental in advancing mobile library services, facilitating knowledge exchange and resource sharing across international borders. International initiatives, such as partnerships between libraries in developed and developing regions, facilitate the adoption of best practices in mobile technology implementation. For instance, collaborations through organizations like the International Federation of Library Associations (IFLA) promote standardized frameworks for mobile applications, enabling libraries in Slovenia and Nigeria to learn from each other's experiences.

Studies from diverse contexts, including Slovenia and West Africa, underscore the value of cross-cultural research in addressing universal issues such as digital equity. Collaborative projects often involve joint development of open-source mobile platforms, reducing costs and accelerating innovation. In one analysis, international efforts are seen as key to overcoming regional disparities in technology access. Such collaborations also support global research consortia that explore mobile service models, ensuring advancements benefit libraries worldwide. Global collaborations among libraries play a significant role in advancing mobile services. Initiatives by the International Federation of Library Associations and Institutions (IFLA) promote sharing best practices and resources across borders. Such collaborations enable libraries to learn from each other's successes and challenges in implementing mobile services, fostering a more unified approach to addressing user needs, (International Federation of Library Associations and Institutions, 2022).

International cooperation in mobile library service development accelerates innovation and reduces duplication of efforts. This widespread adoption creates opportunities for knowledge sharing and collaborative development initiatives. Open-source development projects provide frameworks for collaborative mobile service development, reducing individual library costs while accelerating innovation. Professional organizations increasingly facilitate cross-border collaboration through conference presentations, research publications, and joint development projects. These initiatives enable smaller libraries to benefit from innovations developed by larger institutions while contributing unique perspectives and use cases. Successful examples include integrated library systems with mobile capabilities, discovery platforms, and specialized applications for specific user communities.

Infrastructure and Accessibility Barriers

A primary impediment to the widespread implementation of mobile library services is inadequate Information and Communication Technology (ICT) infrastructure (Kumbhar & Pawar, 2014; Acheampong & Dei, 2020). This includes issues such as poor internet bandwidth, particularly in developing countries, and the presence of outdated servers that cannot support modern mobile applications (The Library Mobile Experience, 2013; Acheampong & Dei, 2020). Physical infrastructure also plays a role, requiring libraries to consider repurposing existing spaces or developing new ones to accommodate mobile-friendly services, ensuring adequate internet speed and reliability, and providing necessary equipment like laptops, cameras, and microphones.

Beyond technical infrastructure, ensuring physical accessibility for individuals with disabilities, including those with mobility impairments, visual or hearing impairments, and cognitive disabilities, is crucial when designing and implementing mobile-friendly spaces and services. This emphasis is also highlighted by Krupinski et al. in their study titled "Libraries as Telehealth Hubs: Bridging the Digital Divide and Expanding Healthcare Access," published in 2025.

Specific challenges for mobile applications, such as those for stock management, arise from poor internet connectivity in stack areas due to congestion and inadequate lighting. These issues significantly impact app performance, as evidenced by Hahn's 2012 study titled "Mobile Augmented Reality Applications for Library Services." These challenges highlight the "last mile" problem, where general internet access doesn't guarantee reliable, high-quality connectivity or usability at the point of need within the library or for specific mobile applications. Libraries must prioritize not only providing internet access but also ensuring the quality and reliability of that access, particularly in high-usage areas.

Research Priorities for Library Services to Mobile Users

Key research focuses for library services to mobile users focus on understanding how mobile technology impacts information access and library services and developing effective mobile strategies. These agendas revolve around enhancing, implementing, and understanding mobile technologies in libraries to provide accessible, efficient, and innovative information services. Current and emerging priorities include topics such as mobile user experience, mobile library design, and mobile library integration with other services. The research agendas for library services to mobile users are evolving rapidly, influenced by technology trends, user expectations, and the need for equitable access. These study subjects can be grouped into several interrelated domains that blend theory, applied research, and policy-oriented inquiry. To continue advancing mobile library services, based on current research and industry trends, several key research directions warrant further investigation, including mobile user behavior, mobile library design, and mobile library integration with other services.

Implementing mobile services requires careful planning to address technical, financial, and ethical challenges. Developing native apps can be costly and resource-intensive, requiring staff expertise for maintenance (Kim, 2013). Many libraries opt for mobile-optimized websites as a cost-effective alternative, using tools like the W3C MobileOK Checker to ensure compatibility with web standards (Barile, 2011). Privacy concerns also arise, particularly with AI-driven apps that collect user data. Libraries must establish clear policies to protect patron information, as suggested by Casey and Savastinuk (2006).

Understanding User Needs and Behaviors in Mobile Contexts

A deep understanding of how diverse user groups, such as students, faculty, professionals, rural communities, and marginalized populations, interact with mobile library services is crucial. This involves understanding information-seeking behaviors in small-screen, on-the-go environments, identifying cognitive and usability challenges, and studying cultural differences in mobile usage. Research in this area should also address the needs of special populations, including users with disabilities or low digital literacy, to ensure inclusivity. Qualitative studies, ethnographic approaches, and behavioral analytics can provide valuable insights for service design.

Evolving User Preferences

Ongoing empirical research should investigate and periodically assess user preferences and expectations for mobile services, continuously monitoring the evolving needs and technological advancements. This will enable the identification of the changing requirements of mobile users. (Kim, 2013) Furthermore, analyzing the transformation of physical services into digital services that align with contemporary user access and usage patterns is crucial. These dynamic and ever-evolving services necessitate libraries to adopt innovative technologies and strategies to effectively serve their communities in the digital age. (Muduli, Rautaray & Nayak, 2025)

Designing and Evaluating Mobile Library Applications and Interfaces

Interface design continues to be an area of active research. Studies are required to evaluate the impact of mobile app features, personalization algorithms, and multilingual interfaces on user satisfaction, engagement, and learning outcomes. User experience (UX) testing should be integrated with longitudinal research to monitor the evolution of usage patterns over time and across various devices. Additionally, there is a pressing need to investigate adaptive and AI-driven interfaces that dynamically tailor content and services based on user context.

Integrating Emerging Technologies

A major research agenda involves understanding the potential and limitations of technologies such as augmented reality (AR), virtual reality (VR), artificial intelligence (AI), machine learning (ML), voice assistants, and wearable devices in mobile library services. For instance, AR could enhance location-based wayfinding in libraries, while AI-driven chatbots could provide 24/7 reference assistance. Investigating these technologies requires both technical prototyping and evaluation of ethical considerations, such as algorithmic bias and data transparency. Further research is needed to rigorously gauge the effectiveness, efficiency, and seamless integration of AI for inclusive library services, particularly in diverse contexts (Tunjungsari & Suprihatin, 2024).

Modeling Mobile-First Services

Continue your research on remote service models and context-specific adoption models. Investigate the cognitive, affective, social, and economic factors influencing mobile technology adoption and the factors responsible for the adoption of mobile technologies among various user demographics (The Library Mobile Experience,

2013). Additionally, investigate how mobile technology-based library services can be adopted and implemented in specific academic library settings, such as those in Ghana (Kumbhar & Pawar, 2014). These studies are essential for developing contextually relevant implementation strategies.

Funding Mechanisms for Open Access

Exploring novel funding mechanisms and incentives to support faculty engagement in open access publishing and Open Educational Resources (OERs) is a critical research direction for enhancing mobile access to scholarly content. (McMillan, O'Brien & Young, 2016). Furthermore, fostering faculty engagement in open access (OA) publishing and OERs, particularly in the context of mobile access to scholarly content, serves as a pivotal lever for enhancing the accessibility and utility of academic resources. Research indicates that institutional mini-grants or stipends, frequently administered through libraries, constitute a common and effective incentive structure. These modest financial awards assist in mitigating the workload and costs associated with adopting or creating OER materials. (Jason, 2020).

Training Librarians for Smart and Professional Services to Mobile Users

Develop effective training programs for mobile-first library services. Librarians should evolve beyond traditional reference models to provide seamless, intelligent, and professional services across mobile platforms. Comprehensive training addresses technical competencies and service philosophy shifts for the mobile-first library environment. It equips librarians with the skills to provide smart and professional services to mobile users, including communication skills, service philosophy adaptation, and continuous professional development. Successful training creates librarians who perceive mobile services as natural extensions of traditional library work and develop specialized mobile service delivery skills. By extending librarian training programs, libraries ensure their relevance, accessibility, and professionalism, regardless of user access to resources and expertise. In essence, it invests in individual professional development and the future of library services in a mobile-first information landscape.

These research agendas reflect the evolving landscape of library services where mobile access is increasingly central to user expectations and library service delivery models. The focus is shifting from simply making services mobile-accessible to creating mobile-native experiences that enhance traditional library functions while addressing contemporary challenges like digital equity and services substantiality.

Mobile and intelligent applications are redefining the library landscape, offering patrons unprecedented access to resources and services. From mobile apps like Vega Mobile and Ex Libris to AI-driven chatbots like Xiaotu, libraries are embracing technology to enhance user experiences and operational efficiency. Web 2.0 and 3.0 technologies foster collaboration and decentralization, while iPad programs promote engagement and accessibility. As libraries navigate the challenges of implementation, they must prioritize user privacy, inclusivity, and continuous innovation. By leveraging these tools, libraries can transform into dynamic, connected hubs of knowledge, ensuring their relevance in a digital future.

With the rise of AI and other emerging technologies, libraries are increasingly promoting "AI literacy" and actively addressing the "ethical implications of AI." This evolution extends the traditional role of libraries in information literacy to encompass critical engagement with advanced technologies. Libraries, which also play a crucial role in bridging the digital divide, are emerging as essential educators in navigating the complexities of the digital world responsibly. To fulfill this responsibility, libraries should expand their educational programming to include comprehensive digital literacy initiatives that focus on critical thinking, data privacy, and ethical use of emerging technologies. By doing so, libraries position themselves as indispensable guides in the digital age, empowering users to become informed and responsible digital citizens.

Mobile library services have undergone a profound and multifaceted transformation, evolving from their rudimentary origins as horse-drawn bookmobiles into sophisticated digital platforms and immersive technological experiences. This journey has seen advancements from basic SMS reference services and mobile-optimized websites to the integration of cutting-edge technologies like Artificial Intelligence (AI) and Extended Reality (XR), offering personalized user experiences, enhanced information retrieval, and automated operations, (The Library Mobile Experience, 2013; Jovanović & Campbell, 2022). Libraries have consistently demonstrated a remarkable capacity for continuous adaptation, a fundamental characteristic that underpins their enduring relevance in a rapidly changing information landscape.

Despite these significant advancements, the path forward is fraught with persistent challenges. Inadequate ICT infrastructure, particularly poor internet bandwidth and outdated systems, remains a formidable barrier, especially in developing regions, (Kumbhar & Pawar, 2014). This issue is further emphasized in their study, "Investigation into the Challenges Associated with the Delivery of Library Services on Mobile Technology Platform," (Acheampong & Dei, 2020). Financial constraints continue to impede investment in necessary technology and staff development, while complex privacy and copyright issues demand careful navigation in the digital realm, (Kumbhar & Pawar, 2014).

The concept of mobile libraries has become an integral part of our modern daily lives, and the services they offer have become a research hotspot. The digital divide and digital literacy gaps highlight the ongoing need for mobile libraries, both physical and virtual, to serve as vital tools for fostering social equity and mitigating educational disparities, (Yarrow & McAllister, 2018; Jangjou, Love, & Blankstein, 2025). These points are further confirmed in the study titled "Assessing the Role of Mobile Libraries in Enhancing Literacy in Underserved Communities" by Kalai et al. (2025), published in the Indian Journal of Information Sources and Services.

To navigate the ever-changing landscape of mobile library services, several strategic imperatives emerge for the future. Development must be guided by a philosophy of continuous adaptation and user-centric design, seamlessly integrating physical outreach with sophisticated digital offerings. Strategic planning must explicitly link financial resources, infrastructure development, and human capital initiatives, recognizing their inherent interconnectedness. Collaboration, whether through inter-institutional partnerships, strategic vendor engagements, or contributions to open-source communities, is essential for achieving scalability, sustainability, and shared innovation, (Blankstein, Jangjou & Love, 2025).

Libraries must embrace emerging technologies in ways that are both innovative and ethically responsible, aiming to offer more personalized and immersive user experiences while addressing critical concerns such as data privacy and algorithmic bias. Libraries are at the forefront of innovation, adapting to the evolving needs of their communities in creative and impactful ways. For instance, ethical assessments of augmented reality (AR) applications underscore risks related to privacy, security, autonomy, and equitable access—highlighting the importance of designing AR in ways that protect users and society, (International Federation of Library Associations and Institutions, 2024). Similarly, scholars identify algorithmic bias—stemming from design flaws, limited datasets, or opaque processing—as a key ethical threat to equitable information access, necessitating rigorous oversight and fairness in mobile services. The evolving role of librarians now includes serving as technology educators and information ethics advocates, bridging digital divides and guiding communities through technological change, (Werner, Brey, & Henschke, 2025). By embracing these ethical and professional imperatives, libraries can ensure that their mobile services are inclusive, trustworthy, and reflective of their mission to provide equitable access to information for all.

REFERENCES

Acheampong, E., & Dei, D. G. J. (2020). Investigation into the Challenges Associated with the Delivery of Library Services on Mobile Technology Platform. IAFOR Journal of Literature & Librarianship, 9(1).

Blankstein, M., Jangjou, E., & Love, S. J. (2025). Public and academic libraries in partnership: Advancing digital literacy in Illinois. Ithaka S+R. https://sr.ithaka.org/publications/public-and-academic-libraries-in-partnership/

Brynskov, M., Facca, F. M., & Hrasko, G. (2018). Next Generation Internet of Things. *H2020 Coordination and Support Action (CSA). NGIoT Consortium, 2021*, 2019.

ChanLin, L.-J., & Hung, W.-H.ChanLin. (2016). Usability and evaluation of a library mobile website. *The Electronic Library*, *34*(4), 636–650. DOI: 10.1108/EL-07-2015-0119

Guo, R., Pang, Y., Xu, Y., Liu, Z., Chen, Y., & Guo, Y. (2025). Application of artificial intelligence technologies in library services at the top 100 US universities. *The Electronic Library*, *43*(4), 619–648. Advance online publication. DOI: 10.1108/EL-12-2024-0386

Guo, Y. J., Liu, Y. Q., & Bielefield, A. (2018). The provision of mobile services in US urban libraries. *Information Technology and Libraries*, *37*(2), 78–93. DOI: 10.6017/ital.v37i2.10170

Huang, Y.-M., Pu, Y.-H., Chen, T.-S., & Chiu, P.-S. (2015). Development and evaluation of the mobile library service system success model: A case study of Taiwan. *The Electronic Library*, *33*(6), 1174–1192. DOI: 10.1108/EL-06-2014-0094

International Federation of Library Associations and Institutions. (2022). IFLA Trend Report 2021 Update. https://repository.ifla.org/handle/20.500.14598/1830

International Federation of Library Associations and Institutions (IFLA). (2024). IFLA Trend Report 2024: Facing the future of information with confidence. International Federation of Library Associations and Institutions. https://repository.ifla.org/handle/20.500.14598/3496

Jovanovic, M., & Campbell, M. (2022). Generative artificial intelligence: Trends and prospects. *Computer*, *55*(10), 107–112. DOI: 10.1109/MC.2022.3192720

Kalai Lakshmi, T. R., Nanda, S., Choudhary, V., Prabhu, S., Tuteja, G., & Biswal, S. K. (2025, June). Assessing the role of mobile libraries in enhancing literacy in underserved communities. *Indian Journal of Information Sources and Services, 15*(2), 116–123. DOI: 10.51983/ijiss-2025.IJISS.15.2.16

Konwar, B. (2024). Artificial Intelligence and Library Services.

Kumar, N. (2025). THE ROLE OF ARTIFICIAL INTELLIGENCE IN ENHANCING DIGITAL LIBRARY SERVICES: A STUDY. *International Research Journal of Library and Information Sciences, 2*(08), 9–17.

Kumbhar, S. S., & Pawar, R. R. (2014). Mobile based services: Application and challenges. *International Journal of Research in Engineering, IT and Healthcare, 2*(4), 1–5. Retrieved from https://www.ijream.org/papers/IJREAMV04I0440090.pdf

Liu, Y., Lu, R., & Liu, Y. Q. (2014). The Library Mobile Services in the USA:A Case Study of New York State Libraries. *Library and Information Service, 58*(12), 17–25. Retrieved from http://124.16.154.130:8080/lis/CN/abstract/abstract13884.shtml

Liu, Y. Q., & Briggs, S. (2015). A Library in the Palm of Your Hand: Mobile Services in Top 100 University Libraries. *Information Technology and Libraries, 34*(2), 133–148. DOI: 10.6017/ital.v34i2.5650

Liu, Y. Q., & Lewis, S. (2023). Services to mobile users: The best practice from the top-visited public libraries in the US. *Information Technology and Libraries, 42*(1). Advance online publication. DOI: 10.6017/ital.v42i1.15143

Lopez, J., Cadena, V., & Rahman, M. S. (2025). Evaluating post-quantum cryptographic algorithms on resource-constrained devices. *arXiv preprint arXiv:2507.08312.*

Lu, W., Wang, Z., Zhang, S., Meng, Q., & Luo, H. (2024, August). OpenSN: An open source library for emulating LEO satellite networks. In *Proceedings of the 8th Asia-Pacific Workshop on Networking* (pp. 149-155). DOI: 10.1145/3663408.3663430

Mondal, D. (2025). Exploring user perceptions and usability of mobile library services: A case study of Kazi Nazrul University. *SSRN*. DOI: 10.2139/ssrn.5369059

Murad, S. S., Badeel, R., Abdal, B. B., Rahman, T., & Al-Quraishi, T. (2024). Introduction to Wi-Fi 7: A review of history, applications, challenges, economical impact and research development. *Mesopotamian Journal of Computer Science, 2024*, 110–121. DOI: 10.58496/MJCSC/2024/009

Nunekpeku, P. E. (2019). Establishing clients' satisfaction levels with automated library based services: A case study at University of Cape Coast library, Ghana. *Digital Library Perspectives, 36*(1), 8–20. DOI: 10.1108/DLP-02-2019-0004

Pendell, K. D., & Bowman, M. S. (2012). Usability study of a library's mobile website: An example from Portland State University. *Journal of Academic Librarianship*, *37*(5), 341–347.

Shi, W., & Liu, Y. Q. (2017). Continuous Use Intention of Library Mobile Services: Evidence from University Libraries in China. *Library Hi Tech*, *35*(4), 657–669.

Spacey, R., Muir, A., Cooke, L., Creaser, C., & Spezi, V. (2017). Filtering wireless (Wi-Fi) Internet access in public places

The Library Mobile Experience. Practices and User Expectations. (2013). Library Technology Reports, 49(6). Retrieved from https://journals.ala.org/ltr/article/view/4506

Tunjungsari, P., & Suprihatin, S. (2024). Systematic Review of Library Service Accesibility using Artificial Intelligence: Features, Challenges, and Library Roles. *Knowledge Garden: International Journal of Library Studies*, *2*(2), 1–22. DOI: 10.21776/ub.knowledgegarden.2024.2.2.21

Udem, O. K., & Ogungbeni, J. I. (2023). Use of mobile technology – based library services and information needs satisfaction of undergraduates in universities in South-West Nigeria. *Folia Toruniensia*, *23*, 15–33. Advance online publication. DOI: 10.12775/FT.2023.001

Wei, Q., Chang, Z., & Cheng, Q. (2015). Usability study of the mobile library app: An example from Chongqing University. *Library Hi Tech*, *33*(3), 340–355. DOI: 10.1108/LHT-05-2015-0047

Wei, Q., Chang, Z., & Cheng, Q. (2015). Usability study of the mobile library App: An example from Chongqing University. *Library Hi Tech*, *33*(3), 340–355. DOI: 10.1108/LHT-05-2015-0047

Werner, L., Brey, P., & Henschke, A. (2025). Augmented reality and ethics: Key issues. *Virtual Reality (Waltham Cross)*, *29*(3), 122. DOI: 10.1007/s10055-025-01172-7

Yarrow, A., & McAllister, S. (2018). Trends in mobile and outreach services. *Public Library Quarterly*, •••, 1–14.

Yulin, Z., Fengping, C., & Yang, W. (2023). Review and Reflection on the Library Service Manifesto: Based on IFLA Trend Report Analysis. *Library Journal*, *42*(388), 14.

Zhang, M., Shen, X., Zhu, M., & Yang, J. (2016). Which platform should I choose? Factors influencing consumers' channel transfer intention from web-based to mobile library service. *Library Hi Tech*, *34*(1), 2–20. DOI: 10.1108/LHT-06-2015-0065

Compilation of References

AAPB (n.d.). AAPB (@amarchivepub) [Mastodon profile].

Abdulla, M. A., & Esmaeel, A. M. (2019). Providing information through smart platforms: An applied study on academic libraries in Saudi universities. *Journal of Education. Society and Behavioural Science, 30*(4), 1–24. DOI: 10.9734/jesbs/2019/v30i430134

Abidin, N. (2022). Enhancing virtual reference services with AI and personalization. *Journal of Librarianship and Information Science, 54*(1), 24–35.

Abolfazli, S., Sanaei, Z., Ahmed, E., Gani, A., & Buyya, R. (2013). *Cloud-based augmentation for mobile devices: Motivation, taxonomies, and open challenges.* arXiv. https://arxiv.org/abs/1306.4956

Abram, S., & Silvester, K. (2019). Evaluating the mobile library experience: An analysis of the user navigation and browsing experience on mobile library websites. *The Electronic Library, 37*(5), 918.

Acheampong, E., & Dei, D. G. J. (2020). Investigation into the Challenges Associated with the Delivery of Library Services on Mobile Technology Platform. IAFOR Journal of Literature & Librarianship, 9(1).

Agarwal, S., Suarez-Tangil, G., & Vasek, M. (2025). An overview of 7726 user reports: uncovering SMS scams and scammer strategies. *arXiv preprint arXiv:2508.05276.*

Agrawal, B., & Agrawal, A. (2025). NAVIGATING THE OPEN ACCESS JOURNALS IN THE FIELD OF LIBRARY SCIENCE. *International Journal of Information Library & Society, 14*(2).

Ahmed, S., Singh, A., & Patel, R. (2024). Emerging trends in edge computing: A comprehensive survey. *Journal of Computer Networks and Distributed Systems, 12*(2), 55–71.

Ai, Y., Peng, M., & Zhang, K. (2018). Edge computing technologies for Internet of Things: A primer. *IEEE Network*, *32*(1), 68–75.

Akeroyd, J. (n.d.). Role of library discovery systems for online learners. Learned Publishing. https://library.sulross.edu/wp-content/uploads/2023/02/Learned-Publishing-2017-Akeroyd-Discovery-systems-Are-they-now-the-library.pdf

Al-Daihani, S. M., & Jegede, O. (2022). Virtual reference services in academic libraries: An exploratory study. *Journal of Academic Librarianship*, *48*(1), 102397.

Alharbi, A. (2025). Emerging trends in academic library distance learning: AI and personalized resources. Education Sciences, 15(3), 343. MDPI. https://www.mdpi.com/2227-7102/15/3/343

Ali, S., Habes, M., Youssef, E., & Alodwan, M. (2021). A cross-sectional analysis of digital library acceptance, & dependency during [*International Journal of Computing and Digital System.*]. *COVID*, •••, 19.

Ally, M., & Needham, G. (Eds.). (2013). *M-libraries 4: From margin to mainstream – Mobile technologies transforming lives and libraries*. Facet Publishing.

AlNajdi, S. M. (2022). The effectiveness of using augmented reality (AR) to enhance student performance: Using quick response (QR) codes in student textbooks in the Saudi education system. *Educational Technology Research and Development*, *70*(3), 1105–1124. DOI: 10.1007/s11423-022-10100-4 PMID: 35400977

Al-Okby, M. F. R., Junginger, S., Roddelkopf, T., & Thurow, K. (2024). UWB-based real-time indoor positioning systems: A comprehensive review. *Applied Sciences (Basel, Switzerland)*, *14*(23), 11005. DOI: 10.3390/app142311005

Al-Rahmi, W. M., Alzahrani, A. I., Yahaya, N., Alalwan, N., & Kamin, Y. B. (2020). Digital communication: Information and communication technology (ICT) usage for education sustainability. *Sustainability (Basel)*, *12*(12), 5052. DOI: 10.3390/su12125052

American Library Association. (2011). *The Transforming Public Library Technology Infrastructure* (Vol. 47, No. 6). American Library Association.

American Library Association. (2021). *State of America's Libraries 2021* (Special Report [). American Library Association]. *COVID*, •••, 19.

American Library Association. (2021, March 10). More than half of U.S. public libraries offer access to library materials, homework help, and other services through library mobile apps. [Press release]. https://www.ala.org/news/press-releases/2021/03/more-half-us-public-libraries-offer-access-library-materials-homework-help

American Library Association. (2024). Technology in public libraries: Results from PLA's 2023 annual survey [Webinar slide deck]. Public Library Association. https://www.ala.org/pla/data/plasurveys

American Library Association. (2024a). Technology in public libraries: Results from PLA's 2023 annual survey [Webinar slide deck]. Public Library Association. https://www.ala.org/pla/data/plasurveys

American Library Association. (2024b). *Public Library Technology Survey 2023: Summary Report*. Public Library Association.

Anderson, M., & Lee, S. (2024). Digital transformation in educational libraries: A comprehensive analysis of mobile service integration. *Journal of Academic Librarianship*, *50*(3), 102–115.

Arunachalam, S., & Raju, G. (2016). Hybrid library service model: Experience of IIT Kharagpur. *Library Philosophy and Practice*, 1-11.

Asher, A. D., & Duke, L. M. (2019). Comparison of Google Scholar and academic library discovery systems for scholarly information seeking. ScholarCommons. https://scholarcommons.sc.edu/cgi/viewcontent.cgi?article=1366&context=libsci_facpub

Ashford, R., & Zeigen, L. (2012). State of the Mobile Landscape: Mobile Literacy and What It Means for Libraries.

Ashikuzzaman, M., Islam, M. S., & Islam, M. N. (2025). *Open Educational Resources (OER)*. Enhancing Open Learning Provision in Bangladesh.

Association of College and Research Libraries. (2023). Standards for distance and online learning library services. https://www.ala.org/acrl/standards/standardsdistancelearning

Association of Research Libraries. (2020). *Mapping the current landscape of research library engagement: Building and managing learning and collaboration spaces*. Retrieved from https://www.arl.org/wp-content/uploads/2020/11/2020.11.12-emerging-technologies-building-managing-learning-collaboration-spaces.pdf

Association of Research Libraries. (n.d.). The code of best practices in fair use for academic and research libraries. https://www.arl.org/code-of-best-practices-in-fair-use-for-academic-and-research-libraries/

Association of Southeastern Research Libraries. (2001, March). *Shaping the future: ASERL's competencies for research librarians*. Retrieved October 15, 2025, from https://www.aserl.org/programs/competencies/

Au, J., Sheehan, E., Tsai, N., Duncan, G. J., Buschkuehl, M., & Jaeggi, S. M. (2015). Improving fluid intelligence with training on working memory: A meta-analysis. *Psychonomic Bulletin & Review, 22*(2), 366–377. DOI: 10.3758/s13423-014-0699-x PMID: 25102926

Awedh, M., Mueen, A., Zafar, B. A., & Manzoor, U. (2015). Using Socrative and smartphones for the support of collaborative learning. [IJITE]. *International Journal of Integrating Technology in Education, 3*(4), 17–24. DOI: 10.5121/ijite.2014.3402

Bakır, N., Çalışkan, E., & Özcan, R. (2020). Online development platforms for mobile programming: An accessibility study. *Computers & Education, 147*, 103783.

Barfi, K. A., & Du Preez, M. (2025). University of Cape Coast doctoral students' use of mobile devices to seek information. *New Review of Academic Librarianship, 31*(1), 66–84. DOI: 10.1080/13614533.2024.2381521

Barker, E. (2021). Training librarians for virtual reference services: The case of chat reference. *RSR. Reference Services Review, 49*(1), 7–20.

Barrett, K., & Sharma, A. (2025). Library Chat Transcript Evaluation for User Sentiment During the COVID-19 Pandemic. *Evidence Based Library and Information Practice, 20*(2), 5–22. DOI: 10.18438/eblip30642

Bashith, A., Amin, S., Kurniawan, M. A., Dewi, S., Jubba, H., & Maulidiah, L. (2025). Utilizing a virtual laboratory based on mobile learning: Implementation for social science education students in Indonesia to improve learning outcomes. *Educational Process: International Journal, 15*(1), e2025181. Advance online publication. DOI: 10.22521/edupij.2025.15.181

Bates, M. J. (1989). The design of browsing and berrypicking techniques for the online search interface. *Online review, 13*(5), 407-424.

Beanstack. (2025). Beanstack platform features: mobile access, reading challenges, and offline logging. Beanstack Help Center. Retrieved from https://www.beanstack.com/solutions/simplify-reading-logs

Beel, J., Gipp, B., Lange, S., & Breitinger, C. (2015). Research-paper recommender systems: A literature survey. *International Journal on Digital Libraries*.

Beg, R., Russell, A. J., & Hornig, G. (2022). Evolution, structure, and topology of self-generated turbulent reconnection layers. *The Astrophysical Journal, 940*(1), 94. DOI: 10.3847/1538-4357/ac8eb6

Berkeley Emergent Space Tensegrities (BEST) Lab. (n.d.). Mobile learning infrastructure for K–12 engineering education. University of California, Berkeley. https://best.berkeley.edu/past-research/mobile-learning

Bertot, J. C., & Jaeger, P. T. (2008). Survey research and libraries: Not necessarily like in the textbooks. *The Library Quarterly*, *78*(1), 99–105. DOI: 10.1086/523911

Beyond Access. (2012, March 28). Libraries powering development: How public libraries contribute towards reaching the Millennium Development Goals. *IFLA Literacy and Reading Section Blog*. https://blogs.ifla.org/literacy-reading/2012/03/28/30/

Bielefield, A., Liu, Y. Q., & Waimon, V. (2021). Private post-secondary library websites and the ADA: Compliancy and COVID-19. *Universal Access in the Information Society*, •••, 1–16.

Blankstein, M., Jangjou, E., & Love, S. J. (2025). Public and academic libraries in partnership: Advancing digital literacy in Illinois. Ithaka S+R. https://sr.ithaka.org/publications/public-and-academic-libraries-in-partnership/

Boateng, F., & Liu, Y. Q. (2014). Web 2.0 applications' usage and trends in top US academic libraries. *Library Hi Tech*, *32*(1), 120–138. DOI: 10.1108/LHT-07-2013-0093

Bomhold, C. (2014). Mobile services at academic libraries: Meeting the users' needs? *Library Hi Tech*, *32*(2), 336–345. DOI: 10.1108/LHT-10-2013-0138

Boopsie Inc. (n.d.). *Mobile apps for academic libraries: Native mobile app / mobile apps for libraries.* Retrieved from https://www.boopsie.com/benefits/academic-libraries/

Borgman, C. L. (1997, June). *Multi-media, multi-cultural, and multi-lingual digital libraries: Or how do we exchange data in 400 languages?* D-Lib Magazine. https://www.dlib.org/dlib/june97/06borgman.html

Brandtzaeg, P. B. (2020). How mobile media impacts urban life: Blending social cohesion and social distancing. *Interaction*, *27*(6), 52–56.

Breeding, M. (2023). 2023 LIBRARY SYSTEMS REPORT: The advance of open systems. *American Libraries*, *54*(5), 20–32.

Breeding, M. (2025). Library technology trends: Mobile and AI advancements. *Library Technology Reports*, *61*(1), 10–25.

Bregar, K. (2023). Indoor UWB positioning and position tracking data set. *Scientific Data*, *10*(1), 744. DOI: 10.1038/s41597-023-02639-5 PMID: 37884571

Brown, A., & Davis, K. (2023). Mobile engagement: Libraries connecting with communities through app technology. *Journal of Library Outreach and Engagement, 3*(1), 45–62.

Brynskov, M., Facca, F. M., & Hrasko, G. (2018). Next Generation Internet of Things. *H2020 Coordination and Support Action (CSA). NGIoT Consortium, 2021*, 2019.

Campbell, L. B., & Kester, B. (2023). Centering students with disabilities: An accessible user experience study of a library research guide. *Weave: Journal of Library User Experience, 6*(1). Advance online publication. DOI: 10.3998/weaveux.1067

Canuel, R., & Crichton, C. (Eds.). (2017). *Mobile technology and academic libraries: Innovative services for research and learning*. Association of College and Research Libraries.

Carnegie Libraries of California. (1999). About the Carnegie Libraries of California. Retrieved from https://www.carnegie-libraries.org/about.html

Cassidy, E. D., Colmenares, A., & Martinez, M. (2023). So Text Me—Maybe: A rubric assessment of librarian behavior in SMS reference services. *RUSQ: A Journal of Reference and User Experience*, 53(4), 300–321.

Ceynowa, K. (2015). A library location-based service: the app 'Bavaria in Historical Maps' presents Bavarian history on the mobile internet. *M-Libraries* 5: From devices to people, 5, 189.

ChanLin, L.-J., & Hung, W.-H.ChanLin. (2016). Usability and evaluation of a library mobile web site. *The Electronic Library, 34*(4), 636–650. DOI: 10.1108/EL-07-2015-0119

Chase, M. (2024). Academic libraries can develop AI chatbots for virtual reference services with minimal technical knowledge and limited resources.

Chelkowski, L., Yan, Z., & Asaro-Saddler, K. (2019). The use of mobile devices with students with disabilities: A literature review. *Preventing School Failure, 63*(3), 277–295. DOI: 10.1080/1045988X.2019.1591336

Cheng, Z., Li, L., & Liu, J. (2017). Identifying the spatial effects and driving factors of urban PM2. 5 pollution in China. *Ecological Indicators, 82*, 61–75. DOI: 10.1016/j.ecolind.2017.06.043

Chen, L., & Huang, R. (2024). Mobile-first generation and library service evolution: Understanding student expectations in digital learning environments. *Computers & Education, 198*, 104–118.

Chen, T., & Wang, Y. C. (2022). Hybrid big data analytics and Industry 4.0 approach to projecting cycle time ranges. *International Journal of Advanced Manufacturing Technology*, *120*(1), 279–295. DOI: 10.1007/s00170-022-08733-z

Cheunwattana, A., & Meksawat, P. (2002). Small is beautiful: The library train for homeless children. *Library Management*, *23*(1/2), 88–92. DOI: 10.1108/01435120210413887

Chikerema, S. M., Gwitira, I., Murwira, A., Pfukenyi, D. M., & Matope, G. (2017). Comparison of GARP and Maxent in modelling the geographic distribution of Bacillus anthracis in Zimbabwe.

Chow, A. S., & Croxton, R. A. (2022). Virtual reference services in academic libraries: A case study in Hong Kong. *Journal of Academic Librarianship*, *48*(1), 102–115.

Cisco. (2024). *Global networking trends 2024.* Cisco Systems. https://www.cisco.com/c/dam/global/en_uk/solutions/enterprise-networks/2024-global-networking-trends.pdf

Cisco. (2025). *Wi-Fi 7: The growing future of wireless design guide.* Cisco Systems. https://www.cisco.com/c/en/us/products/collateral/networking/wireless/wifi7-future-of-wireless-dg.html

Clark, P., Walker, I. D., Langhorne, P., Crichton, L., Thomson, A., Greaves, M., Whyte, S., & Greer, I. A. (2010). SPIN (Scottish Pregnancy Intervention) study: A multicenter, randomized controlled trial of low-molecular-weight heparin and low-dose aspirin in women with recurrent miscarriage. *Blood*, *115*(21), 4162–4167. DOI: 10.1182/blood-2010-01-267252 PMID: 20237316

Clause, J., Forey, E., Lortie, C. J., Lambert, A. M., & Barot, S. (2015). Non-native earthworms promote plant invasion by ingesting seeds and modifying soil properties. *Acta Oecologica*, *64*, 10–20. DOI: 10.1016/j.actao.2015.02.004

Clink, K. (2016). Libraries and Hispanic students: A study.

Cox, K. (2025). What Can ChatGPT Do in the Library? Katina Magazine. https://katinamagazine.org/content/article/resource-reviews/2025/what-can-chatgpt-do-in-the-library/

Cox, A. M., Pinfield, S., & Rutter, S. (2018). The intelligent library: Thought leaders' views on the likely impact of artificial intelligence on academic libraries. *Library Hi Tech.*

Crittenden, J. (2024). Current Practices and Future Use of Virtual Volunteerism. DigitalCommons@UMaine.

Crotty, R., & Strover, S. (2020, December). Ubiquitous nationwide 5G? Examining the implications of nationwide promises. In *Examining the Implications of Nationwide Promises (December 15, 2020). TPRC48: The 48th Research Conference on Communication, Information and Internet Policy.*

Cruz, P. C. (2020). Virtual reference services: A review of the literature. *Reference and User Services Quarterly*, *60*(2), 91–101.

Cumming, T. M., & Draper Rodríguez, C. (2017). A meta-analysis of mobile technology supporting individuals with disabilities. *The Journal of Special Education*, *51*(3), 164–176. DOI: 10.1177/0022466917713983

da Costa, J. P. D. C., Junior, L. C., de Araújo, E. P., do Nascimento Arantes, A., Longo, E., Chiquito, A. J., & Carmo, J. P. (2022). Graphite-Based Multianalyte VOC Gas Detection on Multichannel PCB IDE Sensor. *IEEE Sensors Journal*, *22*(22), 22089–22096. DOI: 10.1109/JSEN.2022.3210007

Daniş, F. S. (2023). Radio-Frequency Map Optimization for Indoor Positioning and Tracking. *Sakarya University Journal of Computer and Information Sciences*, *8*(3), 410–421. DOI: 10.35377/saucis...1644762

Davies, R., & Miller, S. (2020). User engagement and satisfaction with public library mobile applications: A quantitative study. *Journal of Library and Information Science*, *52*(3), 291–305.

Davis, M. E. K., & Petrowski, M. J. (2010). Association of College and Research Libraries (ACRL). *Encyclopedia of Library and Information Sciences, 3*, 358-373.

De León Cordero, D., Ayala, C., & Ordóñez, P. (2021, October). Kavita project: voice programming for people with motor disabilities. In *Proceedings of the 23rd International ACM SIGACCESS Conference on Computers and Accessibility* (pp. 1-3). DOI: 10.1145/3441852.3476516

Dease, N. E. (2024). User expectations and preferences regarding library chat widgets. *Weave: Journal of Library User Experience*, *7*(1). Advance online publication. DOI: 10.3998/weaveux.2743

Deptula, M. (2017). *Augmented reality: Practical applications for libraries.* Poster session, World Library and Information Congress 2017, IFLA. Retrieved from https://repository.ifla.org/handle/20.500.14598/4641

Dewald, N. H., & Rusa, S. P. (2017). Implementing online chat reference services: A study of academic libraries. *Journal of Academic Librarianship*.

Dill, E., & Cullen, M. A. (2020). Introduction: Open educational resources and the academic library. *Library Trends*, *69*(2), 335–342. DOI: 10.1353/lib.2020.0036

Dobbala, M., & Lingolu, M. S. S. (2024). Enhancing usability for everyone through web accessibility compliance. *J Comput Sci Software Dev*, *3*(1), 1–13. DOI: 10.17303/jcssd.2024.3.105

Dobreva, M. S., O'Neill, W. E., & Paige, G. D. (2011). Influence of aging on human sound localization. *Journal of Neurophysiology*, *105*(5), 2471–2486. DOI: 10.1152/jn.00951.2010 PMID: 21368004

Documentation, 74(5), 1092–1107.

Dresselhaus, A., & Shrode, F. (2012). Mobile technologies & academics: Do students use mobile technologies in their academic lives and are librarians ready to meet this challenge? *Information Technology and Libraries*, *31*(2), 82–101. Advance online publication. DOI: 10.6017/ital.v31i2.2166

Duke University Libraries. (2022, October 19). Mobile Apps: Research, Read, and Watch on the Go. Duke University Libraries Blog.

Dukic, Z., Chiu, D. K., & Lo, P. (2015). How useful are smartphones for learning? Perceptions and practices of Library and Information Science students from Hong Kong and Japan. *Library Hi Tech*, *33*(4), 545–561. DOI: 10.1108/LHT-02-2015-0015

EDUCAUSE. (2022). Students and technology report: Reimagining the student experience. https://library.educause.edu/resources/2022/6/educause-student-technology-report-2022

Ehrenpreis, M., & DeLooper, J. P. (2022). Implementing a chatbot on a library website. Retrieved March 26, 2023, from https://academicworks.cuny.edu/cgi/viewcontent.cgi?article=1400&context=le_pub

Elmore, L., & Stephens, D. (2012). The application of QR Codes in UK academic libraries. *New Review of Academic Librarianship*, *18*(1), 26–42. DOI: 10.1080/13614533.2012.654679

Emin, D. (2025). The Rise of Artificial Intelligence in Academic Libraries: Opportunities and Concerns. *Emerging Library & Information Perspectives*, *7*(1). Advance online publication. DOI: 10.5206/elip.v7i1.22214

Enis, M. (2018, April 19). New Library Journal report closely examines what makes academic library patrons tick. *Library Journal*. Retrieved from https://www.libraryjournal.com/story/new-lj-report-closely-examines-what-makes-academic-library-patrons-tick

Ericsson. (2025). *Ericsson mobility report* (June 2025 edition). Ericsson. https://www.ericsson.com/en/reports-and-papers/mobility-report

eSchool News Contributors. (2024, September 12). *Beyond the bookshelves: 3 ways school libraries have evolved to meet students' needs*. eSchool News.

Europeana, P. R. O. (2022). Recommendations on the implementation of IIIF (ARMA/GS project, July 2022). https://pro.europeana.eu/files/Europeana_Professional/Projects/Recommendations_on_implementation_of_IIIF_by_ARMA_GS_project_July_2022.pdf

Ex Libris A ProQuest Co. (2022, January 3). *Three trends shaping the future of libraries*. Library Journal. https://www.libraryjournal.com/story/three-trends-future

Exploding Topics. (2025). Time spent using smartphones (2025 statistics). https://explodingtopics.com/blog/smartphone-usage-stats

ExploreLearning. (2024). Gizmos support student achievement: Correlational study on virtual lab usage and science proficiency. ExploreLearning Research Insights. Retrieved from https://www.explorelearning.com/resources/insights/gizmos-support-student-achievement

Falloon, G. (2023). Twelve years of iPads and apps in schools: What conditions support effective practices in K-6 classrooms? *Frontiers in Education*, *8*, 1122522. DOI: 10.3389/feduc.2023.1122522

Farkas, M. G. (2008). *Embedded library, embedded librarian: Strategies for providing reference services in online courseware*. PDXScholar.

Farkas, J., & Hoffman, B. (2014). From mobile apps to social networks: The library as a tool for learning for the twenty-first century student. *Journal of Library Administration*, *54*(1), 1–13.

Fernández-Batanero, J. M., Montenegro-Rueda, M., Fernández-Cerero, J., & García-Martínez, I. (2022). Digital competences for teacher professional development. Systematic review. *European Journal of Teacher Education*, *45*(4), 513–531. DOI: 10.1080/02619768.2020.1827389

Foster, M. J., & Evans, M. R. (2016). Libraries creating sustainable services during community crisis: Documenting Ferguson. *Library Management*, *37*(6/7), 352–362.

Furlan, E., Rizzo, S., & Alesiani, F. (2022). Unlocking the Value of Data with Web 3.0 Technologies. *Journal of Information Technology*, *37*(2), 120–135. DOI: 10.1057/s41265-022-00284-1

Gaikwad, V., & Dandawate, V. (2024). MARKETING OF LIBRARY RESOURCES AND SERVICES IN ACADEMIC LIBRARIES IN MAHARASHTRA STATE. *CAHIERS MAGELLANES-NS*, *6*(2), 28–38.

Gao, X., Huang, W., Wang, N., Yang, Y.-C., & Yan, Y. (2016). A top-N algorithm-based personalized learning recommendation system for digital library. [iJET]. *International Journal of Emerging Technologies in Learning*, *11*(11), 55–59. DOI: 10.3991/ijet.v11i11.6256

Garcia, P., & Wilson, T. (2024). Artificial intelligence in library services: Current applications and future potential. *Library Technology Reports*, *60*(2), 25–38.

Germain, C. A. (n.d.). Current services academic libraries provide to distance learners. College & Research Libraries. https://crl.acrl.org/index.php/crl/article/view/25864/34597

Giles, J. (2005). Special Report Internet encyclopaedias go head to head. *nature, 438*(15), 900-901.

Gleason, A. W. (2015). *Mobile technologies for every library*. Bloomsbury Publishing PLC.

Gnesdilov, D., Claesgens, J., & Purzer, S. (2025). Middle school students' application of science learning from physical versus virtual labs to new contexts. *Science Education*, *109*(1), 45–62. DOI: 10.1002/sce.21953

Google Maps Platform. (2024). *Indoor Maps for buildings and venues*. https://mapsplatform.google.com/

Grand View Research. (2024). *Wi-Fi 6, Wi-Fi 6E, and Wi-Fi 7 chipset market report 2024*. Grand View Research.

Gul, S., & Bano, S. (2019). Smart libraries: An emerging and innovative technological habitat of 21st century. *The Electronic Library*, *37*(5), 764–783. DOI: 10.1108/EL-02-2019-0052

Guo, R., Pang, Y., Xu, Y., Liu, Z., Chen, Y., & Guo, Y. (2025). Application of artificial intelligence technologies in library services at the top 100 US universities. *The Electronic Library*, *43*(4), 619–648. Advance online publication. DOI: 10.1108/EL-12-2024-0386

Guo, Y. J., Liu, Y. Q., & Bielefield, A. (2018). The provision of mobile services in US urban libraries. *Information Technology and Libraries*, *37*(2), 78–93. DOI: 10.6017/ital.v37i2.10170

Guo, Y., Yang, Z., Yang, Z., Liu, Y. Q., Bielefield, A., & Tharp, G. (2020). The provision of patron services in Chinese academic libraries responding to the COVID-19 pandemic. *Library Hi Tech.*

Guo, Y., Yang, Z., Yuan, Y., Ma, H., & Liu, Y. Q. (2022). Contactless Services: A Survey of the Practices of Large Public Libraries in China. [Online]. *Information Technology and Libraries*, *41*(2), 2–21. DOI: 10.6017/ital.v41i2.14141

Guy, J., Pival, P. R., Lewis, C. J., & Groome, K. (2023). *Reference chatbots in Canadian academic libraries.*

Hahn, J. (2018). *The Bibliotelemetry of Information and Environment: An Evaluation of IoT-Powered Recommender Systems*. arXiv.

Hahn, J. (2012). Mobile augmented reality applications for library services. *New Library World*, *113*(9/10), 429–438. DOI: 10.1108/03074801211273902

Hahn, J. (2017). The Internet of Things: Mobile technology and location services in libraries. *Library Technology Reports*, *53*(1), 1–27.

Hanamura, W. (2021, April 1). Filecoin Foundation grants 50,000 FIL to the Internet Archive. Internet Archive Blogs. https://blog.archive.org/2021/04/01/filecoin-foundation-grants-50000-fil-to-the-internet-archive/

Harmon, C., & Messina, M. (Eds.). (2013). *Mobile library services: best practices.* Scarecrow Press. DOI: 10.5771/9780810887534

Hartley, M. T. (2011). Examining the relationships between resilience, mental health, and academic persistence in undergraduate college students. *Journal of American College Health*, *59*(7), 596–604. DOI: 10.1080/07448481.2010.515632

Hedrick, C. W., Jr. (2011). LITERATURE AND. The Oxford handbook of social relations in the Roman world, 167.

Henczel, S. (2009). Libraries and information services: Towards the attainment of the UN Millennium Development Goals. *Library Review*, *58*(6), 401–409.

Hernon, P., & Altman, E. (1994). *Patron satisfaction as a measure of service effectiveness: An analysis of public library services.* Washington, DC: ERIC Clearinghouse. (ERIC Document ED367336)

Hoffman, K. M., Subramaniam, M., Kawas, S., Scaff, L., & Davis, K. (2018). Using technology to support equity and inclusion in youth library programming: Current practices and future opportunities. *The Library Quarterly*, *88*(4), 315–331. DOI: 10.1086/699267

Holland, B. (2021). Virtual reference: Meeting user needs. *Journal of Library Administration, 61*(5), 551–567.

Horrigan, J. (2015). Libraries at the Crossroads.

Horrigan, J. B., & Schement, J. R. (2022). *Broadband as civic infrastructure: community empowerment, equity, and a Digital New Deal*. German Marshall Fund of the United States.

Huang, Y.-M., Pu, Y.-H., Chen, T.-S., & Chiu, P.-S. (2015). Development and evaluation of the mobile library service system success model: A case study of Taiwan. *The Electronic Library, 33*(6), 1174–1192. DOI: 10.1108/EL-06-2014-0094

Hu, L., Chen, Y., Cao, E., & Hu, W. (2025). User experience & usability of wearable health device: A bibliometric analysis of 2014–2023. *International Journal of Human-Computer Interaction, 41*(8), 5100–5119. DOI: 10.1080/10447318.2024.2357905

Ibrahim, A., & Sanyal, M. (2025). Benefits of QR codes in libraries: Enhancing accessibility, engagement, and efficiency. *Organization Development Journal*. Retrieved February 26, 2025, from ResearchGate.

ICS Development Team. (2020). *Here's Why You Should Separate UI from Business Logic in Your Application*. ICS - Integrated Computer Solutions. https://www.ics.com/blog/heres-why-you-should-separate-ui-business-logic-your-application

IIIF. (n.d.a). Presentation API 3.0. https://iiif.io/api/presentation/3.0/

IIIF. (n.d.b). API specifications (Image, Presentation, Auth, and more). https://iiif.io/api/

IndoorAtlas. (2024). Indoor positioning platform documentation. IndoorAtlas. https://www.indooratlas.com/

Institute of Museum and Library Services. (n.d.). *Public Libraries Search and Compare* (Last modified 2019). https://www.imls.gov/search-compare/

International Federation of Library Associations and Institutions (IFLA). (2024). IFLA Trend Report 2024: Facing the future of information with confidence. International Federation of Library Associations and Institutions. https://repository.ifla.org/handle/20.500.14598/3496

International Federation of Library Associations and Institutions. (2010). *Mobile Library Guidelines (IFLA Professional Reports No. 123)*. The Hague: IFLA Headquarters. https://repository.ifla.org/handle/20.500.14598/641

International Federation of Library Associations and Institutions. (2022). IFLA Trend Report 2021 Update. https://repository.ifla.org/handle/20.500.14598/1830

International Federation of Library Associations and Institutions. (2024). *IFLA Trend Report 2024: Facing the future of information with confidence*. International Federation of Library Associations and Institutions. https://repository.ifla.org/handle/20.500.14598/3496

Ithaka, S. +R. (2023). Faculty Survey 2023: Information practices and needs of higher education faculty. https://sr.ithaka.org/publications/ithaka-sr-us-faculty-survey-2023/

Jabarali, A. K., Sathya Kumar, H., & Barak, Y. S. (2024). Analyzing the COVID-19 pandemic's effects on college students: E-learning experience, challenges and resource awareness. *Journal of Library & Information Services in Distance Learning*, *18*(1-2), 18–37. DOI: 10.1080/1533290X.2024.2332806

Jaeger, P. T., Taylor, N. G., & Gorham, U. (2021). Patron privacy protections in public libraries: IT professionals' points of view. *Journal of Intellectual Freedom and Privacy*, *6*(2), 11–23. https://journals.ala.org/index.php/jifp/article/view/7965

Jiang, T., Luo, G., Wang, Z., & Yu, W. (2024). Research into influencing factors in user experiences of university mobile libraries based on mobile learning mode. *Library Hi Tech*, *42*(2), 564–579. DOI: 10.1108/LHT-11-2021-0423

Jiao, Q., Li, Y., Zhang, J., & Guo, X. (2020). *Privacy-preserving query classification for virtual reference services in libraries.* Program. *Electronic Library and Information Systems*, *54*(1), 90–105.

Jilani, F., Sheikh, A., Shah, H., & Saqlain, M. (2025). Usability evaluation of academic library websites: A systematic literature review (1995–2023). *Information Discovery and Delivery*, *53*(1), 45–63. DOI: 10.1108/IDD-09-2024-0144

Johnson, K., & Davis, M. (2024). Bridging the digital divide: Mobile library services as equity tools in K-12 education. *School Library Research*, *27*(1), 1–18.

Jones, M., & Johnson, L. (2021). The evolution of library mobile applications: Features and user expectations. *Information Technology and Libraries*, *40*(2), 1–18.

Jovanovic, M., & Campbell, M. (2022). Generative artificial intelligence: Trends and prospects. *Computer*, *55*(10), 107–112. DOI: 10.1109/MC.2022.3192720

Jutras, J. (2023). The library's educational role in bridging the digital divide. *Pathfinder: A Canadian Journal for Information Science Students and Early Career Professionals*, 4(1), 19–27. DOI: 10.29173/pathfinder85

Jyoti, J., & Kumar, J. (2024). Mapping the landscape of virtual word of mouth (v-WOM) in the tourism sector: A bibliometric review.

Kajeet. (2024). The state of library device lending programs, 2024. https://www.kajeet.com/en/blog/the-state-of-library-device-lending-programs-2024

Kalai Lakshmi, T. R., Nanda, S., Choudhary, V., Prabhu, S., Tuteja, G., & Biswal, S. K. (2025, June). Assessing the role of mobile libraries in enhancing literacy in underserved communities. *Indian Journal of Information Sources and Services, 15*(2), 116–123. DOI: 10.51983/ijiss-2025.IJISS.15.2.16

Kerr, A., & Pennington, D. R. (2018). Public library mobile apps in Scotland: Views from the local authorities and the public. *Library Hi Tech, 36*(2), 237–251.

Khaled, A. (2024, May). Smart Location-based Services (Smart-LBS): Platform for Smart Space-Independent LBS. In *2024 IEEE International Conference on Electro Information Technology* (eIT) (pp. 643-650). IEEE. DOI: 10.1109/eIT60633.2024.10609850

Khan, A. U. (2024). Beyond bookshelves: How 5G/6G technology will reshape libraries. *Library & Information Science Research, 46*(3), 210–225.

Kim, Y., & Park, J. H. (2021). Factors influencing the intention of knowledge sharing in public Libraries: Based on self-determination theory. *Journal of the Korean BIBLIA Society for library and Information Science, 32*(1), 247-265.

Kim, B. (2013). *The library mobile experience: Practices and user expectations.* American Library Association.

Kim, B. (2013). The present and future of the library mobile experience. *Library Technology Reports, 49*(6), 15–28.

Kim, H.-J. (2018). *The present and future of the library mobile experience.* Library Technology Reports., Retrieved from https://journals.ala.org/ltr/article/view/4506/5285

Kim, J., & Park, H. (2021). The effects of mobile library app design on user satisfaction and loyalty. *International Journal of Information Management, 58*, 102315.

Kim, Y. M., & Abbas, J. (2010). Adoption of Library 2.0 functionalities by academic libraries and users: A knowledge management perspective. *Journal of Academic Librarianship, 36*(3), 211–218.

Kleftodimos, A., & Evangelidis, G. (2023). The effect of physical and virtual inquiry-based experiments on students' attitudes and learning. *Journal of Science Education and Technology, 33*(2), 234–248.

Konwar, B. (2024). Artificial Intelligence and Library Services.

Kopak, R., Low, D., & Keltie, M. (2020). Improving e-resource problem resolution workflows in academic libraries. *Library Resources & Technical Services, 64*(4), 177–193. https://journals.ala.org/index.php/lrts/article/view/7511

Kroski, E. (2008). Library mobile initiatives. Library Technology Reports, 44(5), 29–40. American Library Association. https://journals.ala.org/ltr/article/view/4894/5902

Kuhlthau, C. C. (2005). Information search process. *Hong Kong, China, 7*(2005), 226.

Kuhlthau, C. C., Maniotes, L. K., & Caspari, A. K. (2015). *Guided inquiry: Learning in the 21st century* (2nd ed.). Libraries Unlimited. DOI: 10.5040/9798400660603

Kulkanjanapiban, P., Silwattananusarn, T., & Lambovska, M. (2025). Research on AI-driven innovations and services in academic libraries: A bibliometric and systematic literature review. *Journal of Data and Information Science, 10*(3), 1–51. DOI: 10.2478/jdis-2025-0036

Kumar, S. (2025). *Embedded librarian online courses: Academic library examples.* College & Research Libraries News. https://crln.acrl.org/index.php/crlnews/article/view/24813/32645

Kumar, A., & Singh, P. (2023). Mobile technology adoption in academic libraries: A systematic review of implementation strategies. *Library Philosophy and Practice, 2023*, 1–22.

Kumar, N. (2025). THE ROLE OF ARTIFICIAL INTELLIGENCE IN ENHANCING DIGITAL LIBRARY SERVICES: A STUDY. *International Research Journal of Library and Information Sciences, 2*(08), 9–17.

Kumbhar, S. S., & Pawar, R. R. (2014). Mobile based services: Application and challenges. *International Journal of Research in Engineering, IT and Healthcare, 2*(4), 1–5. Retrieved from https://www.ijream.org/papers/IJREAMV04I0440090.pdf

Labster. (n.d.). How Labster used research to imagine, build, test, and improve its platform. Retrieved from https://www.labster.com/blog/research-story

Lamb, A. (2012). An investigation of academics' perception, uptake and use of ebooks at the University of Waikato (Doctoral dissertation, Open Access Te Herenga Waka-Victoria University of Wellington).

Lappalainen, Y., & Narayanan, N. (2023). Aisha: A custom AI library chatbot using the ChatGPT API. *Journal of Web Librarianship, 17*(3), 37–58.

Lee, J., & Kim, S. (2023). Enhancing library user engagement through gamification and augmented reality in mobile applications. *Library Hi Tech*, *41*(2), 273–289.

Lee, K. B. (2011). Developing Mobile Collaborative Learning Applications for Mobile Users. *International Journal of Interactive Mobile Technologies*, *5*(4), 42. DOI: 10.3991/ijim.v5i4.1823

Lee, Y., & Kim, H. Y. (2019). Consumer need for mobile app atmospherics and its relationships to shopper responses. *Journal of Retailing and Consumer Services*, *51*, 437–442. DOI: 10.1016/j.jretconser.2017.10.016

LeMire, S., Gilbert, S., Graves, S., & Faultry-Okonkwo, T. (2017). *Selfie as Guide: Using Mobile Devices to Promote Active Learning and Student Engagement*. Mobile Technology and Academic Libraries: Innovative Services for Research and Learning, 55-71.

LeMire, S., Gilbert, S., Graves, S., & Faultry-Okonkwo, T. (2017). *Selfie as Guide: Using Mobile Devices to Promote Active Learning and Student Engagement. Mobile Technology and Academic Libraries: Innovative Services for Research and Learning, 55-71. American Library Association. (2024). Public Library Technology Survey 2023: Summary Report*. Public Library Association.

Lenormand, M. (2023). Understanding the diversity of mobile service usage in cities. arXiv Preprint. https://arxiv.org/abs/2311.06269

Lenstra, N., Oguz, F., & Duvall, C. S. (2020). Library services to an aging population: A nation-wide study in the United States. *Journal of Librarianship and Information Science*, *52*(3), 738–751. DOI: 10.1177/0961000619871596

Li, Y. (2024). Factors influencing the behavioral intention to use mobile library services: An empirical study of university students in China. *AU-GSB e-Journal*, 17(2), 45–57. https://auojs.au.edu/index.php/Scholar/article/view/7474

Library of Congress. (2024). Digital Strategy at the Library of Congress. Retrieved from https://www.loc.gov/digital-strategy/

LibraryH3lp. (2024). *Text a Librarian SMS reference service*. Retrieved from https://libraryh3lp.com/text-a-librarian

Li, I. W., Jackson, D., & Koshy, P. (2025). Student's reported satisfaction at university: The role of personal characteristics and secondary school background. *Higher Education*, *89*(6), 1513–1531.

Lin, X., Sun, Y., & Zhang, Y., & MR, K. (2023, February). Application of AI in library digital reading promotion service. In *2023 IEEE International Conference on Integrated Circuits and Communication Systems (ICICACS)* (pp. 1-6). IEEE. DOI: 10.1109/ICICACS57338.2023.10100096

Lisedunetwork. (2024, September 25). Remote access to library resources. https://www.lisedunetwork.com/remote-access-to-library-resources/

Liu, Y. Q. (2023). Digital library of high school virtual laboratories: Empowering STEM education through mobile access. *CSU Grant 2023: Southern Connecticut State University.* Franklin, D. (2013). *A practical guide to gender diversity for computer science faculty*. Morgan & Claypool Publishers.

Liu, Y. Q., & Wang, Y. (2023). Research of Library's Smart Service, Derived from Recommender Systems and Knowledge Discoveries (Traditional Chinese Edition) (p. 163). Nook Digital LLC.

Liu, Y., Lu, R., & Liu, Y. Q. (2014). The Library Mobile Services in the USA:A Case Study of New York State Libraries. *Library and Information Service, 58*(12), 17–25. Retrieved from http://124.16.154.130:8080/lis/CN/abstract/abstract13884.shtml

Liu, L., Su, X., Akram, U., & Abrar, M. (2020). The user acceptance behavior to mobile digital libraries. [IJEIS]. *International Journal of Enterprise Information Systems, 16*(2), 38–53.

Liu, Y. Q. (2007). Utilizing Recommender Systems in Digital Libraries, Current Study and Tech Issues. *Library and Information Service, 51*(12), 6–18.

Liu, Y. Q. (2023). *Digital library of high school virtual laboratories: Empowering STEM education through mobile access. CSU Grant 2023.* Southern Connecticut State University.

Liu, Y. Q., Bielefield, A., & McKay, P. (2019). Are urban public libraries websites accessible to Americans with disabilities? *Universal Access in the Information Society, 18*(1), 191–206. DOI: 10.1007/s10209-017-0571-7

Liu, Y. Q., & Briggs, S. (2015). A Library in the Palm of Your Hand: Mobile Services in Top 100 University Libraries. *Information Technology and Libraries, 34*(2), 133–148. DOI: 10.6017/ital.v34i2.5650

Liu, Y. Q., & Lewis, S. (2023). Services to mobile users: The best practice from the top-visited public libraries in the US. *Information Technology and Libraries, 42*(1). Advance online publication. DOI: 10.6017/ital.v42i1.15143

Liu, Y. Q., & Wnuk, A. (2009). The Impact of Digital Resource and Service Use on Urban Residents in New England Public Libraries—A Survey Report. *Public Library Quarterly*, *28*(1), 4–23. DOI: 10.1080/01616840802675242

Lo, P. C., & Stark, A. J. (2020). Examining the relationship between social inclusion and mobile libraries in the age of Internet connectivity: A qualitative study of mobile librarians around the globe. Journal of Librarianship and Information Science. Advance online publication. An, J., Zhu, X., Wan, K., Xiang, Z., Shi, Z., An, J., & Huang, W. (2024). Older adults' self-perception, technology anxiety, and intention to use digital public services. BMC Public Health

Löff, J., Hoffmann, R. B., Pieper, R., Griebler, D., & Fernandes, L. G. (2022). Dsparlib: A c++ template library for distributed stream parallelism. *International Journal of Parallel Programming*, *50*(5), 454–485.

Loh, C. E., & Koh, Y. Q. (2025). *Mobile libraries bridging literacy divides and fostering social cohesion: A PRISMA review of mobile library initiatives.* Office of Education Research, National Institute of Education., https://repository.nie.edu.sg/bitstreams/2b6c0ace-4c8b-4a2e-b0256758e272e422/download

Lo, P. C., He, M., & Liu, Y. Q. (2019). Social inclusion and social capital of the Shanghai Library as a community place for self-improvement. *Library Hi Tech*, *37*(4), 811–828. DOI: 10.1108/LHT-04-2018-0056

Lo, P. C., & Stark, A. J. (2020). *Examining the relationship between social inclusion and mobile libraries in the age of Internet connectivity: A qualitative study of mobile librarians* around the globe. *Journal of Librarianship and Information Science*. Advance online publication.

Lopez, J., Cadena, V., & Rahman, M. S. (2025). Evaluating post-quantum cryptographic algorithms on resource-constrained devices. *arXiv preprint arXiv:2507.08312*.

Lotts, M., & Graves, S. (2011). Using iPads for roving reference services. *College & Research Libraries News*, *72*(4), 217–220. DOI: 10.5860/crln.72.4.8544

Luo, L. (2016). Professional preparation for "Text a Librarian": What are the requisite competencies? RUSQ: A Journal of Reference & User Experience, 52(1), 44-65.

Luo, L. (2020). Digital reference service: A review of literature and a case study of academic libraries in China. *Information Development*, *36*(6), 1224–1237.

Luo, L. (2021). Artificial intelligence and library services: A review of research and practices. *Journal of Academic Librarianship*, *47*(6), 102390.

Luo, L., Zhang, T., & Xu, H. (2022). The promise and potential of chatbots in academic libraries. *College & Research Libraries, 83*(2), 193–208.

Lu, W., Wang, Z., Zhang, S., Meng, Q., & Luo, H. (2024, August). OpenSN: An open source library for emulating LEO satellite networks. In *Proceedings of the 8th Asia-Pacific Workshop on Networking* (pp. 149-155). DOI: 10.1145/3663408.3663430

Magazine, S. Q. (2025). Mobile phone usage statistics 2025: What the latest data tells us about global trends. SQ Magazine. https://sqmagazine.co.uk/mobile-phone-usage-statistics/

Mahlangu, L., Botha, A., & Buys, T. (2021). Designing an AI chatbot to support library services at a South African university. *Library Hi Tech, 39*(1), 101–119.

Mao, Y., You, C., Zhang, J., Huang, K., & Letaief, K. B. (2017). A survey on mobile edge computing: The communication perspective. arXiv. https://arxiv.org/abs/1701.01090

Marshall, J. G., Fitzgerald, D., Busby, L., & Heaton, G. (1993). A study of library service and user education needs in a problem-based medical curriculum. *Bulletin of the Medical Library Association, 81*(3), 316–322. https://pubmed.ncbi.nlm.nih.gov/8374587/ PMID: 8374589

Martinez, C., & Brown, J. (2024). Comprehensive mobile service portfolios in academic libraries: Usage patterns and student preferences. *College & Research Libraries, 85*(2), 178–195.

Masenya, J., & Mashabela, G. (2025). Going mobile for smart public libraries: Factors influencing the adoption of smartphones in accessing EIRs. *South African Journal of Library and Information Science, 91*(2), 1–12.

Matteson, J. (2024). A systematic review of research on live chat service. *RUSQ: A Journal of Reference and User Experience*, 60(2).

Mehta & Wang, X. (2020). COVID-19 and digital library services–a case study of a university library. *Digital Library Perspectives.*

Meyers-Martin, C. (2024). It's all about the chat and COVID-19: Virtual one-on-one research consultations surveyed. *Journal of Academic Librarianship, 50*(4), 102904. DOI: 10.1016/j.acalib.2024.102904

Milano, F., da Rocha, H., Laracca, M., Ferrigno, L., Espírito Santo, A., Salvado, J., & Paciello, V. (2024). BLE-based indoor localization: Analysis of some solutions for performance improvement. *Sensors (Basel), 24*(2), 376. DOI: 10.3390/s24020376 PMID: 38257468

Mocanu, R. (2024). Text messaging for schools: Improving communication with SMS. Textmagic. https://www.textmagic.com/blog/sms-solutions-schools/

Moen, W. E., & McClure, C. R. (n.d.). *Virtual reality and the academic library of the future*. ResearchGate. https://www.researchgate.net/publication/359601186_Virtual_reality_and_the_academic_library_of_the_future

Mohammed, M. S., & Zidan, K. A. (2023). Enhancing attendance tracking using animated QR codes: A case study. *Indonesian Journal of Electrical Engineering and Computer Science*, *31*(3), 1716. DOI: 10.11591/ijeecs.v31.i3.pp1716-1723

Mondal, D. (2025). Exploring user perceptions and usability of mobile library services: A case study of Kazi Nazrul University. *SSRN*. DOI: 10.2139/ssrn.5369059

Moorefield-Lang, H., Meier, C., & Miller, R. (2014). *Tablet computers in school libraries and classrooms*. American Library Association.

Moorefield-Lang, H., Meier, C., & Miller, R. (2014). *Tablet computers in school libraries and classrooms*. American Library Association. Gray, B., & Thomas, D. P. (2025). Student perceptions of primary school library support: Investigating reading, research, and technology use. *Journal of Librarianship and Information Science*, •••, 09610006251322983.

Mounir, K., & Shafeek, F. (2003). *History of mobile libraries*.

Moya, S., & Camacho, M. (2023). Developing a framework for mobile learning adoption and sustainable development. Technology. *Knowledge and Learning*, *28*(2), 727–744. DOI: 10.1007/s10758-021-09537-y

Murad, S. S., Badeel, R., Abdal, B. B., Rahman, T., & Al-Quraishi, T. (2024). Introduction to Wi-Fi 7: A review of history, applications, challenges, economical impact and research development. *Mesopotamian Journal of Computer Science*, *2024*, 110–121. DOI: 10.58496/MJCSC/2024/009

Mustapha, A., Ibrahim, F., Garba, K. D., & Muhammed, M. (2024, December). *Transforming academic libraries: The impact of mobile technology*. ResearchGate. https://www.researchgate.net/publication/390121750_Transforming_Academic_Libraries_The_Impact_of_Mobile_Technology

Nahum-Shani, I., Smith, S. N., Spring, B. J., Collins, L. M., Witkiewitz, K., Tewari, A., & Murphy, S. A. (2016). Just-in-time adaptive interventions (JITAIs) in mobile health: Key components and design principles for ongoing health behavior support. *Annals of Behavioral Medicine*, •••, 1–17. PMID: 27663578

National University. (2025, July 15). What are the challenges of online learning for students? National University Blog. https://www.nu.edu/blog/challenges-of-distance-learning-for-students/

Nawaz, N., & Saldeen, M. A. (2020). ARTIFICIAL INTELLIGENCE CHATBOTS FOR LIBRARY REFERENCE SERVICES. *Journal of Management Information & Decision Sciences*, 23.

Nicholas, D., Rowlands, I., Clark, D., & Williams, P. (2009). Student information-seeking behaviour in a digital academic environment. *The Journal of Documentation*, *65*(1), 106–130. DOI: 10.1108/00220410910926149

Nichols, M. (2020). Developing a virtual reference service with limited resources. *New Review of Academic Librarianship*, *26*(1-2), 158–172.

Nunekpeku, P. E. (2019). Establishing clients' satisfaction levels with automated library based services: A case study at University of Cape Coast library, Ghana. *Digital Library Perspectives*, *36*(1), 8–20. DOI: 10.1108/DLP-02-2019-0004

O'Muircheartaigh, C., & Palmberg, I. (2023). Public libraries as community health partners: Promoting wellness and health literacy among adults. *Public Library Quarterly*, *42*(4), 345–362.

Ocran, T. K., Underwood, P. G., & Arthur, P. A. (2020). Strategies for successful implementation of mobile phone library services. *Journal of Academic Librarianship*, *46*(5), 102174. DOI: 10.1016/j.acalib.2020.102174

Ogbebor, O. A. (2024, December 19). Artificial intelligence transforms academic libraries into dynamic, cutting-edge solutions. International Journal of Library and Information Science. Academic Journals. https://academicjournals.org/journal/IJLIS/article-full-text-pdf/52BB02772891

Ogbole, G. I., Adepoju, A. E., Ibrahim, A., Togunwa, T. O., & Nkeakam, F. A. (2023). MRI training in Africa. *Annals of Ibadan Postgraduate Medicine*, *21*(2), 75–80. PMID: 38298339

Okonkwo, I., & Ejikeme, C. (2023). Specialized digital collections accessible through mobile platforms serve students with diverse learning needs. *Library Hi Tech*, *41*(1), 234–251. DOI: 10.1108/LHT-01-2023-0014

Oliveira, S. M. (n.d.). Information seeking behavior of online students: Academic library perspectives [Preprint]. Preprint Press JHU. https://preprint.press.jhu.edu/portal/sites/default/files/oliveira.pdf

Ontario Council of University Libraries. (2024). *Scholars Portal Ask A Librarian service annual report 2024.*

Osika, B., & Kaufman, C. (2012). 'MOBILIZING'COMMUNITY COLLEGE LIBRARIES: Trends and Solutions. *Searcher (Medford, N.J.), 20*(9).

Oulasvirta, A. (2008, September). Designing mobile awareness cues. In *Proceedings of the 10th international conference on Human computer interaction with mobile devices and services* (pp. 43-52).

OverDrive, Inc. (2023). Libby: The library reading app. OverDrive, Inc. https://www.overdrive.com/apps/libby

OverDrive, Inc. (2024). *Libraries achieve record-breaking circulation of digital media in 2023.* OverDrive company news.

Owens, E., Arce, V., Del Bosque, D., Fowler, R., & Sheffield, S. (2024). Beyond the Library: The Role of Academic Libraries' Chat Reference in Answering Campus Questions. *The Reference Librarian, 65*(1-2), 1–24. DOI: 10.1080/02763877.2024.2334216

Owolabi, K. A., Abdullahi, Z. S., Agboola, I. O., & Ogungbade, R. M. (2024). Professional competence and user satisfaction in academic libraries in Lagos State, Nigeria. *Chinese Journal of Library and Information Science, 17*(2), 85–99. https://www.cjolis.org/index.php/cjolis/article/view/70

Owusu-Ansah, F. E. (2015). Sharing in the life of the person with disability: A Ghanaian perspective. *African Journal of Disability, 4*(1), 1–8. DOI: 10.4102/ajod.v4i1.185 PMID: 28730035

Parry, D. (2023, May 4). iSchool researchers release Autism-Ready Libraries Toolkit for libraries. University of Washington Information School.

Partners, S. T. L. (2024). *Mobile edge computing: Industry overview and applications.* https://stlpartners.com/articles/edge-computing/mobile-edge-computing

Patel, A. K., & Singh, K. (2022). A Bibliometric Study of Global Publications on the Effect of Social Media in Libraries from 1991 to 2021. *Management of Modern Libraries in New Normal*, 216-226.

Patel, N., & Williams, R. (2023). Strategic implementation of mobile technologies in school libraries: A framework for success. *School Library Journal, 69*(8), 32–37.

Paterson, L., & Low, B. (2011). Student attitudes towards mobile library services for smartphones. *Library Hi Tech, 29*(3), 412–423. DOI: 10.1108/07378831111174387

Pauline, A. (2011). User satisfaction with academic libraries services: Academic staff and students' perspectives. *International Journal of Library and Information Science*, *3*(10), 209–216.

Pendell, K. D., & Bowman, M. S. (2012). Usability study of a library's mobile website: An example from Portland State University. *Journal of Academic Librarianship*, *37*(5), 341–347.

Penner, K. (2024). iPad and Digital Sheet Music Delivery: Supporting Student Development and Encouraging Digital Score Use Through Library-Circulated Hardware. *Fontes Artis Musicae*, *71*(3), 179–189. DOI: 10.1353/fam.2024.a939059

Pereira, J. L. S., Andrade, R. L., Vilela, L. F., Oliveira, M. M., Marques, L. N., Inácio, J. V., Aguiar, M. D., Lima, R. N., Pinton, L. O., Morosoli, A. C. C., Reale, L. P. A., Rocha, J. P. M., Maciel, G. I. F., Moreira, R. V., & Cerqueira, R. L. (2025). Uso de wearables para monitoramento contínuo de pacientes com doenças cardíacas. *Caderno Pedagógico*, *22*(1), e13284–e13284. DOI: 10.54033/cadpedv22n1-091

Perritt, H. H. (2002). *Americans with disabilities act handbook*. Wolters Kluwer.

Peterson, G., Griffin, J., LaFrance, J., & Li, J. (2017). Smartphone participation in web surveys: Choosing between the potential for coverage, nonresponse, and measurement error. *Total survey error in practice*, 203-233.

Peters, T. A., & Bell, L. A. (Eds.). (2013). *The handheld library: Mobile technology and the librarian*. Bloomsbury Publishing USA. DOI: 10.5040/9798400661679

Pew Research Center. (2021). *Mobile fact sheet*. Pew Research Center. https://www.pewresearch.org/internet/fact-sheet/mobile/

Pew Research Center. (2022, August 10). Teens, social media and technology 2022. Pew Research Center.

Pew Research Center. (2025, July 9). *Teens and Internet, Device Access Fact Sheet*. Retrieved from Pew Research Center.

Phillips, A. L. (2024, November 1). Embracing the metaverse: A survey of virtual reality and augmented reality practices at the United States' top one hundred university libraries. College & Research Libraries. https://crl.acrl.org/index.php/crl/article/view/25664/34458

Pike, G. (2021). Virtual reference services in academic libraries: Current issues and future directions. *Journal of Academic Librarianship*, *47*(4), 102398.

Pinto, M., & Segura, A. (2025). Toward a conceptual framework on mobile information literacy in higher education. *Journal of Academic Librarianship*, *51*(3), 103051. DOI: 10.1016/j.acalib.2025.103051

Popenici, S. A. D., & Kerr, S. (2024). Virtual and augmented reality in Science, Technology, Engineering, and Mathematics (STEM) education: An umbrella review. *Information (Basel)*, *15*(9), 515. DOI: 10.3390/info15090515

Potnis, D. (2015). Wilson's information-seeking behavior models (1981, 1996, 1999). In *Information seeking behavior and technology adoption: Theories and trends* (pp. 94–112). IGI Global Scientific Publishing. DOI: 10.4018/978-1-4666-8156-9.ch006

Providence College. (2017). *A novel application: Using mobile technology to connect physical and virtual reference collections*. Digital Commons @ Providence. https://digitalcommons.providence.edu/facstaff_pubs/49/

Providence Public Library. (n.d.). Mobile learning. In *Propagating Promising Practices*. Mobile Learning.

Public Library Association. (2024, July). 2023 Public Library Technology Survey: Summary report [PDF]. American Library Association. Retrieved from https://www.ala.org/sites/default/files/2024-07/PLA_Tech_Survey_Report_2024.pdf

Public Library Association. (n.d.). DigitalLearn.org: Online digital literacy courses [Website]. Retrieved August 16, 2025, from https://www.ala.org/pla/initiatives/digitalliteracy

Putro, N. H. P. S., & Lee, J. (2017). Reading interest in a digital age. *Reading Psychology*, *38*(8), 778–807. DOI: 10.1080/02702711.2017.1341966

Qayyum, M. A., Mahmood, K., & Alqahtani, F. (2022). Students' perceptions of library virtual reference services: A case of an academic library. *Journal of Librarianship and Information Science*, *54*(2), 529–542.

Quaye Affum, M. (2022). Assessment of the impact of mobile libraries on traditional libraries. *International Journal of Computing. Programming and Database Management*, *3*(1), 45–48.

Radford, M. L., & Gary, J. (2021). Ethics of AI and chatbots in reference services. *RSR. Reference Services Review*, *49*(3), 353–371.

Rahman, S., & Clarke, D. (2023). Mobile library services and educational equity: Addressing access barriers through technology. *Information Technology and Libraries*, *42*(3), 45–58.

Ram Prakash, R. (2023). Use of mobile technology in library systems. *NeuroQuantology : An Interdisciplinary Journal of Neuroscience and Quantum Physics, 21*(13), 24–30. DOI: 10.14704/nq.2023.21.13.NQ88302

Reference & User Services Association, Multilingual Materials Subcommittee. (1990, June). Guidelines for Multilingual Materials Development and Library Services (Reference & Adult Services Division Board of Directors, American Library Association). Retrieved from https://www.ala.org/rusa/sections/rss/rsssection/rsscomm/spanishspeaking/guidelinesmultilingual

Reference and User Services Association. (2017). *Guidelines for Implementing and Maintaining Virtual Reference Services*. American Library Association.

Restianty, A., Sumartias, S., Hadisiwi, P., & Hafiar, H. (2024). Digital applications as assistive technology for students with disabilities: Promoting inclusive practices through configurable and multisensory interfaces. Asia-Pacific Journal of Special Education Vancouver Public Library Foundation. (2024, January 12). Funding priorities: Programs and collections for children and youth – Sensory Storytimes. Retrieved from https://vplf.ca/category/funding_priorities/programs_and_collections_for_children_and_youth/

Reyes, R. L., Isleta, K. P., Regala, J. D., & Bialba, D. M. R. (2024). Enhancing experiential science learning with virtual labs: A narrative account of merits, challenges, and implementation strategies amid COVID-19. *Journal of Computer Assisted Learning*, *40*(6), 3167–3186. DOI: 10.1111/jcal.13061

Roberts, E., & Kim, H. (2024). Inclusive design principles in mobile library applications: Supporting diverse learning needs. *Universal Access in the Information Society*, *23*(2), 289–302.

Rodriguez, S., Szydlowski, N., & Cai, J. (2025, March). Library-led AI: Building a library chatbot as service and strategy (Workshop report). San José State University Library. https://www.ala.org/sites/default/files/2025-03/Library-LedAI.pdf

Roy, L. (2025). Using data analytics to identify information gaps and create inclusive library programs. Journal of Web Librarianship, 19(2), 123–138. Taylor & Francis Online. https://www.tandfonline.com/doi/full/10.1080/1941126X.2025.2497729

Ruan, J., & Wang, S.-B. (2016, December). Study on innovation of smart library service model in the era of big data. In *Proceedings of the 2016 4th International Conference on Electrical & Electronics Engineering and Computer Science (ICEEECS 2016)* (pp. 1077-1081). Atlantis Press. DOI: 10.2991/iceeecs-16.2016.207

Safdar, M., Ur Rehman, S., Yousaf, A., & Ashiq, M. (2023). A systematic literature review on the use of mobile phones to access library services and resources: Challenges and benefits. *Global Knowledge, Memory and Communication.* Advance online publication. DOI: 10.1108/GKMC-04-2023-0122

Saragossi, J., Costello, L., & Kasten, K. (2024). Mobile applications in academic libraries. Library Resources & Technical Services, 62(4), 198–... DOI: 10.5860/lrts.62n4.198

Saragossi, J., Kasten, K., & Costello, L. (2018). Mobile Applications in Academic Libraries.

Saravani, S., & Haddow, G. (2017). A theory of mobile library service delivery. *Journal of Librarianship and Information Science*, *49*(2), 131–143. DOI: 10.1177/0961000615595854

Schroeder, T., Dodds, L., Georgiou, A., Gewald, H., & Siette, J. (2023). Older adults and new technology: Mapping review of the factors associated with older adults' intention to adopt digital technologies. *JMIR Aging*, *6*, e44564. DOI: 10.2196/44564 PMID: 37191976

Scottish Library & Information Council. (2018, November 29). *Research to examine the value and impact of mobile library services.* https://www.cilips.org.uk/research-to-examine-the-value-and-impact-of-mobile-library-services/

Scott, S. M., & Ward, L. (2025). The use of AI in academic libraries. In *Examining AI Disruption in Educational Settings* (pp. 59–100). IGI Global. DOI: 10.4018/979-8-3693-7873-1.ch003

Seeholzer, J., & Salem, J. A.Jr. (2011). Library on the go: A focus group study of the mobile Web and the academic library. *College & Research Libraries*, *72*(1), 9–20. DOI: 10.5860/crl-65r1

Senior, A., & Ward, L. (2019). Measuring patron satisfaction with reference service: A rapid review. *Evidence Based Library and Information Practice*, *14*(2), 76–101. https://journals.library.ualberta.ca/eblip/index.php/EBLIP/article/view/29556

Shariq Imran, A., Uddin, I., Fayyaz, M., & Sajjad, M. (2023). Pattern-based hybrid book recommendation system using semantic relationships. *Scientific Reports*, •••, 13.

Shaw, M., & Matusiak, K. K. (2022). User perceptions of library chatbots: A case study. *Information Technology and Libraries*, *41*(4).

Shi, W., & Liu, Y. Q. (2017). Continuous Use Intention of Library Mobile Services: Evidence from University Libraries in China. *Library Hi Tech, 35*(4), 657–669.

Sidorko, P. E. (2009). Virtually there, almost: Educational and informational possibilities in virtual worlds. *Library Management, 30*(6/7), 404–418. DOI: 10.1108/01435120910982104

Simmonds, P. L., & Andaleeb, S. S. (2001). Usage of academic libraries: The role of service quality, resources, and user characteristics.

Singh, A. K., & Narayan, R. (2021). Progressive Web Apps: A Smart Way to Build Mobile-Web Apps for Academic Libraries. Applying Mobile Technologies in Transformation of Library Services, 91-117.

Singh, B. P., & Madhusudhan, M. (2023). Mobile apps–based applications in libraries and information centers: A systematic review of the literature and future research agendas. *International Journal of Librarianship, 8*(3), 83–102. DOI: 10.23974/ijol.2023.vol8.3.294

Singh, S., & Madhusudhan, M. (2023). Mobile applications in libraries: A systematic review of global trends (2015–2022). *The Electronic Library, 41*(3), 476–493. DOI: 10.1108/EL-02-2022-0058

Smart Learning Environments. (2025). Factors influencing virtual lab adoption in marginalized rural schools: Insights from South Africa. Smart Learning Environments, 12, Article 11. https://doi.org

Smith, J. (2020, April 14). Mobile Libraries: Culture on the Go. The Unwritten Record. National Archives at College Park. https://unwritten-Morales, M., Knowles, E. C., & Bourg, C. (2020). Multilingualism, neoliberalism, and language ideologies in libraries. In the Library with the Lead Pipe. https://www.inthelibrarywiththeleadpipe.org/2020/multilingualism-in-libraries/

Smith, G. (2015). The future of library mobile websites: A usability study. *Library Hi Tech, 33*(4), 574–587.

Smith, P., Garcia, R., & Lee, S. (2022). Integrating digital resources through library apps: A user-centered perspective. *Digital Library Perspectives, 38*(3), 157–175.

SolarSPELL. (2024). SolarSPELL: Portable solar-powered offline digital libraries empowering underserved K-12 learners. Wikipedia. Retrieved from https://en.wikipedia.org/wiki/SolarSPELL

Sowole, I. A. (1995). Collection development at the University of Agriculture library, Abeokuta,(UNAAB) Nigeria. *The International Information & Library Review, 27*(4), 383–391. DOI: 10.1080/10572317.1995.10762383

Spacey, R., Muir, A., Cooke, L., Creaser, C., & Spezi, V. (2017). Filtering wireless (Wi-Fi) Internet access in public places

Speidel, D., & Liu, Y. Q. (2007). Are Recommender Systems Good for Libraries? *Library and Information Service, 51*(12), 19–25.

Spires, T. (2008). Handheld librarians: A survey of librarian and library patron use of wireless handheld devices. *Internet Reference Services Quarterly, 13*(4), 287–309. DOI: 10.1080/10875300802326327

Springshare. (2023, October 10). Introducing the LibAnswers NPS® Module: Your new favorite feedback tool. Springshare Blog.

Springshare. (2025). LibAnswers product overview.

Stănescu, G., & Oprea, S. V. (2025). Recent Trends and Insights in Semantic Web and Ontology-Driven Knowledge Representation Across Disciplines Using Topic Modeling. *Electronics (Basel), 14*(7), 1313. DOI: 10.3390/electronics14071313

StatCounter. (2016) *Mobile and Tablet Internet Usage Exceeds Desktop for First Time Worldwide.* https://gs.statcounter.com/press/mobile-and-tablet- internet-usage-exceeds-desktop-for-first-time-worldwide.

Statista. (2025). *Mobile subscriptions worldwide by region 2005-2023.* Retrieved March 9, 2025, from https://www.statista.com

Stattia, A., & Villegas, S. (2020). The use of mobile learning in Grades K–12: A literature review of current trends and practices. *Peabody Journal of Education, 95*(2), 139–147. DOI: 10.1080/0161956X.2020.1745613

Straits Research. (2025). *Progressive Web Apps: Market size and global outlook 2025.* Straits Research.

Subramaniam, M. (2020). *Understanding digital equity and inclusion among vulnerable adult populations: A mixed-methods approach.* Portland State University Digital Equity Research Report. Retrieved from https://pdxscholar.library.pdx.edu/digital_equity/

Sulisworo, D., Maryani, I., Mastul, A.-R. H., & Kusumaningtyas, D. A. (2024). Incorporating gamified elements into mobile learning to enhance critical thinking skills. *Jurnal Pendidikan Progresif, 14*(1), 237–248. DOI: 10.23960/jpp.v14.i1.202418

Svaler, T. B. (2024). On making libraries and museums more accessible for autistic people. *IFLA Journal*, *56*(2), 418–430. DOI: 10.1177/03400352231202516

Tafesse, W. (2024). Digital overload, coping mechanisms, and student well-being: Evidence from higher education. *SAGE Open*, *14*(2). Advance online publication. DOI: 10.1177/21582440241236087

Tampa Bay Library Consortium. (2025). *Ask A Librarian fee structure and sustainability plan. Tampa Bay Library Consortium.* https://tblc.org/ask-a-librarian-fee-structure/

Tang, F., & Østvold, B. M. (2023). Transparency in app analytics: Analyzing the collection of user interaction data. manuscript (preprint).

Tao, D., McCarthy, P. G., Krieger, M. M., & Webb, A. B. (2009). The mobile reference service: A case study of an onsite reference service program at the School of Public Health. *Journal of the Medical Library Association: JMLA*, *97*(1), 6–12. DOI: 10.3163/1536-5050.97.1.006 PMID: 19159004

TechRT. (2025). Smartphone statistics 2025. https://techrt.com/smartphone-statistics/

The Digital Librarian. (2025). AI in libraries: Opportunities and ethical considerations. Retrieved from https://thedigitallibrarian.org/ai-in-libraries

The Library Mobile Experience. Practices and User Expectations. (2013). Library Technology Reports, 49(6). Retrieved from https://journals.ala.org/ltr/article/view/4506

Theiss, D. (2022). Distance and online learners and library anxiety: An exploration into the causes, impact, and recommendations for practice. *Journal of Library & Information Services in Distance Learning*, *16*(2), 152–167. DOI: 10.1080/1533290X.2022.2112799

Thomas, L. C. (2013). Mobile present, mobile future. *The Handheld Library: Mobile Technology and the Librarian*, 13.

Thompson, A., Rodriguez, M., & Chang, L. (2024). ICT infrastructure requirements for mobile library service implementation: A multi-institutional study. *DESIDOC Journal of Library and Information Technology*, *44*(1), 12–25.

Thompson, B., Cook, C., & Thompson, R. L. (2002). Reliability and structure of LibQUAL+ scores: Measuring perceived library service quality. *portal. Portal (Baltimore, Md.)*, *2*(1), 3–12.

Thurlow, I., Duke, A., & Davies, J. (2006). Applying semantic web technology in a digital library. In *Demos and Posters of the 3rd European Semantic Web Conference, available at: www. eswc.*

Timotheou, S., Angelopoulos, C. M., & Kitsios, F. (2022). Digital transformation in cultural and educational institutions: A systematic literature review. *Information (Basel)*, *13*(12), 568. DOI: 10.3390/info13120568

Torres-Pérez, P., Méndez-Rodríguez, E., & Orduna-Malea, E. (2016). Mobile web adoption in top ranked university libraries: A preliminary study. *Journal of Academic Librarianship*, *42*(4), 329–339. DOI: 10.1016/j.acalib.2016.05.011

Tsirulnikov, K., Suart, C., Abdullah, R., Vulcu, F., & Mullarkey, C. E. (2023). Game on: Immersive virtual laboratory simulation improves student learning outcomes & motivation. *FEBS Open Bio*, *13*(3), 396–407. Advance online publication. DOI: 10.1002/2211-5463.13567 PMID: 36723273

Tunjungsari, P., & Suprihatin, S. (2024). Systematic Review of Library Service Accesibility using Artificial Intelligence: Features, Challenges, and Library Roles. *Knowledge Garden: International Journal of Library Studies*, *2*(2), 1–22. DOI: 10.21776/ub.knowledgegarden.2024.2.2.21

Turner, A. (2021). *How Many Smartphones Are in the World? 2020.* URL: https:// www. bankmycell. com/blog/how-many-phones-are-in-the-world [accessed 2021-05-21].

Twomey, B., Johnson, A., & Estes, C. (2024). *It takes a village: a distributed training model for AI-Based chatbots.*

Ubogu, A., & Chukwusa, O. (2024). Mobile library services revisited: Applications, prospects, and challenges in Nigerian academic libraries. *Delta Journal of Computing. Communications & Media Technologies*, *1*(1), 1–12.

Udem, O. K., & Ogungbeni, J. I. (2023). Use of mobile technology – based library services and information needs satisfaction of undergraduates in universities in South-West Nigeria. *Folia Toruniensia*, *23*, 15–33. Advance online publication. DOI: 10.12775/FT.2023.001

University at Buffalo Health Sciences Library. (2013). Providing mobile reference and instruction services using iPads. *Journal of the Medical Library Association*, 101*(3). https://pmc.ncbi.nlm.nih.gov/articles/PMC4733882/

University at Buffalo Health Sciences Library. (2013). Providing mobile reference and instruction services using iPads. *Journal of the Medical Library Association: JMLA*, *101*(3). https://pmc.ncbi.nlm.nih.gov/articles/PMC4733882/

University of Texas Libraries. (2024, November). *Building a Bot: An Exploration of AI to Assist Librarians*. TexLibris. https://texlibris.lib.utexas.edu/2024/11/building-a-bot-an-exploration-of-ai-to-assist-librarians/?utm_source=chatgpt.com

Urrea, G., & Yoo, E. (2023). The role of volunteer experience on performance on online volunteering platforms. *Production and Operations Management*, *32*(9), 3051–3071. DOI: 10.1111/poms.13879

Valente, L., Feijó, B., Leite, J. C. S. D. P., & Clua, E. (2018). A method to assess pervasive qualities in mobile games. *Personal and Ubiquitous Computing*, *22*(4), 647–670. DOI: 10.1007/s00779-017-1107-0

Varlamis, I., Sardianos, C., Chronis, C., Dimitrakopoulos, G., Himeur, Y., Alsalemi, A., Bensaali, F., & Amira, A. (2023). Using big data and federated learning for generating energy efficiency recommendations. *International Journal of Data Science and Analytics*, *16*(3), 353–369. DOI: 10.1007/s41060-022-00331-2

Vasantha Kumar, P. J. R. (2022). Impact of mobile technology in library services. *Indian Journal of Library Science and Information Technology*, *7*(1), 40–43. DOI: 10.18231/j.ijlsit.2022.008

Vasishta, P., Dhingra, N., & Vasishta, S. (2025). Application of artificial intelligence in libraries: A bibliometric analysis and visualisation of research activities. *Library Hi Tech*, *43*(2/3), 693–710.

Villagrasa, S., Fonseca, D., Redondo, E., & Martí, N. (2024). A systematic review of accessibility and inclusive design in virtual laboratories. *Frontiers in Education*, *9*, 1351711. Advance online publication. DOI: 10.3389/feduc.2024.1351711

Vogus, B. (2019). Open educational resources (OER) and textbooks: Opportunities for academic libraries. *Public Services Quarterly*, *15*(3), 242–247. DOI: 10.1080/15228959.2019.1629858

W3C. (2018). ActivityPub (W3C Recommendation). https://www.w3.org/TR/activitypub/

Wagwu, V., Okpala, A. E., Maxwell, C. E., & Oladokun, B. D. (2024). AI-Driven Chatbots for Library Services towards Transforming Patrons' Engagement in Nigerian University Libraries. Communicate. *Journal of Library and Information Science*, *26*(1), 89–98. https://www.cjolis.org/index.php/cjolis/article/view/78

Walsh, A. (2012). *Using mobile technology to deliver library services a handbook*. Facet. DOI: 10.29085/9781856048996

Walsh, A. (2012). *Using mobile technology to deliver library services: a handbook*. Facet Publishing.

Wang, Y., Dong, X., & Liu, Y. Q. (2018). Service Mode of Smart Recommendation Based on Knowledge Discovery. *Research on Library Science* Kukiolczynski, K. T. & Liu, Y. Q. (2021). Public Library Events: Who, What, Where, When, and Why? *Public Library Quarterly*, 1-22. Fung, L. W., Zhao, J., Yan, V. K., Blais, J. E., Chan, J. C., Li, S. T., ... & Chan, E. W. (2022). COVID-19 vaccination preferences of university students and staff in Hong Kong. *JAMA Network Open*, *5*(5), e2212681–e2212681.

Warren, S. J. (2021). A systematic review of virtual reference services in academic libraries. *College & Research Libraries*, *82*(4), 472–495.

We Are Social. (2025). Digital 2025 Global Overview Report. https://wearesocial.com/wp-content/uploads/2025/02/GDR-2025-v2.pdf

Wei, Q., Chang, Z., & Cheng, Q. (2015). Usability study of the mobile library app: An example from Chongqing University. *Library Hi Tech*, *33*(3), 340–355. DOI: 10.1108/LHT-05-2015-0047

Weiser, M. (1991). The Computer for the 21 st Century. *Scientific American*, *265*(3), 94–105. DOI: 10.1038/scientificamerican0991-94 PMID: 1675486

Werner, L., Brey, P., & Henschke, A. (2025). Augmented reality and ethics: Key issues. *Virtual Reality (Waltham Cross)*, *29*(3), 122. DOI: 10.1007/s10055-025-01172-7

Whitchurch, C. (2011). QR codes and library engagement. *Bulletin of the American Society for Information Science*, *38*(1), 14–17. DOI: 10.1002/bult.2011.1720380107

Williams, S., Marks, L., & Kwan, Y. (2021). The Potential Role of Blockchain in Libraries: A Systematic Review. *. *Library Hi Tech*, *39*(3), 520–535. DOI: 10.1108/LHT-06-2021-0114

WorldMetrics Research. (2024). *Cellphones in school statistics: Market data report 2024*. WorldMetrics.

Xie, I., Lee, J. H., Lee, Y., Wang, S., & Babu, R. (2023). Accessibility and usability of digital libraries for blind and visually impaired users: A comparative study of mobile apps and mobile websites. *Information Research*, 28(3), paper 937. https://informationr.net/infres/article/view/337

Xu, M., Chen, Y., & Zhao, L. (2024). Digital literacy education for older adults: A scoping review. *Proceedings of the Association for Information Science and Technology*, *61*(1), 1081. DOI: 10.1002/pra2.1081

Yadukrishnan, T. A., Kumar, A., & Ihejirika, K. T. (2023). More user engagement, better for libraries: Proposing a framework for social media integration in libraries. *Journal of Librarianship and Information Science*, *55*(4), 1045–1058.

Yan, R., Zhao, X., & Mazumdar, S. (2023) conducted a systematic literature review titled "Chatbots in libraries: A systematic literature review," published in Education for Information (Volume 39, Issue 4).

Yang, S., Yang, J., Yang, C., Yan, W., & Li, P. (2025). Human Activity Recognition using WiFi Signal Features and Efficient Residual Packet Attention Network. *IEEE Sensors Letters*, *9*(4), 1–4. DOI: 10.1109/LSENS.2025.3551337

Yanuar, R., Ibrahim, M., & Jemain, A. A. (2020). Factors influencing user behavior intention to use mobile library application: A theoretical and empirical research based on grounded theory. *International Journal of Advanced Science and Technology*, *29*(4), 11442–11453.

Yarra Libraries. (n.d.). Connected Community: iPad lending program. https://library.yarracity.vic.gov.au/using-the-library/ipad-lending-program

Yarrow, A., & McAllister, S. (2018). Trends in mobile and outreach services. *Public Library Quarterly*, •••, 1–14.

Yoon, H. Y. (2016). User acceptance of mobile library applications in academic libraries: An application of the technology acceptance model. *Journal of Academic Librarianship*, *42*(6), 687–693. DOI: 10.1016/j.acalib.2016.08.003

Yorio, K. (2023). School Librarians Explore Possibilities of ChatGPT. School Library Journal. https://www.schoollibraryjournal.com/story/School-Librarians-Explore-Possibilities-of-ChatGPT

Young Adult Library Services Association. (2015, March 4). Teen programming guidelines. American Library Association.

Young, S. W. H., Chao, Z., & Chandler, A. (2019). User experience methods and maturity in academic libraries. *Information Technology and Libraries*, *38*(3), 68–88. DOI: 10.6017/ital.v39i1.11787

Yuan, N., Yu, Q., & Liu, W. (2025, August). The impact of digital literacy on learning outcomes among college students: The mediating effect of digital atmosphere, self-efficacy for digital technology and digital learning. [). Frontiers.]. *Frontiers in Education*, *10*, 1641687. DOI: 10.3389/feduc.2025.1641687

Yulin, Z., Fengping, C., & Yang, W. (2023). Review and Reflection on the Library Service Manifesto: Based on IFLA Trend Report Analysis. *Library Journal*, *42*(388), 14.

Zhang, M., Shen, X., Zhu, M., & Yang, J. (2016). Which platform should I choose? Factors influencing consumers' channel transfer intention from web-based to mobile library service. *Library Hi Tech*.

Zhang, N., & Liu, Y. (2024). Design and implementation of virtual laboratories for higher education sustainability: A case study of Nankai University. *Frontiers in Education*, *8*, 1322263. Advance online publication. DOI: 10.3389/feduc.2023.1322263

Zhang, W., Liu, B., & Tsai, S. B. (2022). Analysis and Research on Digital Reading Platform of Multimedia Library by Big Data Computing in Internet Era. *Wireless Communications and Mobile Computing*, *2022*(1), 2022. DOI: 10.1155/2022/5939138

Zhao, F., Liu, Y. Q., & Wang, Y. (2017). Research on the Current Situation and Countermeasures of the Mobile Information Service in Chinese Academic. *Library and Information Service, 61*(21), 49-55. https://www.lis.ac.cn/CN/10.13266/j.issn.0252-3116.2017.21.006

Zhao, Y., Xie, D., Zhou, R., Wang, N., & Yang, B. (2022). Evaluating users' emotional experience in mobile libraries: An emotional model based on the Pleasure-Arousal-Dominance emotion model and the Five Factor Model. *Frontiers in Psychology*, *13*, 942198. Advance online publication. DOI: 10.3389/fpsyg.2022.942198 PMID: 35874402

Zhong, L., Cao, J., & Xue, F. (2024). The paradox of convenience: How information overload in mHealth apps leads to medical service overuse. *Frontiers in Public Health*, *12*, 1408998. DOI: 10.3389/fpubh.2024.1408998 PMID: 39668954

Zhu, X., Whitaker, E., Cho, M., & Zhang, M. (2025). (forthcoming). Student engagement in academic libraries: A conceptual framework. *College & Research Libraries*, *86*(1). Advance online publication. DOI: 10.5860/crl.86.2.275

Zhu, Y., & Hose, K. (2025). A survey on knowledge-aware news recommender systems. *Semantic Web*, *15*(1), 18–25.

Zickuhr, K. (2012, December 31). *Mobile connections to libraries*. Pew Research Center. https://www.pewresearch.org/internet/2012/12/31/mobile-connections-to-libraries-2

About the Author

Yan Quan Liu, Ph.D., is a Professor of Information and Library Science at Southern Connecticut State University. He holds a Ph.D. in Information Studies from the University of Wisconsin-Madison and is an author of over one hundred papers in the areas of information and library science. The Fulbright Scholar (https://fulbrightscholars.org/grantee/yan-quan-liu). The member of Sigma Xi, The Scientific Research Honor Society.

Index

A

C

D

P

R

S

U

W

www.ingramcontent.com/pod-product-compliance
Lightning Source LLC
LaVergne TN
LVHW080906300926
849162LV00003B/571